A Handbook of Soil Terminology, Correlation and Classification

Edited by

Pavel Krasilnikov, Juan-José Ibáñez Martí,
Richard Arnold, Serghei Shoba

Routledge
Taylor & Francis Group

LONDON AND NEW YORK

First published 2009 by Earthscan in the UK and USA

2 Park Square, Milton Park, Abingdon, Oxon OX14 4RN
711 Third Avenue, New York, NY 10017, USA

Routledge is an imprint of the Taylor & Francis Group, an informa business

First issued in paperback 2016

ISBN 978-1-138-96549-2 (pbk)
ISBN 978-1-84407-683-3 (hbk)

Typeset by JS Typesetting Ltd, Porthcawl, Mid Glamorgan
Cover design by Susanne Harris

A catalogue record for this book is available from the British Library

Library of Congress Cataloging-in-Publication Data
A handbook of soil terminology, correlation and classification / edited by Pavel Krasilnikov ... [et al.]. – 1st ed.
       p. cm.
   Includes bibliographical references and index.

   1. Soils–Classification. 2. Soil science–Terminology. I. Krasil'nikov, P. V. (Pavel Vladimirovich)
   S592.16.H36 2009
   631.4'4–dc22

                                                    2009017344

Contents

List of Figures and Tables

Figures

Tables

List of Acronyms and Abbreviations

AFES   Association for Soil Investigation (France)
CPCS   Commission on Pedology and Soil Cartography (France)
CRT-CRG   Cooperative Research Group on Chinese Soil Taxonomy (Chinese)
ECZ   Earth Critical Zone
FAO   UN Food and Agriculture Organization
RSG   reference soil group
UNESCO   United Nations Educational, Scientific and Cultural Organization
USDA   United States Department of Agriculture
WRB   World Reference Base for Soil Resources
YBP   years before present (Chinese)

Introduction

The cosmos it seems is a vast unending system of phenomenal energy and the Earth is a unique subsystem of the cosmos not known to be duplicated. The numerous subsystems of the Earth contribute to this variability, and life, as we perceive it, comprises a myriad of subsystems the evolution of which has given rise to humans whose systems and subsystems constantly amaze us. In simplistic terms life on terrestrial portions of Earth requires air, water, shelter and sources of energy in order to survive.

Survival depends on an ability to classify variability, to envision and interpret relationships among domains and objects within domains, and to respond to known or perceived consequences of those relationships. Thus we are innately conditioned and trained to classify, correlate, interpret and store or discard information about variability.

As humans we are exposed to variability, either without choice or through our curiosity, and this causes us to classify these massive sources of information into manageable segments for our existing mental capacities. We mentally put objects, visual clues, sounds and so forth into generalized groups or domains for which we develop archetypes that enable us to quickly distinguish and segregate different populations. The objects in these domains can then be separated into smaller clusters with which we associate specific properties (either imagined or measured). This is the process of correlation that enables us to establish relationships among and between members of populations. For many classifications and correlations of object variability our mental processes seem to be unconscious rather than deliberate activities.

Common interpretations of relationships refer to whether we believe they are cause-and-effect, maybe cause-and-effect or unknown (empirical). We tend to prefer the cause-and-effect ones because they provide us with possible explanations of what is happening in the milieu of variability that surrounds us.

Do we save the information about our interpretations or discard it? As can be seen, this is another classification issue. If we have a spoken language we save them as stories to pass on to others; if we have a written language we write down the details of the stories. If, as individuals, we choose to discard our interpretations of relationships they go into our huge areas of ignorance.

We can imagine that when our ancestors were hunters and gatherers that they learned, often by trial and error, where the better locations were to find the animals they could kill for food, where the seeds and fruits were more abundant and during which seasons, where it was possible to find or make shelters to protect them from the environment and adversaries, and where the water was safe to drink. As they clustered as families and small communities they could diversify their activities by passing on information about survival in their stories. Others then, in their curiosity, learned how to do many other things that enriched the lives of one and all, and cultures and civilizations evolved.

When communities were formed there was a need to locate where water and food could be made available, and the need to classify locations was critical to their development. Eventually the earth in which plants and trees of interest existed was examined in more detail, maybe only the surface layer, but systems of classification were developed and the information passed on from generation to generation. Management practices, such as ploughing and irrigation, were found to be useful, as was the addition of organic matter and manure to maintain and increase biomass yields. Soil classification obviously had a strong foundation in agricultural pursuits, and also in the construction of shelter and the provision of clean potable water.

With the advent of communities and of tribes and nations that were not mainly nomadic, there arose the need for some system of land ownership. And so land quality (soil quality) and location became important issues when dealing with the terrestrial ecosystems of the planet. How many hundreds of soil/land classifications may have existed we probably will never know as now many languages, both spoken and written, have been lost.

This book is about some of the stories of humanity's struggle to develop and interpret relationships of the earthy materials at or near the surface of our planet. Although we provide many terms that have been, and are, used to describe kinds of soils, many are not in organized formats of modern classification systems but give an indication of meaningful aids in some cultures. Most of the chapters of this book refer to national systems of soil classification developed since the advent of modern soil science in the late 1800s.

Several systems have been designed to be international in scope, such as the US Soil Taxonomy and the French Reference System. An attempt to produce a collaborative soil map of the world was undertaken by the Food and Agriculture Organization of the United Nations (UN FAO) in the late 1960s, and its legend became a type of correlation tool to help people relate soils of one place to those of another. Follow-up attempts to be able to correlate major generalized concepts of soils throughout the world have culminated in the World Reference Base for Soil Resources (WRB) (now in a new 2006 version). We cannot translate any system directly into another system because of definitional differences in concepts, in physical and chemical measurements and in organizational formats of the many soil classification systems currently in existence. What we have tried to do is create an awareness of the challenges that the scientific soil science community is facing at the present time.

For a long time soil classification was the focus of attention in pedology: it reflected, as a mirror, the struggle of pedogenetic and geographic concepts in organizing information about soil profiles and their spatial representation on maps. Currently soil classification has moved from the nucleus to the margin of the attention of soil science community as environmental issues of terrestrial ecosystems have gained prominence. In the last decades, developments in digital mapping now facilitate combining various information layers, somewhat replacing traditional soil classification-based maps. Even in soil genesis and soil geography studies researchers commonly speak in terms of pedogenetic processes and particular soil characteristics rather than the use of formal soil names. For many purposes mathematical ad hoc classifications work better than more general basic classifications. Does it mean that soil classification is 'dead and buried'? We think not. Soil classification fulfils a necessary function for communication, education and mapping. Even in everyday communication we need proper names, and science is not an exception. Soil education would be incomplete and flawed, if we spoke about soil properties only, without a holistic view on natural soil groups. In other disciplines, for example in geology, the development of digital geochemical maps did not result in the abandonment of traditional geological survey. Finally, soil science has already accumulated thousands of maps and documents developed on the basis of soil classification, which cannot be ignored.

The book draws on an earlier title (*Soil Terminology and Correlation*) published in Russia in 1999 and which was edited by Professor S. Shoba. Though this work has been completely rewritten for the new book, it includes some concepts proposed in the earlier version. The book consists of three main parts. The first part, written by Professor R. W. Arnold, Dr J.-J. Ibáñez Martí and Dr P. Krasilnikov, provides an introduction to the theoretical bases of soil classifications. The second part gives an overview of the existing soil classifications. This part was written mainly by Drs P. Krasilnikov and R. W. Arnold; Chapter 9 was written together with Dr P. Schad, Chapter 16 with Professor E. Michéli and Chapter 26 with Dr A. Hernández Jiménez. The book includes all the published classification systems actually in use that have a system of soil diagnostics and are suited for large-scale soil mapping. Out of use classifications, soil maps' legends without a system of soil diagnostics or conceptual groupings of soils are regarded as underdeveloped or incomplete classifications and are discussed in a special chapter. The third part discusses folk soil classifications and includes a list of documented indigenous soil names. This part was prepared by Drs P. Krasilnikov, J. Tabor and R. W. Arnold.

In the second part we attempt to correlate the terms of the national classifications with the WRB (IUSS Working Group WRB, 2006). This classification was designed especially for soil correlation, not to replace national taxonomies, but to serve as an umbrella system, facilitating understanding between the specialists of different countries. In this volume we understand correlation as a method for visualization of soil units. Generally the limits of soil classes in different classifications are overlapping, and one class in a national classification may correspond to various WRB groups, especially if conceptually different criteria are used for defining the classes. In these cases we tried to limit correlation to a rational minimum of groups. Otherwise the reader may be confused, finding that some classes correspond to dozens of WRB groups; the concept of the class would be washed off, and the reader would not get any idea about the concept of the soil unit. In the list of the WRB groups corresponding to every unit of national soil classifications we put the group with major overlapping first, followed by less important analogies, and the groups with minor overlap are ignored. In certain cases the concepts of the units in national soil classifications required the use of modifiers that are not included in the recommended list for the particular reference group; in these cases the modifiers are listed in italics.

The reader should be warned that this book is not a dictionary that can help to convert the names of a particular soil into another classification system. In fact, such a conversion is generally impossible. This volume may be helpful for understanding soil maps and research papers prepared in the other classification, but it cannot be used for translating the maps from one system to another.

It is our belief that many of the gaps and difficulties in trying to interpret correlations among soil systems are, in fact, opportunities that can serve as the basis for the stories of the future. Some of these are highlighted below.

The editors

Part 1
The Theoretical Bases of Soil Classifications

P. Krasilnikov, J.-J. Ibáñez and R. Arnold

1
Introduction to Classifications with an Emphasis on Soil Taxonomies

Theory of classifications in natural sciences

Classification is a vast theme, which includes concepts ranging from basic perceptions and logic to routine procedures of separation and clustering. We start to classify objects in early childhood, and keep doing it all of our life; in fact, classification is one of the more important human tools for understanding the world (Kemp and Tenenbaum, 2008). In statistics, classification is a method for grouping objects (Good, 1965); though relatively simple in its basis, the procedures have many complications and the subject is studied, discussed and improved.

Basic natural classifications, that is, special groupings of particular objects into classes, are used by scientific communities as a common language for a specific branch of science. Cormack (1971) shows that not every separation of entities into groups should be called classification; when the separation is completely artificial and does not imply any criteria to distinguish the groups, the separation can be called dissection. For example, administrative divisions of a territory mainly have no basis other than a historical one.

Most classifications are based on ideas of the discreteness of entities. However, it is not evident for most earth sciences. In these cases the domains are commonly continua in space (like lithology or soils), or the objects have a continual gradient of properties. The continua/discrete dilemma is especially important for soil classifications, and will be discussed in more detail in the next chapter. The human mind can separate continua as well as group different objects into classes. It should be noted that this property of the human mind is typical not only for scientific thinking, but also for everyday living.

With the recognition of objects of interest that are similar in many features, they become parts of a larger grouping, a domain that separates them from all other objects. Long before Carl Linnaeus (1707–1778), humans distinguished different living beings and named them; even soils were recognized for their suitability for growing crops. The great achievement of Linnaeus was that he transformed the existing nominal classifications into taxonomies. He used a hierarchical structure – observable characters, based on a binomial nomenclature, strictly defined the borders of taxa – and worked out a uniform Latin terminology. The levels were kingdom, phylum, class, family, genus, species and variety. Before the appearance of modern scientific classifications many archetypes existed (in the sense of Aristotle – 'a prototype, the main form, with possible deviations' (Shreyder, 1993)) for many natural objects.

A unified theory of natural scientific classifications does not exist. Numerous publications concentrate on particular aspects of natural classification, or on the application of classifications in specific branches of science. We will outline some theoretical bases, structures and terminology of modern soil classification. For better understanding of soil classification we specify some important definitions and concepts. We will consider any division of a population of entities, based on their intrinsic properties or functions, as a classification. The classifications used by scientists will be regarded as scientific ones since we have no criteria to separate 'scientific' from 'non-scientific' other than the acceptance by the scientific community (Kuhn, 1970). Scientific classifications can be separated into natural or basic ones, and applied ones. Basic ones classify the whole set of the objects studied by a certain branch of science on the basis of their internal relations, and applied classifications classify according to values of a few properties or attributes. These latter ones are more appropriately termed technical interpretive groupings.

There is some confusion with the terms 'taxonomy' and 'classification'. For some, taxonomy is the theoretical study of classification, including its bases, principles, procedures and rules (Simpson, 1961, p11), or in more general terms, the study of the general principles of scientific classification. Dobrovolsky and Trofimov (1996) consider taxonomy to be the theory of classification. For Mayr (1969) taxonomy is a retrieval information system. Classification is often understood as placing objects into a hierarchical arrangement of categories and classes. Two objects are placed in the same class because they share one or more attributes of the defining properties of the class (Minelli, 1993, p5). Minelli (1993, p6) further states: 'there is not a general agreement as to the definition of systematics, and of related words such as taxonomy and classification'. We shall use the terms as synonyms.

Objectives of natural scientific classifications

As mentioned previously the main objective of classification is a perception of the world. That is, we attempt to search for order in the universe. In natural sciences, the purposes of systematization of existing knowledge and language function are of major importance. Classification nomenclature is necessary to enable specialists to understand each other. There is a problem in having a taxonomic unit serve two purposes, that is, the arrangement of data, convenient for search and retrieval, and the discovery of the internal system relations (Rozova, 1993). So-called genetic classifications, which are traditionally considered to be the most fundamental, are based on parameters of the basic object that in turn determine many other properties. Quite often the primary parameter is a complicated theoretical construction, and in practice it is impossible to diagnose it directly. Because of this conflict there is concern about the extent to which natural classifications reflect the world and to what extent they reflect our minds and theories. It has been noted (Cline, 1949) that classifications are products of the human mind and are subject to all of its limitations.

Systematics and classification

Soil classifications are often compared with biological ones. The improved Linnaean taxonomy seems to be an example of order and precision compared with numerous imprecise soil classifications. As an argument against hierarchical soil classifications Swanson (1993) mentioned that biological classification has a basis (evolution) and soil classification does not. Evolution implies a relation between the entities, whereas soil classification mainly deals with creating provisional classes, including formal grouping using numerical methods, and establishing 'pseudorelations' between them. Albrecht et al (2005) referred to the first type as 'systematics' and the second type as 'classification'. These authors indicate that systematics is a fundamental scientific and deductive ordering of objects into systematic units and suggest two examples of systematics: genealogical trees and evolutionary schemes.

As Gorochov (2001) noted, the idea about a certain natural classification of organisms originated from theological assumption of the plan of creation, which could be discovered or reconstructed. Thus in scientific systematics, the idea was that the evolution of organisms or other entities (e.g. ethnic groups, languages or soils) should be reconstructed to group them correctly. However, it does not mean that systematics has priority or is more scientific. Even in biology there are doubts that a classification based on the evolutionary relation between organisms (phylogenetic taxonomy, or cladistics) is the best option for grouping taxa, as cladistics leads to separation of similar life forms at the base of different branches of a phylogenetic tree, and very different forms at the final parts of these branches must be united with primitive initial ones (Gorochov, 2001).

Soil classifications commonly represent a mixture of the above approaches, being intermediate between 'systematics' and 'classification' (Albrecht et al, 2005). Because there are numerous soil classifications, they show a broad spectrum of attempts to combine convenient grouping and genetic principles. The concern about convenience and genetic principles relates to the diversity of approaches in different scientific schools and also to the pressure for practical applications, primarily agriculture. This circumstance to some extent may limit the classification activity of soil scientists to the pages of scientific journals and monographs. In some countries the situation resulted in bilingual terminology in soil science; in Australia until recently two classifications existed, one for scientists, and one for other users (Moore et al, 1983).

Theory-based (systematics) and properties-oriented (classification) approaches of soil grouping vary in different taxonomic systems from almost completely speculative constructions to numerical ad hoc empirical ones. The latter type usually constitutes interpretive groupings in which the units are pragmatic groups rather than natural soil classes – for example, suitability for irrigation, steep or stony land, areas of high salinity and so forth. Such groupings are not covered in this book. There is a broad spectrum of approaches to grouping soils, yet most combine convenient groupings and genetic principles. Many of the variations occur because of the desire to provide information for practical uses of the soil resources, such as agriculture or forestry (Ibáñez and Boixareda, 2002; Ibáñez et al, 2005). Theory-based and property-oriented groups of soils vary in different schemes from almost completely speculative to numerical ad hoc empirical groupings. The more effective soil classification systems combine these approaches; however, the basic theoretical concepts used to classify soils vary among the scientific schools of thought.

Theoretical bases of soil classifications

The conceptual basis for modern soil classification is genesis (Schelling, 1970) most commonly being morphogenetic. Since the early works of Dokuchaev, there has been belief in the dependence of soil properties on soil processes, which depend in turn on soil-forming factors. Now it is expressed as: factors of soil formation → internal soil system functioning → specific pedogenic processes → soil properties and features → external soil functions (Targulian and Krasilnikov, 2007). The concept is practical as it is used in soil mapping. The limits of soil polygons are usually placed on the borders of areas with uniform soil-forming factors (the same relief, parent materials, hydrology, etc.) (Hudson, 1992). Because mapping has been the main area of application of soil classification, it seems logical that the soils with the same soil-forming factors, pedogenic processes, properties and external functions should be grouped into the same classes. However, the dependence of soil properties on soil-forming factors is not linear, and as most soils are polygenetic they vary in space reflecting also the soil-forming conditions of the past which may only partially be reconstructed (Phillips et al, 1996).

Because the direction and intensity of similar processes vary in time and space it is useful to consider the concepts of 'divergence' and 'convergence' (Rozanov, 1977). Divergence means that soils formed under similar conditions in different places commonly exhibit variable properties (Johnson et al, 1990; Phillips, 1993; Ibáñez et al, 1994) due to local factors. Convergence means that different pedogenic processes under different environmental conditions might lead to similar soil properties and morphology. For example, such processes as podzolization, clay eluviation and surface gleying generally lead to the formation of a bleached, clay-depleted surface horizon. Thus one cannot expect a complete correspondence of soil-forming factors, pedogenic processes and soil properties. External functions are less suitable for classifying soils; however, they are important for most interpretive groupings. For example, the most important soil function, its productivity, depends mostly on the availability of nutrients and toxicants in the surface horizon, which is dynamic, spatially variable, and easily modified by human impact.

Early soil classifications used mainly factor-based approaches. At that time it was believed that the factors→processes→properties relation was linear and simple. Empirical data were usually insufficient for establishing the ranges of properties for every soil group. This approach survived almost to the end of the 20th century; for example, it was used in the Classification and Diagnostics of Soils of the USSR (Egorov et al, 1977). In that classification, the emphasis on soil-forming factors resulted in certain contradictions with empirical data when different soils were placed in the same taxa if they were found to be in the same climatic zone. For example, the soils without a bleached surface horizon in the 'podzolic zone' were called 'cryptopodzolic' or 'weakly differentiated podzolic soils'. Though most specialists currently reject a strictly factor-oriented approach, some soil classifications now consider surrogates of the climate factor to be measurable internal properties of soils. The US Soil Taxonomy (Soil Survey Staff, 1999) and Chinese Soil Taxonomic Classification (Gong Zitong, 1994) use internal water and temperature regimes as quantitative diagnostic properties at high levels of taxonomic hierarchies.

Another conceptual approach was proposed by Kubiëna (1953), who based his classification on soil evolution. Chronosequences of soils were established on the basis of empirical data and theoretical concepts about the stages of soil development. This scheme is widely used in German and other classifications. Soils are grouped according to the kind and degree of development of their profiles, considering that more developed soils should have more complex profiles. The concept is not universally accepted as many deep steppe soils, for example, might be considered weakly developed because they have a simple A/C profile. It is important to note, that the evolutionary grouping of entities in soil science is distinct from that in biology (cladistics). Soils are grouped according to the stage of their development (horizontal grouping) and not on the basis of the same 'evolutionary branch' (vertical grouping). For example, shallow young soils on limestone, *rendzinas*, are not grouped with ancient limestone-derived residual soils, *terra rossa*. The evolutionary approach does not modify significantly basic soil grouping compared with other approaches and can be regarded as a convenient arrangement of soils, useful mainly for educational purposes to indicate possible pathways of soil development.

Most existing soil classifications use soil morphology and properties as marks or evidence of definitions used for soil groupings. The soils are still grouped according to concepts of their genesis, and observable and measurable criteria are used to provide objective identification for the placement of the objects in the classification system. Quite often the term 'genetic classification' is misunderstood in soil science. Every classification based on grouping soil profiles as combinations of horizons can be considered genetic. The diagnostics may be done on the basis of surrogates of soil-forming factors or on the basis of the properties of the soil profile itself, but the concept is the same. The profile is the result of soil formation and evolution, and the sequence and properties of the horizons are explainable and interconnected.

Summarizing the information on the theoretical bases of soil classification, we conclude that practically all the approaches of various scientific schools have the same foundation, and most soil groups have rather similar abstract central concepts. The differences in theoretical approaches result mainly in the methods of defining and measuring diagnostics.

Characterizing soil classifications

It is important to discern for taxonomies their formal structures (graphs topology), from rationale, diagnostic criteria and nomenclature. For instance, two taxonomies may have similar topological structures, but different rationale (theory) and/or diagnostic criteria, and the products could be quite different. Classifications can be grouped and characterized both qualitatively and quantitatively. From a basic structural point of view, four main types exist: nominal systems, tables, reference bases and hierarchical taxonomies (Shoba, 2002). In nominal systems every class of the objects of interest is unique, and the objects are not grouped into taxa of higher levels. Nominal systems are typical of most indigenous biological and soil classifications (Tabor, 1992). A system of soil series once used in the US and some other countries was also nominal. These systems are the most basic and are proposed at the first stages of the growth of classification (Mosterín, 1984). For scientific purposes nominal systems are not comprehensive and in most cases during the process of accumulating empirical data they are transformed into taxonomies (Mosterín, 1984; Kemp and Tenenbaum 2008). For example, the system of soil series in the US was incorporated into Soil Taxonomy (Soil Survey Staff, 1999) as each series was redefined to be within the limits imposed by all higher categories. In some small countries, like Guatemala, the soil series are still listed in alphabetical order (Beinroth, 1978). Soil classes might also be grouped into a table, similar to the periodic table of elements. In soil science an example is the scheme used in the South African Republic (Soil Classification Working Group, 1977) where the table columns include topsoil classes, and the lines consist of the sequences of mineral horizons. Examples of reference bases are the new French classification (AFES, 1998) and the latest version of the World Reference Base for Soil Resources (WRB) (IUSS Working Group WRB, 2006). In the French system, soil classes are regarded as points or regions in n-dimensional space of properties. These classes are optionally grouped, but the groupings are not part of the system. If some specification is needed, an explanatory modifier is added to the name of a reference group. The WRB has a distinct hierarchical structure although it was declared to be a reference base. Hierarchical taxonomy is a more complex case with classes grouped into bigger ones in the higher levels. The difference between these types is not very great although the formats are specific. In earlier versions of the US Soil Taxonomy the family level had the structure of a reference base where soil subgroups were supplemented with modifiers giving specific information about texture, mineralogy and other properties, and the series level was a nominal system. Today both categories have been modified to be consistent with categorical definitions.

To better catalogue classifications it is useful to know their internal structure. Anthropologists and psychologists studying folk classification provided profound insight about them. The seminal work of Berlin et al (1973) showed that such classifications have six taxonomic levels: kingdom, life form, intermediate, generic, specific and varietal. The system was adapted by Williams and Ortiz-Solorio (1981) for soil science with five taxonomic levels: kingdom, collective, generic, specific and varietal. Every classification would have at least two levels – the kingdom, which is the universe of objects (soils), and the generic level that includes the entities of that universe. The number of taxonomic levels (categories) may be called the depth of classification (Holman, 1992, 2005). Systems may be characterized by the kinds of levels – that is, collective, specific or varietal – as several may be present in a system.

Another characteristic is the breadth of classification, that is, the number of objects at the lowest taxonomic level. For most folk classifications the breadth hardly surpasses 1000 objects (Holman, 2005), but for scientific classifications the number is unknown but may reach tens or hundreds of thousands. Using this approach we note that tables and reference bases are particular cases of hierarchical taxonomies. Only nominative systems are flat classifications with two ranks: kingdom and generic. Tables have three ranks: kingdom, collective and generic. The hierarchy of tables is flexible since both columns and lines can be considered as classes. The reference bases have two obligatory levels – kingdom and generic – and specific and varietal ranks are included if modifiers are used.

Tables and reference bases are more closed than nominal systems. Any new object can be introduced easily into a nominal system, but for tables and reference bases it is more complicated because an internal relationship must be determined for any new object. For example, in a dimeric table a new object should be included with an existing class or fill an empty cell. In some cases it is necessary to add new columns or lines to accommodate additional entities. Systems with clear hierarchies permit limiting taxon sizes, thus they are easy to use and to search for objects. Reference bases may be less convenient for users but they are more flexible and permit including new taxa with minimal changes in the whole system. In the next chapters we will point out some advantages and disadvantages of many soil classifications. The ideal system does not yet, and may never, exist.

References

AFES (1998) *A Sound Reference Base for Soils (The 'Referentiel pedologique': text in English)*, INRA, Paris, 322pp

Albrecht, C., Jahn, R. and Huwe, B. (2005) 'Bodensystematik und Bodenklassifikation. Teil I: Grundbegriffe', *Journal of Plant Nutrition and Soil Science*, vol 168, no 1, pp7–20 (Proceedings of a Seminar held at CIAT, Cali, Colombia, 10–14 February 1974)

Beinroth, F. H. (1978) 'Relationship between U.S. soil taxonomy, the Brasilian soil classification system and FAO/UNESCO soil units', in E. Bornemisza and A. Alvarado (eds) *Soil Management in Tropical America*, Gordon Press, Raleigh, NC, pp92–108

Berlin, B., Breedlove, D. E. and Raven, P. H. (1973) 'General principles of classification and nomenclature in folk botany', *American Anthropologist*, vol 75, no 2, pp214–242

Cline, M. G. (1949) 'Basic principles of soil classification', *Soil Science*, vol 67, no 1, pp81–91

Cormack, R. M. (1971) 'A review of classification', *Journal of Royal Statistical Society*, Ser. A (General), vol 134, no 3, pp321–342

Dobrovolsky, G. V. and Trofimov, S. Ya. (1996) *Soil Systematic and Classification (History and Modern State)*, Moscow State University Publishers, Moscow, 78pp (in Russian)

Egorov, V. V., Fridland, V. M., Ivanova, E. N., Rozov, N. N., Nosin, V. A. and Fraev, T. A. (1977) *Classification and Diagnostics of Soils of USSR*, Kolos Press, Moscow, 221pp (in Russian)

Gong Zitong (ed) (1994) *Chinese Soil Taxonomic Classification (First proposal)*, Institute of Soil Science, Academia Sinica, Nanjing, 93pp

Good, I. J. (1965) 'Categorization of classification', in *Mathematics and Computer Science in Medicine and Biology*, HMSO, London, pp115–128

Gorochov, A. V. (2001) 'On some theoretical aspects of taxonomy (remarks by the practical taxonomist)', *Acta Geologica Leopoldensia*, vol 24, nos 52–53, pp57–71

Holman, E. W. (1992) 'Statistical properties of large published classifications', *Journal of Classification*, vol 9, no 3, pp187–210

Holman, E. W. (2005) 'Domain-specific and general properties of folk classifications', *Journal of Ethnobiology*, vol 25, no 1, pp71–91

Hudson, B. D. (1992) 'The soil survey as paradigm-based science', *Soil Science Society of America Journal*, vol 56, no 4, pp836–841

Ibáñez, J. J. and Boixadera, J. (2002) 'The search for a new paradigm in pedology: A driving force for new approaches to soil classification', in E. Micheli, F. Nachtergaele, R. J. A. Jones and L. Montanarella (eds) *Soil Classification 2001*, EU JRC, Hungarian Soil Sci. Soc., UN Food and Agricultural Organization, Italy, pp93–110

Ibáñez, J. J., Pérez-González, A., Jiménez-Ballesta, R., Saldaña, A. and Gallardo, J. (1994) 'Evolution of fluvial dissection landscapes in Mediterranean environments. Quantitative estimates and geomorphological, pedological and phytocenotic repercussions', *Zeitung Geomorphol*, vol 38, pp105–119

Ibáñez, J. J., Ruiz-Ramos, M., Zinck, J. A. and Brú, A. (2005) 'Classical pedology questioned and defended', *Eurasian Soil Science*, vol 38, suppl 1, pp75–80

IUSS Working Group WRB (2006) *World Reference Base for Soil Resources*, 2nd edition, World Soil Resources Reports no 103, UN Food and Agriculture Organization, Rome, 128pp

Johnson, D. L., Keller, E. A. and Rockwell, T. K. (1990) 'Dynamic pedogenesis: New views on some key soil concepts and a model for interpreting Quaternary soils', *Quaternary Research*, vol 33, no 3, pp306–319

Kemp, C. and Tenenbaum, B. (2008) 'Discovery of structural form', *Proceedings of the National Academy of Sciences*, vol 105, no 31, pp10687–10692

Kubiëna, W. L. (1953) *Bestimmungsbuch und Systematik der Boden Europas*, Ferdinand Enke Verlag, Stuttgart

Kuhn, T. S. (1970) *The Structure of Scientific Revolution*, University of Chicago Press, Chicago

Mayr, E. (1969) *Principles of Systematic Zoology*, McGraw-Hill, New York, 428pp

Minelli, A. (1993) *Biological Systematics. The State of the Art*, Chapman & Hall, London, 387pp

Moore, A. W., Isbell, R. F. and Northcote, K. H. (1983) 'Classification of Australian soils', in *Soils: An Australian Viewpoint*, CSIRO, Melbourne and Academic Press, London, pp253–266

Mosterín, J. (1984) *Conceptos y Teorías en la Ciencia*, Alianza Universitaria, Madrid, 220pp

Phillips, J. D. (1993) 'Stability implications of the state factor model of soils as a nonlinear dynamical system', *Geoderma*, vol 58, no 1, pp1–15

Phillips, J. D., Perry, D. C., Garbee, A. R., Carey, K., Stein, D., Morde, M. B. and Sheehy, J. A. (1996) 'Deterministic uncertainty and complex pedogenesis in some Pleistocene dune soils', *Geoderma*, vol 73, no 1, pp47–164

Rozanov, B. G. (1977) *Soil Cover of the Globe*, Moscow State University Publishers, Moscow, 248pp (in Russian)

Rozova, R. R. (1993) 'Methodological analysis of the classification problem', in *The Theory and Methods of Biological Classifications*, Nauka Press, Moscow, pp6–17 (in Russian)

Schelling, J. (1970) 'Soil genesis, soil classification and soil survey', *Geoderma*, vol 4, no 2, pp165–193

Shoba, S. A. (ed) (2002) *Soil Terminology and Correlation*, 2nd edition, Centre of the Russian Academy of Sciences, Petrozavodsk, 320pp

Shreyder, Yu. A. (1993) 'Systematics, typology, classification', in *The Theory and Methods of Biological Classifications*, Nauka Press, Moscow, pp90–100 (in Russian)

Simpson, G. G. (1961) *Principles of Animal Taxonomy*, Columbia University Press, New York, 247pp

Soil Classification Working Group (1977) *Soil Classification: A Binomial System for South Africa*, Science Bulletin no 390, Department of Agriculture Technical Survey, Pretoria, 150pp

Soil Survey Staff (1999) *Soil Taxonomy: A Basic System of Soil Classification for Making and Interpreting Soil Surveys*, 2nd edition, USDA-NRCS, Lincoln, NB, United States Government Printing Office, Washington DC, 696pp

Swanson, D. K. (1993) 'Comments on "The soil survey as paradigm-based science"', *Soil Science Society of America Journal*, vol 57, no 5, p1164

Tabor, J. A. (1992) 'Ethnopedological surveys – soil surveys that incorporate local systems of land classification', *Soil Survey Horizons*, vol 33, no 1, pp1–5

Targulian, V. O. and Krasilnikov, P. V. (2007) 'Soil system and pedogenic processes: Self-organization, time scales, and environmental significance', *Catena*, vol 71, no 3, pp373–381

Williams, B. J. and Ortiz-Solorio, C. A. (1981) 'Middle American folk soil taxonomy', *Annals of Association of American Geographers*, vol 71, no 3, pp335–358

2
Soil Classifications:
Their Peculiarity, Diversity and
Correlation

Defining the objects of soil classification

The discussion on the categories in soil classification has a long history (Dmitriev, 1991). Initially, pedologists considered each soil profile to be both representative and typical of its own class. However, the extension of soil survey resulted in recognizing the great heterogeneity of soil morphology and properties. Thus, a number of concepts were proposed to define the objects of soil classification (Arnold, 1983). The first version of the US Soil Taxonomy defined the soil individual as a small landscape called a polypedon and its sampling unit was a pedon, which was considered to be the object of classification (Buol et al, 1977). Many discussions dealt with the definition of the uniformity and size of the pedon (Dmitriev, 1991); however, Sokolov (1978) noted that most specialists ignored these features and considered the object of classification to be the profile itself. In more recent publications we can find the opinion, based on cognitive psychology, that it is not the natural object which is classified, but our idea of this object (Haskett, 1995). A similar thought is outlined in the modern French soil classification (AFES, 1998). French scientists proposed to separate the soil mantle (a real three-dimensional natural body), soil profile (vertical change of a certain property in a soil pit; e.g. a morphological profile, a humidity profile, etc.) and solum, which is the concept of a soil as a sequence of pedogenically developed horizons. The object of classification is solum, a sequence of horizons (or ideas of horizons). This approach seems logical, because in a strict sense, except in our mind, there are neither soil classes nor soil profiles. In reality there is a substrate, with some properties changing in vertical and lateral directions that we divide into horizons by overlaying the structure of our mind on the real object. The fact that the projection is subjective does not make it 'less scientific', if we use a unified technique for distinguishing horizons and soil classification.

Another problem is the continuum nature and spatial variability of soils (Heuvelink and Webster, 2001). In contrast to biological classifications, which classify mostly discrete objects, soil classification divides continual soil cover (the pedosphere) into discrete taxonomic units. The situation is complex both in pedology and other natural sciences.

Soil – a discrete/continuum dilemma

If the pedosphere were a true continuum, all combinations of soil properties could be possible. Yaalon (2003) showed that soil types or pedotaxa could vary both in a continuum and in a discrete way because of their soil-forming factors dependence. Usually, the human mind processes information through discreteness or 'reification' (Mosterín, 1984). For this reason, the first step in classification science is to recognize the objects of concern and place them into discrete classes with either strict or fuzzy boundaries. Such classes may be limited by mental constructs, spatial entities, temporal processes and so forth. Whether soil variability is conveyed in discrete or fuzzy units, both represent idealized end members of a continuum with innumerable transitions between 'pure continuous' and 'pure discrete' units (Levy-Leblond, 1996). They are purpose dependent and lead to different representations of the natural structure of the soil cover (Ibáñez and Boixadera, 2002; Ibáñez et al, 2005). Kay (1975) analysed the application of the fuzzy set theory and showed that it is not the best way to solve the category-as-prototype problem. Fuzzy set theory has been not very successful in biotaxonomic practice; however, it can be employed in 'ad hoc' classifications for certain interpretive products. The same can be said about soil map representations when a certain degree of fuzziness that occurs between different polygons is considered.

The dilemma of the continuum can be seen in Borhr's Principle of Complementarity 'Wave-Particle' (duality). According to this principle, the soil cover can be viewed as a continuum field (pedosphere) that comprises numerous aggregates of 'artificial' or 'natural' entities (pedotaxa). Conceptually, a continuum is not incompatible with discrete units (Fridland, 1976). It is possible to analyse the structure of the soil cover using complementary approaches, the same way that plant cover is analysed by combining phytosociological typology and gradient analysis of the plant landscape as a continuum.

According to Yaalon (2003), of the five soil-forming factors acting as driving forces for the nature of soils, parent material and topography frequently change abruptly over small distances. He estimated from his own experience that the occurrence of sharp soil boundaries for most medium- and large-scale maps would be between one-quarter and one-half of the soil boundary lines drawn on them. Likewise, vertical lithologic discontinuities are frequent in the soil cover. Their frequency is about 33 per cent in US soil series, based on a sample of 1000 soil series descriptions (Schaetzl, 1998).

Practically all the natural classifications deal with somewhat fuzzy objects. Rocks and minerals are spatially heterogeneous, and have numerous transitional species. Novel approaches in biology show the importance of horizontal gene flow and other biological mechanisms, making the separation of biotaxa into hard classes subjective and questionable (Ghiselin, 1974; Sattler, 1986). Thus, biological individuals are discrete only in appearance. As a consequence, many biotaxonomists accept that species are the true individual, while the biological entities of a given species are mere organisms of that species (Ghiselin, 1974). It is interesting to note that a number of disadvantages (abundance of transitional taxa, definition of taxa at the same level by different criteria, or at a different level by the same criteria), which are often considered to be typical only for soil classifications, are mentioned among the problems of classifications in biology.

Soil archetypes

Since Dokuchaiev most pedologists accept that soil is a natural body. However, insofar as soil is also a continuum (e.g. Dmitriev, 1991), many pedologists think that pedotaxa or soil types are artificial entities. Do soil taxonomists break the soil continuum artificially into classes? This is an old dilemma, termed the Naturalia/Artifitialia, which concerns philosophical and ontological controversies more than scientific ones. It started with a controversy between Buffon and Linnaeus in the 18th century (see Ibáñez and Boixadera (2002) and references therein).

Gorochov (2001) noted that since we do not know the overall organizations of nature, we might produce various alternative natural classifications, or consider them all as artificial ones. His observation refers mainly to the manner of organizing the structure of classification and does not mean that classes can be absolutely provisional. In appears that any classification is intuitively based on the concept of archetypes; that is, certain entities have similar characteristics. Most natural classifications grew from pre-scientific ones, mostly non-verbal concepts of archetypes. The archetype concept of Aristotle implied the existence of ideal images of entities. As far as we know, archetypes do exist in our mind, or in the 'collective mind' of humankind (or scientific community). They can be defined as some ideal 'central images' of objects, or 'prototypes'. It is believed that archetypes are basic for natural classifications, and the diagnostics for recognition and placement are a technical matter, which has secondary priority.

At the initial stage of the development of modern soil classification, soil types in the sense of V. V. Dokuchaev and his successors were archetypes. The names of soil types were mainly borrowed from folk soil classifications: the words *chernozem, solod, solonetz, rhendzina* were used by Russian, Ukrainian and Polish peasants for ages. The use of indigenous soil names reinforced the use of the archetypes in scientific soil classifications. Any scientific classification system based only on formal logic could be regarded as an artificial one as it cannot be verified through correlation with the real world. For biological and pedological classifications a necessary stage is to compare them with our artificial classes and/or with pre-scientific indigenous knowledge (Militarev,

1993). Folk soils groupings are not accepted directly in scientific systems, as certain limitations of indigenous classifications in soil science exist.

Subconscious perception of soil as a body, which can be objectively separated into existing archetypes, is based on several features of the soil cover. First, in real space different soils are associated and more or less distinct boundaries exist between these soils. Second, in a given landscape a limited number of soils usually exist. Some of these soils are dominant and are considered archetypes; others cover smaller areas and often are considered to be transitional taxa. Additional research commonly shows that in other landscapes these 'transitional' soils are dominant and previously distinguished 'archetypes' can be described as neighbours of the 'transitional' class. As a result soil archetypes seem to be very unstable, depending on the soils of a given landscape, and partially account for the diversity of soil classifications in the world. Historically in soil classification the archetypes and their diagnostics are primary, and additional diagnostics (based on factors, morphology of the profile or anything else) are secondary because they are constructed to separate the archetypes in a more detailed and precise manner. The principle difference among different national classifications is associated with the different ways of separation of the soil continuum. For example, most soil classifications consider *solonetz* as a stable soil archetype, but the US taxonomy (Soil Survey Staff, 1999) divides this archetype into many great groups belonging to two different orders, Alfisols and Mollisols, depending on the ratio of organic matter accumulation and argillic horizon development. The US Soil Taxonomy, regarding archetypes, is a special case and will be discussed below. Peat soils, which are separated in most classifications on the highest level, are divided into different taxa in the Australian (Isbell, 2002) and US taxonomies (Soil Survey Staff, 1999). In Australia tidal peat soils are separated from Organosols and placed in the Hidrosol order, and in the US peat soils with permafrost are placed not in the Histosol order, but in the Histel suborder of the Gelisol order. Many other examples occur.

Historically, soil archetypes were established at the highest level of soil taxonomy, and with the accumulation of diagnostic information, they were subdivided at lower levels. The development of soil classification resulted in the growth of classifications both vertically and horizontally. At the basic (archetype) level the number of taxa was increasing and in some classifications the number of types (great groups) was already difficult to manage; thus upper hierarchical levels were constructed, such as divisions in Russia, and orders and suborders in the US.

Apart from the growth of soil classifications 'from top to bottom', parallel systems developed in a number of national scientific schools. In the beginning of the 20th century M. Whitney and colleagues developed in the US a system of soil series as lower level landscape units with individual ranges of features (Simonson, 1989). Soil series could be regarded as a population of profiles with more or less fixed properties corresponding somewhat to species in biology. The same system was also used in the United Kingdom and New Zealand. In practice, it was a parallel classification for soil survey with archetypes at the lowest taxonomy level. Practical soil surveyors used series for both large- and medium-scale soil mapping without interest in upper levels of taxonomy until the number of series increased to several thousand and was difficult to manage. The fact that the system of soil series did not fit well into Great Soil Groups, which in turn did not fit well in a global perspective of archetypes, was one of the reasons for the creation of new soil taxonomy in the US (Banfield, 1984). Classifications based on soil series could not be easily transformed into a morphogenetic structure of global soil taxonomy. Instead, these classifications had two archetypes levels: one at the 'bottom' of a hierarchy and the other in the upper part. In the US Soil Taxonomy (Soil Survey Staff, 1999) there were archetypes at the series and great group levels. This dilemma was later resolved in the US by redefining series to be within family limits. Some 20,000 series were redefined, thereby losing many of the geographic–parent material relations on which they had originated.

Babel tower

As we can see, more than one natural soil classification can exist – that is, there is no unique 'true' classification to be discovered. The same is true for other natural disciplines, but unlike in biology or geology, the possible existence of different classifications realized in soil science occurred because of a lack of a strong guiding international interest to develop a global system. The existence of numerous national soil classifications is a serious problem of perception of soil science by other specialists. To some extent it is related to the differences in soil cover in different countries that leads to distinguishing different archetypes as a basis for classification. Modern biology and geology originated in medieval time in Europe, and later were distributed all over the world in a 'semi-mature' state. Soil science was distributed in a rudimentary state, and was often developed independently in different countries. Sokolov (1978) noted that the lack of a uniform classification resulted from the fact that soil science was relatively young and similar to an 'infant disease' that would be overcome in the near future. Some researchers proposed the US Soil Taxonomy as a world classification; others hoped that the Soil Map of the World legend by FAO-UNESCO (or, later, the World Reference Base for Soil Resources (WRB)) would replace national classifications. However, the development of soil classifications in the last decade of the 20th century dashed these hopes. National schools did not try to integrate, but intensified activities to update and revise their classifications. In these years new versions of classifications were proposed in New Zealand (Hewitt, 1998), China (Gong Zitong, 1994), Australia (Isbell, 1996), Russia (Shishov et al, 1997), France (AFES, 1998) and Brazil (EMBRAPA, 1999). However, what resulted was the development of improved quantitative diagnostics to support the designation of units and their classification in hierarchical systems.

The situation is very unfortunate. One of the strongest arguments against the acceptance of uniform soil taxonomy was that the change of classification would make obsolete all the existing soil maps made with older national systems. However, an introduction of a new national classification system, usually quite different from the older one, would lead to the same problem. Experience in Russia showed that the introduction of a new national system of soil classification caused much confusion for students, and, sometimes, a complete rejection of pedology. We believe that soil classification harmonization and, finally, acceptance of a uniform classification is a priority task in pedology.

Hierarchy in soil taxonomies

One of the most important features of any classification is its hierarchical organization, especially the definition of the categories and the soil properties used to satisfy those definitions. It helps us understand why some soil properties are used at the upper levels and others at lower levels. If instead, it is proposed that all soil properties have the same value and importance, then in an ideal case the classes grouped according to attribute A should be separated according to attribute B from the next lower level, and attribute C would separate at the next lower level, and so forth (Figure 2.1a). If these properties are equally important, then an alternative variant of taxonomy is possible: for example, using C at the highest level (Figure 2.1b). A hierarchy of importance of soil features to be used as indicators of the category definitions is based on the empirical experience of a classifier. As a result the hierarchy of soil taxonomy is subjective and varies in different scientific schools. For example: in Russian soil classification (Shishov et al, 2004) alluvial soils are distinguished on the highest level (the level of 'trunks'); in the WRB (IUSS Working Group WRB, 2006) on the level of reference groups (the same as for other soils); and in the US Soil Taxonomy (Soil Survey Staff, 1999) only on the third, great group level. Fortunately, the hierarchy of classes is not completely arbitrary in soil science: the importance of soil properties for agriculture and environment is also considered. However, the opinions on the value of particular soil properties for practical interpretations may also vary among the experts.

The situation can be even more complex because it may be impossible to use uniform diagnostic properties for all classes even at the same categorical level of taxonomy (Gennadiev and Gerasimova, 1980). If taxonomy has categories (levels) defined as generalized abstract concepts then diagnostic soil features may satisfy the definitions at different levels. Temperature and moisture regimes in the US Soil Taxonomy are examples of definitive characteristics used at different levels depending on the other properties present in the soil. Properties important for some taxa may be of less importance for others (or be absent in other taxa). An example is soil grouped according to fine earth texture; for organic soils this is not meaningful, but a surrogate of fibre composition may be used. The list of modifiers used for soil denomination within the Histosol reference group of the WRB (IUSS Working Group WRB, 2006) is very different from the modifiers for mineral soils. Also, particular lists of modifiers are found in the reference groups of Anthrosols, Crysols, Solonchaks and some others. The same criterion may be applied at different

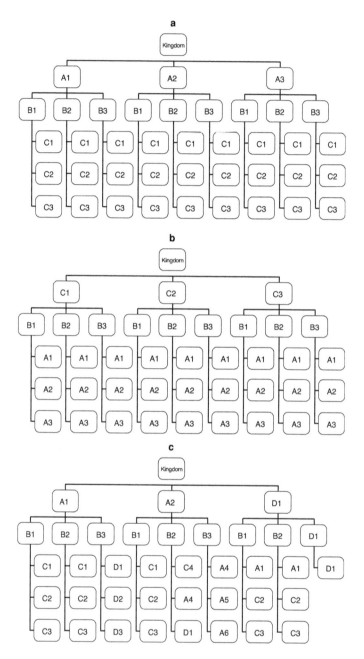

Figure 2.1 *The hierarchy of the criteria used in taxonomies: a – an ideal arrangement of hierarchy, considering the property A as the most important one; b – alternative arrangement of taxa listed in (a); c – real taxa arrangement, where the same criteria might be used on different hierarchical levels (in the case of soil classification D, for example, might mean the presence of organic material layer)*

levels of soil taxonomies, depending on its expression or on the presence of other attributes (Figure 2.1c). For example, the presence of organic material (*histic* horizon) in the WRB is the criterion for Histosols reference group, if the depth of organic layer is greater than 50cm, and for *histic* modifiers in most groups, if the depth of organic layer is less than 50cm (IUSS Working Group WRB, 2006). The presence of a clay-enriched *argic* horizon is used as a diagnostic criterion for Albeluvisols, Alisols, Acrisols, Lixisols and Luvisols reference groups, and at the modifier level in reference groups that appear before Albeluvisols in the key (IUSS Working Group WRB, 2006). In fact, the keys, which are declared to be instruments for soil diagnostics, also determine the hierarchy of the value of different soil attributes, and form a kind of a 'hidden structure' of soil taxonomy.

Thus, hierarchies of soil taxonomies are subjective, expert-dependent structures, which facilitate the search and recall of objects within the system rather than being a reflection of any real organization of entities into natural groups.

Soil terminology

The terminology used in soil classifications may be divided into two groups: traditional (indigenous and common folk terms) and artificial terms. The principle of including folk soil names (*podzol, chernozem, gley,* etc.), as well as stylised terms (*krasnozem, burozem*), in scientific classifications was used by Dokuchaev and his followers. The early Russian pedologists followed the example of Carl Linnaeus by recognizing existing archetypes, but unlike biology, no scientific names were proposed; instead, the folk terms were borrowed directly. Dokuchaev did not collect these terms himself, rather he used soil names from existing publications such as the *Materials on Statistics of Russia* which contained numerous folk names for soils. He understood that folk names could not be converted directly into scientific terms (Krasilnikov, 1999) but should be determined more strictly because in folk tradition different soils could be grouped under the same name, or the same soil was named differently in other localities. As a result, scientific terms that have originated from folk terminology often differ significantly in their meaning from the original concept. For example, the peasants used the name 'podzol' for any soil with a bleached horizon close to the surface. Dokuchaev made his first description of a 'podzol' in Middle Russia; nowadays these soils are called Planosols or Stagnosols. Later the name 'podzol' turned out to be confusing as it was used in different scientific schools in rather different ways. Wilde (1953) showed that extrapolation of local traditional terms, and creation of quasi-traditional soil names, can often be misleading: for example, the term 'brown-podzolic soil' means literally 'a brown soil with a colour of ash'. Wilde proposed using local vernacular names (*regoor, badobe, smonitza*) for newly discovered soil taxa instead of compiling old 'folk' names derived from distant localities having different environmental conditions. Traditional terminology has the advantage that most soil information users can easily understand these terms: the names are meaningful, and reflect some important characteristics of soils. However, some of these names may be misleading: for example, initially Russian folk soil name 'belozem' (literally meaning 'white earth') was used for soils with a white (*albic*) superficial horizon, but later Gerasimov (1979) used the same name for whitish gypsiferous soils of Armenia. In that case the evident indication of a certain attribute (colour) caused the replacement of the initial meaning by a new object, which had nothing in common with the 'old' except a whitish colour. In general, various old soil names of Russian and German schools, derived from folk soil terminology, were distributed over the world and finally lost their initial meaning, causing mistakes and confusion in correlation. Vernacular names are most informative when used in the language of their origin, but the information is often lost when the term is used outside its language media. For example, the name 'kuroboku' (literally, 'inky black') used in Japan for dark-coloured volcanic soils is meaningless for those who do not know Japanese.

The other option for constructing scientific soil terminology was to apply completely new artificial names. It was first proposed by Guy Smith (Banfield, 1984) while preparing the 7th Approximation, a new American soil classification system. Guy Smith considered that old traditional soil names were confusing, and with the help of philologists developed a completely new system of soil terminology; a wide group of philologists participated in the development of soil nomenclature that was mnemonic (Heller, 1963). In addition, the levels in the taxonomy are recognizable by the number of syllables of the base words and the 'ic' ending of modifiers. The idea was brilliant and could work very well if the system remained the only artificial nomenclature. Unfortunately soon a number of 'clones' of the US classification terminology appeared, and now some of the artificial terms cause almost the same confusion as traditional ones. For example, the name Histosols is used both in the US Soil Taxonomy and in the classifications of Cuba, China and in World Reference Base; the problem is that the definitions and diagnostic criteria for this group vary in different classifications. Attempts to avoid confusion by modifying slightly the names, like in Australian classification (Isbell, 2002) (*Vertosols* instead of *Vertisols*, *Podosols* instead of *Spodosols*), only increased the chaos.

There are some technical problems when introducing an artificial system of terminology instead of a traditional one. For a simple user of soil information it is difficult to perceive the rejection of traditional soil names. In the US the situation favoured this change because both farmers and soil surveyors still deal with the system of soil series, which relate to recognizable landscapes and parent materials. But in the countries where the system of soil series is not in use, the users have to use the same scientific classification terms. Any change of terms leads to changes in large-scale maps and all attendant documentation.

Some classifications use a mixture of traditional and artificial terminology; the WRB (IUSS Working Group WRB, 2006) is such an example, due mainly to its compiled origin. Some other classifications partially use traditional terminology but may apply artificial terms if a new group needs to be distinguished. These mixed systems share the advantages, but also the disadvantages, of both traditional and artificial terminology.

It is difficult to say which kind of terminology is better. Artificial terminology causes less confusion, and gives an impression of a 'more scientific' approach; consequently most classifications now mainly use artificial soil names.

Comparing soil classifications

In this book we provide short summaries of soil classifications. To present the structure of various soil classifications we use a system proposed by Berlin et al (1973) and adapted for pedology by Williams and Ortiz-Solorio (1981). Table 2.1 suggests a schematic structure of an ideal taxonomic classification.

Table 2.1 *The structural characteristics of soil classifications*

Level	Taxon name	Taxon characteristics	Borders between classes	Diagnostics	Terminology
0	Soils	Kingdom			
1		Collective			
2		Generic			
3		Specific			
4		Varietal			

Zero level means the kingdom – soils in this case. It assumes that a definition of soil has been made, thereby excluding all things not soil. This level is usually assumed and therefore not discussed; however, some confusion arises if the definition of soils, or profiles, and their limits are not provided.

The collective level means groups of archetypes of soils gathered together according to some specified properties. It has been common to define this abstract level as soils having properties thought to be associated with major pathways of soil development. In some cases it relates to kind of material, such as organic or volcanic; in others it related to kinds of pedogenic horizons. The collective level commonly serves the task of more rapid search of objects in classification and may form the basis for developing keys for identification. In some cases this grouping also indicates the reason for the selection of certain diagnostics and provides evidence of the importance assigned by the designers of the classification. For example, in Russian soil classification (1997) three trunks of sinlithogenic, postlithogenic and organogenic soils are made just to show the importance of the ratio of the processes of sedimentation and soil formation. In some classifications there are no collective levels, and in some others there are two or more.

The generic level includes the archetypes of major groups of soils. Sometimes it is difficult to understand which of the levels is generic. A possible criterion is that on this level one can imagine the profile, at least in general. Most classifications have only one generic level. Two generic levels are present in the classifications using the concept of soil series; in the process of formation of these classifications two levels of archetypes were formed simultaneously, one for landscapes, one for soil profiles.

The specific level includes soils differing from the 'central image' by the presence of additional horizons or properties. For the Russian classification the levels of subtype and genus are specific; for the US Soil Taxonomy it is the level of subgroups. The varietal level shows quantitative variations in some properties. For the Russian classification it is the level of soil class; for the US Soil Taxonomy it is the subgroup level.

The varietal level shows quantitative variations in some properties. For the Russian classification it is the soil class, for the US it is the family and series levels.

We mention the criteria used to divide classes (mainly morphology, chemical properties, climatic/soil regimes characteristics, or a combination of various methods), the borders of classes (strict or fuzzy) and terminology used at every level. The combination of structural, diagnostic and terminology features composes a kind of 'passport' of soil classification, and permits comparison of different taxonomic systems.

The correlation of soil classifications

This book, apart from the characteristics of national soil classifications, has another objective: it is a tool to assist the correlation of national classifications with the World Reference Base of Soil Resources (IUSS Working Group WRB, 2006).

The WRB was selected for correlating national classifications because it was initially constructed as a system for international correlation and is one of the most widely used classification systems for medium- and small-scale soil maps especially since the European Union accepted WRB as a basic system for soil mapping. The WRB is now used throughout the world, theoretically including all the world soils, and the system is relatively simple, which is an advantage for correlation purposes.

Sometimes correlation is understood as a translation of the terms of different classifications, a kind of a dictionary, where every object in one classification has a direct analogy in the other. However, our previous experience with correlating soil classifications (Shoba, 2002) showed that direct analogies cannot be found. The main problems of soil correlations were summarized by Krasilnikov (2002); later Schad (2008) showed that not only different classifications, but even different versions of the same classification do not permit simple translation of soil names.

Direct soil terminology translation is often not possible, due to differences in diagnostics and in archetypes. Quantitative diagnostics relate to methods of measurement as well as the soil features being considered. Depth, thickness, chemical limits, colours and so forth result in overlaps between taxa of one system relative to another. The difference in diagnostics may result in the following: a soil taxon of a national classification is broader than a WRB group; a soil taxon of a national classification is narrower than a WRB group; or soil taxa of a national classification and the WRB classes only partially intersect. The difference in the limits of classes may be due to: (a) one classification uses fuzzy qualitative criteria, and the other uses quantitative criteria; (b) different qualitative criteria are used in two classifications; and (c) both classifications use quantitative criteria, but these criteria are different.

Different approaches to diagnostics of Podzols in WRB and Russian classifications may illustrate the first case. Russian 'podzol' type seems to be broader, because only fuzzy morphological criteria (bleached horizon above iron and/or humus-enriched one) are used, whereas the WRB requires certain depth and chemical requirements for a *spodic* horizon. However, some WRB Podzols are not within the Russian type 'podzol', because in Russian classification the soils with Fe, Al and humus illuviation, lacking a bleached superficial horizon, are included into other groups (Shishov et al, 2004). Thus, we have a partial intersection of the WRB and Russian 'podzol' classes. The other case is that different criteria are used for soil diagnostics. For example, Solonetz soils both in the WRB and older Russian classification (Egorov et al, 1977) were distinguished according to sodium content in the sodic (solonetzic) horizon. However, the new Russian classification (Shishov et al, 2004) uses the kinetics of soil swelling as the main criteria, because in southern Russia there are solonetz soils that do not meet the requirements for sodium content; their physical properties are the same, but sodium has been washed out. In that case some of these solonetz correlate with the Solonetz group in the WRB, but some should be correlated with Hyposodic Luvisols. The most striking example is the use of factors of soil formation as diagnostic criteria in some national classifications. Older Russian classification used such terms as 'meadow', 'forest' or 'desert' soil; these terms are difficult to correlate with the WRB, which is based on profile properties only. The same is true for the US Soil Taxonomy, which uses temperature and water regime criteria on the highest level of hierarchy. It is impossible to reflect the difference between, for example, Usterts and Xererts, correlating them with the WRB. The difference in quantitative criteria between classifications can be illustrated by an example of organic soils. Histosols in the WRB, as well as peat soils in the new Russian classification, should have an organic horizon thicker than 50cm, whereas in old Russian classification this limit was 30cm, and in Soil Taxonomy 40cm (60cm for fibric material). Thus, the WRB Histosols group appears to be in some cases broader and in some cases narrower than Histosols of the US Soil Taxonomy. The problem is aggravated by the fact that some terms used in these classifications are similar, thus increasing misunderstanding.

The most complicated situation is found if soil archetypes are different in different classifications; the situation occurs due to both historical reasons and the subjective opinion of classifiers. Sometimes old archetypes are broken apart in some classifications, and the borders between fragments differ in national classifications. For example, 'sod-podzolic soils' of Russian classification correspond partially to Albeluvisols in the WRB system, partially to Planosols, and partially to Albic Luvisols or Alisols. It is not a problem of diagnostics: the classes in these classifications have different 'central concepts', and different definitions. Also anthropogenic soils are difficult to correlate because human-transformed soils have not always been included in some classifications: the profiles of Technosols and Anthrosols (IUSS Working Group WRB, 2006), for example, are scattered among Mollisols, Inceptisols or Entisols in the US Soil Taxonomy (Soil Survey Staff, 1999). In some systems urban soils, badlands and mine spoils are recognized as miscellaneous land types rather than as soil bodies.

Difficulties exist with correlation of soils at different levels of a hierarchy, that is, if a soil group is recognized as an archetype in one classification (and appears on the generic level), and in the other it is not recognized as an archetype (and, thus, appears on the specific or varietal level). In the latter case the historically established archetypes are often divided between two or more archetypes. For example, *terra rossa* soils in the WRB (IUSS Working Group WRB, 2006) are included as parts of Luvisols, Nitisols or Cambisols reference groups. Rendzinas are classified as Rendzic Leptosols and Rendzic Phaeozems, and Braunerde are particular cases of Cambisols or Umbrisols.

Thus, correlation of soil names in different classifications cannot be regarded as a simple 'translation', and cannot be used, for example, for conversion of soil maps into another system of soil classification. In this case the primary field data should be used, and the profiles should be reclassified in the other classification.

Some may ask why soil correlation should be done at all? We consider that soil correlation has the main purpose of providing a general idea about the soils in unfamiliar classifications. For example, if we indicate that the central concept of *Lou* soils in China corresponds to Terric Anthrosols in the WRB (instead of giving the whole definition used in the Chinese classification) it saves time and effort, and gives a reasonable idea of these soils although the diagnostic criteria are slightly different in the two classifications. We are trying to follow this main aim of soil correlation; consequently only one or two corresponding WRB names are suggested for taxa of national classifications, even though some intersect with other WRB classes. For example, the order Dermosols in Australian soil classification (Isbell, 2002) includes very different soils, which may be correlated with seven WRB reference groups (Cambisols, Chernozems, Kastanozems, Phaeozems, Nitisols, Umbrisols and Durisols). However, we try to avoid such a broad correlation, which can mean that 'everything correlates with everything', and provide the major equivalents in the WRB.

References

AFES (1998) *A Sound Reference Base for Soils (The 'Referentiel pedologique': text in English)*, INRA, Paris, 322pp

Arnold, R. W. (1983) 'Concepts of soils and pedology', in L. P. Wilding, N. E. Smeck and G.F. Hall (eds) *Pedogenesis and Soil Taxonomy*, vol I, *Concepts and Interactions*, Elsevier Scientific Publishers, Amsterdam, pp1–21

Banfield, J. F. (ed) (1984) 'Guy D. Smith discusses soil taxonomy', SSSA, Madison, WI, 42pp

Berlin, B., Breedlove, D. E. and Raven, P. H. (1973) 'General principles of classification and nomenclature in folk botany', *American Anthropologist*, vol 75, no 2, pp214–242

Buol, S. W., McCracken, R. J. and Hole, F. D. (1977) *Soil Genesis and Classification*, 7th edition, Iowa State University Press, IA, 360pp

Dmitriev, E. A. (1991) 'What classifies soil classification?', *Pochvovedenie*, no 2, pp122–133 (in Russian)

Egorov, V. V., Fridland, V. M., Ivanova, E. N., Rozov, N. N., Nosin, V. A. and Fraev, T. A. (1977) *Classification and Diagnostics of Soils of USSR*, Kolos Press, Moscow, 221pp (in Russian)

EMBRAPA (1999) *Sistema Brasileiro de Clasificação de solos*, Embrapa Produção de Informação, Brasília – Embrapa Solos, Rio de Janeiro, 412pp

Fridland, V. M. (1976) *Patterns of Soil Cover*, translation from Russian 1972, D. H. Yaalon (ed), IPST, Jerusalem and Wiley, Chichester, 291pp

Gennadiev, A. N. and Gerasimova, M. I. (1980) 'On some trends in the actual soil classification in the USA', *Pochvovedenie*, no 9, pp3–12 (in Russian)

Gerasimov, I. P. (1979) 'Belozems', in I. P. Gerasimov (ed) *Genetic Soil Types of the Subtropics of Trans-Caucasian Region*, Nauka Press, Moscow, pp251–255 (in Russian)

Ghiselin, M. (1974) 'A radical solution to the species problem', *Systematic Zoology*, vol 23, no 4, pp536–544

Gong Zitong (ed) (1994) *Chinese Soil Taxonomic Classification (First proposal)*, Institute of Soil Science, Academia Sinica, Nanjing, 93pp

Gorochov, A. V. (2001) 'On some theoretical aspects of taxonomy (remarks by the practical taxonomist)', *Acta Geologica Leopoldensia*, vol 24, nos 52–53, pp57–71

Haskett, J. D. (1995) 'The philosophical basis of soil classification and its evolution', *Soil Science Society of America Journal*, vol 59, no 1, pp179–184

Heller, J. L. (1963) 'The nomenclature of soils, or what's in a name?', *Soil Science Society of America Proceedings*, vol 27, no 2, pp216–220

Heuvelink, G. B. M. and Webster, R. (2001) 'Modelling soil variation: Past, present and future', *Geoderma*, vol 100, nos 3–4, pp269–301

Hewitt, A. E. (1998) *New Zealand Soil Classification*, 2nd edition, Maanaki Whenua–Landcare New Zealand Ltd, Dunedin, Landcare Research Science Series No 1, Lincoln, Canterbury, New Zealand, 122pp

Ibáñez, J. J. and Boixadera, J. (2002) 'The search for a new paradigm in pedology: A driving force for new approaches to soil classification', in E. Micheli, F. Nachtergaele, R. J. A. Jones and L. Montanarella (eds) *Soil Classification 2001*, EU JRC, Hungarian Soil Sci. Soc., UN Food and Agriculture Organization, Italy, pp93–110

Ibáñez, J. J., Ruiz-Ramos, M., Zinck, J. A. and Brú, A. (2005) 'Classical pedology questioned and defended', *Eurasian Soil Science*, vol 38, suppl no 1, pp75–80

Isbell, R.F. (1996) *Australian Soil Classification*, CSIRO Land & Water, Canberra, 160pp

Isbell, R. F. (2002) *Australian Soil Classification*, revised edition, CSIRO Land & Water, Canberra, 144pp

IUSS Working Group WRB (2006) *World Reference Base for Soil Resources*, 2nd edition, World Soil Resources Reports no 103, UN Food and Agriculture Organization, Rome, 128pp

Kay, P. (1975) 'A model-theoretic approach to folk taxonomy', *Social Science Information*, vol 14, no 2, pp151–166

Krasilnikov, P. V. (1999) 'Early studies on folk soil terminology', *Eurasian Soil Science*, vol 32, no 10, pp1147–1150

Krasilnikov, P. V. (2002) 'An experience in correlating World Reference Base for Soil Resources with national soil classifications', in *Transactions of 17th World Congress of Soil Science*, 14–21 August 2002, Bangkok, Thailand, CD-ROM, pp2031-1–2031-10

Levy-Leblond, J. M. (1996) *L'exercice de la pensée et la pratique de la science*, Éditions Gallimard, Paris, 353pp

Militarev, V. Yu. (1993) 'Principles of the theory of classifications in natural sciences', in *The Theory and Methods of Biological Classifications*, Nauka Press, Moscow, pp101–115 (in Russian)

Mosterín, J. (1984) *Conceptos y Teorías en la Ciencia*, Alianza Universitaria, Madrid, 220pp

Sattler, R. (1986) *Biophilosophy. Analytic and Holistic Perspectives*, Springer-Verlag, Berlin, 344pp

Schad, P. (2008) 'New wine in old wineskins: Why soil maps cannot simply be "translated" from WRB 1998 into WRB 2006', in W. H. Blum, M. H. Gerzabek and M. Vodrazka (eds) *EUROSOIL 2008, Book of Abstracts*, BOKU, Vienna, Austria, p120

Schaetzl, R. J. (1998) 'Lithologic discontinuities in some soils on drumlins: Theory, detection and application', *Soil Science*, vol 163, no 5, pp570–590

Shishov, L. L., Tonkonogov, V. D. and Lebedeva, I. I. (1997) *Classification of Soils of Russia*, Dokuchaev Soil Science Institute, Moscow, 236pp

Shishov, L. L., Tonkonogov, V. D., Lebedeva, I. I. and Gerasimova, M. I. (2004) *Classification and Diagnostics of Soils of Russia*, Oykumena, Smolensk, 342pp (in Russian)

Shoba, S. A. (ed) (2002) *Soil Terminology and Correlation*, 2nd edition, Centre of the Russian Academy of Sciences, Petrozavodsk, 320pp

Simonson, R.W. (1989) *Historical Highlights of Soil Survey and Soil Classification with Emphasis on the United States, 1899–1970*, International Soil Reference and Information Centre, Wageningen, Technical Paper no 18, 83pp

Soil Survey Staff (1999) *Soil Taxonomy: A Basic System of Soil Classification for Making and Interpreting Soil Surveys*, 2nd edition, USDA-NRCS, Lincoln, NB, United States Government Printing Office, Washington DC, 696pp

Sokolov, I. A. (1978) 'On basic soil classification', *Pochvovedenie*, no 8, pp103–107 (in Russian)

Wilde, S. A. (1953) 'Soil science and semantics', *Journal of Soil Science*, vol 4, no 4, pp1–4

Williams, B. J. and Ortiz-Solorio, C. A. (1981) 'Middle American folk soil taxonomy', *Annals of Association of American Geographers*, vol 71, no 3, pp335–358

Yaalon, D. H. (2003) 'Are the soils spatially a continuum?', *Pedometron*, vol 14, pp3–4

3
The Structures of Soil Taxonomies

Some basic principles

In the previous chapter we discussed the structure of classifications in terms of the levels of taxonomy and the criteria used for their diagnostics. An alternative approach to a basic definitional structure of a classification involves mathematical modelling using topology, fractal theory and some other instruments. Before analysing such structures of hierarchical classifications some basic principles of graph theory will be explained.

In topological terms hierarchies imply tree structures. A tree structure is a way of representing a hierarchy structure in a graphical form that has been called a 'tree structure' because the graph looks a bit like an inverted above-ground tree. In graph theory, a tree is a connected acyclic structure (or sometimes, a connected directed structure, in which every vertex has a degree of 0 or 1). An acyclic graph, which is not necessarily connected, is sometimes called a forest because it resembles a group of trees. Every finite tree structure has a member that has no superior (e.g. the soil universe or pedosphere). This member is called the root node. The lines connecting elements are called 'branches', the elements themselves are called 'nodes'. Nodes without other branching nodes are called 'end-nodes' (the lowest hierarchical level of a given hierarchy, such as species).

Background

Willis and Yule (1922) carried out one of the early mathematical analyses of the structure of biological classifications. These authors showed that the branching of biological classification from the top to the bottom of their hierarchical trees conforms to a hollow (or Willis) curve, with a very large number of taxa having one or very few subtaxa, and a small number of taxa with many subtaxa. It is noticeable that field biodiversity inventories also conform to this curve type (Walters, 1986; Dial and Marluff, 1989; Burlando, 1990, 1993). Further mathematical analysis shows that Willis curves often fit to power, geometric or lognormal distributions, and also broken stick and Weibull models (Ibáñez and Ruiz-Ramos, 2006; Ibáñez et al, 2006). Ibáñez et al (1990, 1995, 1998, 2005a, b) show that soil inventories conform both to the Willis curve and the above-mentioned linear regression models. Furthermore, Ibáñez and Ruiz-Ramos (2006) show that the US Soil Taxonomy also fits the same mentioned mathematical patterns.

In the next subsections we use material published mainly by Ibáñez and Ruiz-Ramos (2006) and Ibáñez et al (2006, 2009) where the mathematical structures of pedological and biological classifications have been studied and compared.

Taxa-size distributions

Clayton (1972) showed that 'taxa-size distribution' in a given biological taxonomy (number of subtaxa in the taxa) fits well to a power law formalized as:

$$N = kS^{-D} \tag{1}$$

where N is the frequency of genera, S is the number of species per genus (genus size), k is the group-inherent constant and D is the exponent (fractal dimension), corresponding to the slope of the regression line in a log–log plot. However, taxa-size frequencies can also fit other distributions with similar fits and confidence degrees (Ibáñez and Ruiz-Ramos, 2006), thus, in practice it is difficult to detect the best fit of a given taxa-size distribution. In general, power laws fit well to most data sets in both pedological and biological taxonomies. The same is true with lognormal and sometimes Weibull distributions. In any case simultaneous fits to power laws and lognormal distributions are in agreement with previous studies of biological classifications (Burlando, 1990, 1993; Minelli et al, 1991; Minelli, 1993).

Thus biological and pedological classifications seem to have the same structure from the point of view of statistical distribution models. Burlando (1990, 1993), Minelli et al (1991) and Minelli (1993) believe that the fit to a power law is a signature of underlying fractal structures of biological classifications.

To analyse and compare biological and pedological taxonomies, Ibáñez and Ruiz-Ramos (2006) selected the Tylenchina suborder (Siddiqi, 2000), a multi-species taxa of plant-parasitic soil nematodes, and the US Soil Taxonomy (Soil Survey Staff, 1999). For each taxonomic construct, the size of each taxon (logarithm of the number of individuals) was plotted as a function of taxonomic category. Regression lines were also drawn for the whole taxonomic set, as well as for each nematode superfamily and soil order. Some of the higher categories of both classifications have few taxa, and statistical comparisons were not useful and are not reported.

Figures 3.1 and 3.2 present the branching systems of Tylenchina suborder and the US Soil Taxonomy from top to bottom. It is clear that similar results for pedological and biological classification are obtained (Minelli et al, 1991).

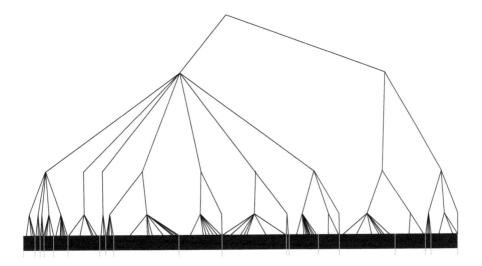

Figure 3.1 *The branching system of Tylenchina suborder classification (reproduced with permission of Springer Publishing Company)*

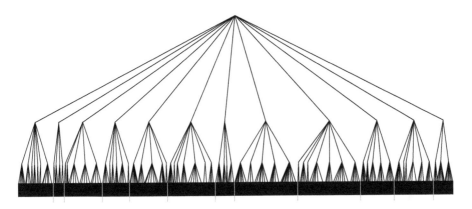

Figure 3.2 *The branching system of the US Soil Taxonomy (reproduced with permission of Springer Publishing Company)*

In addition to whole taxonomic sets, several distributions of subtaxa in taxa at higher levels were considered to test if the same statistical pattern could be detected at different levels of the taxonomies. The US Soil Taxonomy has a slightly more regular structure than the biological one for the Tylenchina suborder of nematodes; nevertheless the results demonstrate similar branching systems based on taxa/per taxon relationships in all the cases (Ibáñez and Ruiz-Ramos, 2006). It is noticeable that the former began with 12 soil orders and the latter with only two superfamilies. There was not a scientifically sound fit to any distribution model of the higher categories in the biological system because of the small sample size.

In Figures 3.3 and 3.4 are Willis curve plots of the two classifications (Ibáñez and Ruiz-Ramos, 2006). It is clear that, in view of their respective idiosyncrasies or the different 'social' taxonomic practices in both disciplines, the US Soil Taxonomy has a more compact shape and its taxa sizes have a smaller range of values. The opposite is true for the Tylenchina suborder where a larger number of very small taxa were found (number of subtaxa per taxon

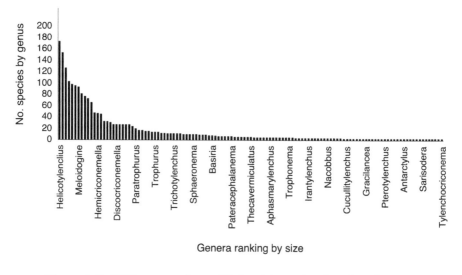

Figure 3.3 *Willis curve plots of Tylenchina suborder classification (reproduced with permission of Springer Publishing Company)*

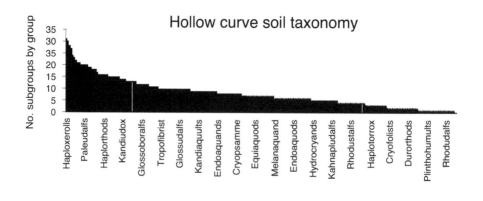

Figure 3.4 *Willis curve plots of the US Soil Taxonomy (reproduced with permission of Springer Publishing Company)*

between 1 and 2). In this respect it is important to keep in mind that, while soil classifications are a result of the consensus among experts (closed systems), the biological systems are open and grow continuously over time with the inputs of the whole scientific community involved in the detection of new taxa (Walters, 1986; O'Donnell et al, 1995).

Ibáñez and Ruiz-Ramos (2006) also showed linear functions for the size of each taxon versus the taxonomic rank of each category (hierarchical level). Regressions done for whole taxonomic sets (Figures 3.5 and 3.6), and for each nematode superfamily and soil order, followed similar patterns at these different classification levels and had R^2 values higher than 0.89. The US Soil Taxonomy regression coefficients were slightly higher that those for the Tylenchina suborder. This might be a consequence of the consensual origin of the US Soil Taxonomy versus the open nature of biological classifications that are built up by individual ('experts') additions over time of new taxa and some rearrangement of the old ones.

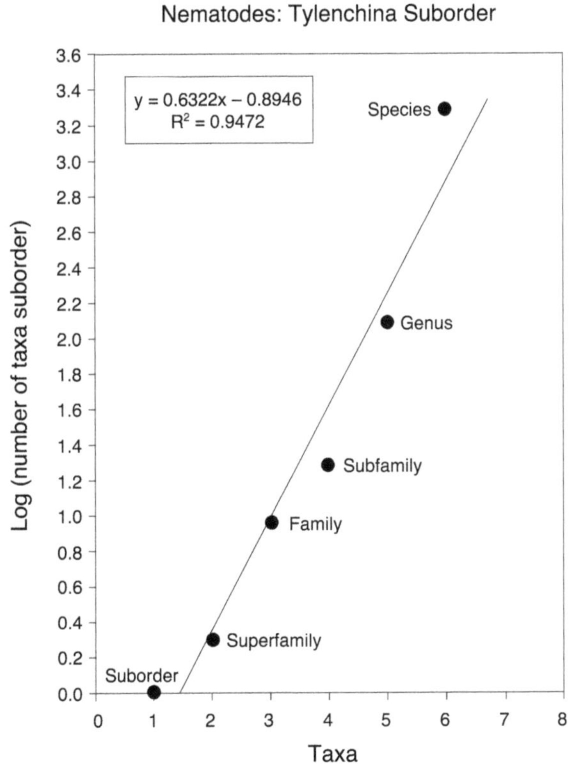

Figure 3.5 *Regressions for the size of each taxon (logarithm of the number of individuals) versus the taxonomic rank of each category of Tylenchina suborder classification (reproduced with permission of Springer Publishing Company)*

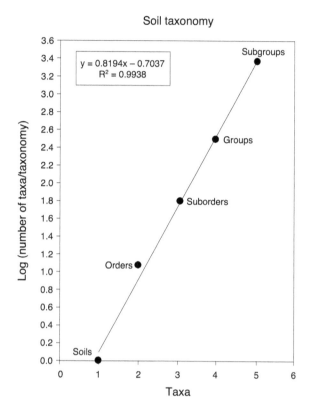

Figure 3.6 *Regressions for the size of each taxon (logarithm of the number of individuals) versus the taxonomic rank of each category of the US Soil Taxonomy (reproduced with permission of Springer Publishing Company)*

Bifurcation ratios

Bifurcation ratios (proposed by Strahler (1957) for studying the geometry of drainage basins) of both taxonomic systems were calculated to compare the branching systems tested. These ratios measure the number of branches (subtaxa) or 'growth' from a given taxa (node), for the taxa of all hierarchical categories.

The bifurcation ratios for both classification systems (Ibáñez and Ruiz-Ramos, 2006) have a wide range of variation, as is common in natural systems (e.g. drainage basins) and in different biological classifications (Burlando, 1990, 1993; Minelli et al, 1991; Minelli, 1993, pp195–197). Results showed that the biological taxonomy tested has a higher branching system (is much more profusely branched with more subtaxa per taxa) than the pedologic one at the end-nodes. The branching of Soil Taxonomy for soils of the US is illustrated in Figure 3.4.

Entropy analysis, Mayr criteria and Maximum Entropy Principle

Mayr (1969, p236) encouraged practising biological taxonomists to create equal sized taxonomic units. According to this author, 'excessively large' taxa, as well as an excessive number of small taxa, reduce the usefulness of these constructs as 'information retrieval systems'. Currently, biological classifications produce high variability of taxon sizes, and according to the Mayr criteria are not compatible with an optimal efficiency in transmission of information for human minds that prefer equal size groups. However, current classification schemes seem, 'at first glance' to refute his suggestion because Willis curves seem to be ubiquitous. Ibáñez and Ruiz-Ramos (2006) demonstrated that the distribution of taxa in the branching system of biological and pedological taxonomies is much more equitable than Mayr (1969) thought, as explained below.

In a more formal way, the Principle of Maximum Entropy (MaxEnt Principle) states that: 'the least biased and most likely probability assignment is that which maximizes the total entropy subject to the constraints imposed on the system' (Jaynes, 1957 cit. in Pastor-Satorras and Wagensberg, 1998). The MaxEnt Principle allows a novel statistical characterization of the taxonomic constructs irrespective of whether or not they are fractal objects (Ibáñez et al, 2006). This is illustrated with the evenness index: the total entropy H' divided by maximum entropy H_{max} of a given dataset that can be estimated, as is done in the biodiversity literature (see Ibáñez et al, 1995 and 1998 and references therein). This index may be used as a measure of the Mayr criteria and the MaxEnt Principle. Its values range between 0 for the less equitable possible distributions and 1 for the most equitable distributions where all taxa have the same number of subtaxa. H_{max} is the entropy value for the cases where all classes have the same number of objects. This value may be applied as an indicator of the number of subtaxa included in a given taxa and their respective sizes. H' is the calculated entropy value. The H_{max} value, calculated for biological and soil classification mentioned above, shows that the Soil Taxonomy has a more regular branching system than Tylenchina suborder. H' has higher values in the tested soil classification than in the biological one, suggesting that taxa sizes are more regularly distributed in US Soil Taxonomy; however, the most noticeable result is that 'evenness' in both taxonomies was very high (close to one), indicating maximum efficiency (Ibáñez and Ruiz-Ramos, 2006). Thus, in terms of entropy, biological and soil taxonomies appear to be 'nice information systems'.

The fractal and/or multifractal structure of hierarchical taxonomies

Fractal trees imply hierarchies and are thought to be the best way to maximize the economy of the flow of matter and energy (Pastor-Satorras and Wagensberg, 1998) according to the MaxEnt Principle (Jaynes, 1957). Ibáñez et al (2006,

2009) hypothesized that the human mind might follow the same MaxEnt Principle to gain efficiencies of information flow (classifications in this case).

Burlando (1990), Burlando et al (1993), Minelli et al (1991) and Minelli (1993) examined biological hierarchical taxonomies for fractal structures by fitting taxonomies of several target biotaxa to lognormal and power laws. Their published results showed that biological taxonomies conform to power laws, and also to lognormal distributions. Thus, these authors concluded that biological classifications and phylogeny have fractal structures. However, since biological taxonomies also fit lognormal distributions (which do not conform to fractal structures) the hypothesis is weak.

Guo et al (2003) state that the US Soil Taxonomy (Soil Survey Staff, 1999) structure conforms to a lognormal distribution. However, Ibáñez and Ruiz-Ramos (2006), using the same tools and criteria as Burlando and Minelli, show that the structure of the US Soil Taxonomy fits better to a power law than to a lognormal distribution. In summary these analyses seem to demonstrate that both classifications have similar structures from different mathematical perspectives.

A multifractal analysis of the data used by Ibáñez and Ruiz-Ramos (2006) indicates that both classifications fit well to multifractal distributions, when the biggest taxa are not considered. The underlying reasons are thought to be agronomic, geographic and cognitive biases. Thus, for example, it is noticeable that in biotaxonomies the biggest taxa sizes correspond to the species of more economic interest: plant-parasitic nematodes that, in general, produce the pests most harmful to crop production, or the most cosmopolitan ones. In a similar way the biggest taxa sizes in the US Soil Taxonomy are some 'haplo' great groups indicative of a cognitive bias. This classification also has an underlying agronomic bias (Ibáñez et al, 2009). Mollisols, the soil order of great interest from an agronomic point of view, contain many large-size taxa. It seems that the human mind tends to devise hierarchical taxonomies as full fractal trees but different biases divert them to quasi-multifractal structures.

The fractal mind of pedologists

There has not been much work in science dealing with the organizational, political and scientific layering of database structures or of classifications and surveys of natural resources. Currently, information infrastructures are of paramount importance in the new wave of how science should work to solve societal demands. And yet there is still disagreement among scientists whether taxonomies are invented (human-made constructs) or are discovered ('natural' structures) independent of the discipline involved. It is thought that it would be helpful to study the nature of taxonomies from different points of view to find how classifications and data collection (surveys) are linked. It is generally accepted that much institutional work on soil classification systems has been nationally biased, especially in terms of practical land management demands. Ibáñez et al (2009) tried to demonstrate empirically that the multifractal nature of the US Soil Taxonomy is strongly linked with conventional soil survey practices using the same mathematical tools as above. The working hypothesis originates from a seminal paper of Beckett and Bie (1978) who showed that many soil maps have attributes that conform to power law distributions including: (i) map scale–area surveyed; (ii) standard line density–scale dependency; (iii) sampling density–mapped area; (iv) hierarchic taxonomic level used according to the scale map; (v) minimum polygon size fits the functions to the map scale; (vi) soil survey effort depending on the map scale; and (vii) boundary density–scale map relationships.

Experts in cartography know the rules that they apply; however, in the bibliographic material we find few references to implications of human cognitive bias in this practice, or to mathematical relationships between classifications and map scale. Although not many documents explicitly mention this latter topic (Dent and Young, 1981; Forbes et al, 1982; Dent, 1985; Avery, 1987; Soil Survey Division Staff, 1993), in general, these publications are in agreement with the plots shown by Beckett and Bie (1978).

Quadratic functions have been recognized for a long time in cartography; however, this is not so evident for power laws. These statistical distribution models may be formalized as follows:

Power Law: $y = ax^b$ (1)
Quadratic Fit: $y = a+bx+cx^2$ (2)
Linear Fit: $y = a+bx$ (3)

It is clear that depending on the scale and purpose of a map, the various classes and subclasses to be represented will be expressed in various degrees of detail. In thematic maps, such as the soil ones, the principal subject is represented in detail by including a large range of subclasses. Thus, the definition of the classes of features represented by the symbols is just as much part of the map accuracy as putting them in the right place on the map. Defined classes and subclasses of map features should be consistent and used systematically. Pedologists attempt to do this by applying taxonomy and other guidelines when surveying soil landscapes. Dent and Young (1981) associate map scales with different types of soil cartographic units and the US Soil Survey Manual (Soil Survey Division Staff, 1993, p48) has a key for identifying kinds of soil surveys and their common taxonomic components.

In general, all land thematic maps need a topographic base map; thus it was once common for natural resource maps (e.g. soils) to utilize existing topographic base maps and there emerged a clear relation between maps and classifications to cover the area of a given country in a systematic pattern. The conclusions obtained by Ibáñez et al (2009) show that (i) there is a clear mathematical relationship between the US Soil Taxonomy (as an example of pedological classification) and soil maps to follow a scale invariant pattern (a cascade of power laws); (ii) the taxonomic information portrayed in soil maps tends also to be scale invariant; (iii) the size–number frequency of soil map delineations follow the same pattern as in (i) and (ii); and (iv) other types of natural resources maps, such as vegetation ones, also follow the above-mentioned relationships.

In view of these results, several soil surveys in the US were examined using the different categories (order, suborder, great group, subgroup, family and series) in Soil Taxonomy to determine if the above power relationships are regular. The results were conclusive: in each soil map the pedosphere continuum was broken iteratively in less inclusive categories according to a power law. Obviously, the coefficient values varied, as different areas have different pedodiversities; however, the spatial pattern was the same in all cases.

We conclude that pedologists have fractal minds and that cognitive, geographic and utilitarian biases divert the results to multifractal products. It seems that nature works in a similar pattern. It is logical that our cognitive apparatus works similar to nature even though many patterns are difficult to study and interpret. Corroboration of hypotheses regarding scale invariant relationships indicates that our mind works in a similar way in the making of taxonomies and thematic map representations. That is, the two activities are strongly linked in the minds of soil surveyors and soil taxonomists. Both products follow some type of scale invariance structures such as fractal or multifractal ones, suggesting an innate way to process information. Corroboration of hypotheses regarding size–scale distributions indicate that our cognitive apparatus is in resonance with the abundance distribution models detected in nature in biodiversity and pedodiversity studies.

Hidden structures in soil classifications

As mentioned in Chapter 2, a hierarchical taxonomy is just one of the possible ways to arrange soil classes. Some important classification systems, such as the new French soil classification (AFES, 1998) and the World Reference Base for Soil Resources (WRB) (IUSS Working Group WRB, 2006) are stated to be reference bases with no or little hierarchy. Though these classifications will be discussed in more detail later, it is important to know if structural rules may be applied to these 'non-hierarchical' classifications. Some authors (e.g. Swanson, 1999) note that apart from an 'honest' hierarchy, some soil classifications may have a 'hidden structure', for example, included in the keys for soil diagnostics. At a first glance, a key arranges the soils or some soil attributes according to a scheme to facilitate an orderly separation of the classes at each categorical level. For instance, in the US Soil Taxonomy (Soil Survey Staff, 1999) the first soil order in the key is Gelisols (permafrost soils), followed by Histosols (organic soils), then Spodosols (acid strongly leached soils) and so forth. This sequence means that permafrost is used to make an early separation which reduces the scope of defining all other classes at this level by adding exclusion statements. The presence of a thick organic layer separates another unique group of soils, again saving space to recognize the remaining taxa. All the soil orders are on the same level of the hierarchy. The structure of the hierarchy can be observed by following the keys down to each lower category which reveals the rationale (although seldom stated) of diagnostics used at each level of the hierarchy. In biology keys are used for species determination, with no intention to range the species along with their value. It would be better say that the ranging of the 'weight' of soil attributes in classifications determines the horizontal structure of soil taxonomies, that is, the internal relations between soils at the same taxonomic level.

However, another type of 'hidden hierarchy' exists. The most illustrative example of a hidden structure may be found in the last edition of the WRB (IUSS Working Group WRB, 2006). For teaching purposes it is common to group the reference groups into ten environmental conditions; consequently this has the appearance of a higher category in a hierarchy. Apart from two evident and declared levels of a hierarchy, namely reference soil groups (RSGs) and combinations of modifiers with RSG names, it has a strict sequence of the use of modifiers. Thus, every subsequent modifier may be regarded as a subgroup of a soil, characterized by the previous modifier. In fact, the situation is more complex, because each modifier implies also the existence of soils with opposite properties: for example, the modifier *leptic* means that these soils are relatively shallow, less than 1m in depth, but it also means that deeper soils exist, though they do not receive a special modifier. It can be reflected by non-existent, 'chimerical' taxa which lack the attributes indicated by a modifier, but may have any attributes associated with the 'lower' modifiers. Moreover, several modifiers may be mutually exclusive and, thus, be at the same hierarchical level; 'chimerical modifiers' may also be present at the same level. For instance, many reference groups have *folic* (having a thick dry organic horizon) and *histic* (having a thick water-saturated organic horizon) as the first modifiers in the list. It means that soils at the first level are divided into those having either dry organic topsoil or a moist organic topsoil (peat), and those having no organic horizon. In the real reference groups the structure is also complicated with the restrictions in some 'upper' modifiers which do not permit the use of certain 'lower' modifiers because they are repetitive or contradictory to the 'upper' ones. A fragment of the reconstructed hidden hierarchy of Vertisols is presented at Figure 3.7.

The French soil classification (AFES, 1998) has three declared hierarchical levels; from top down they are great groups of references, references and soil types. The upper level is optional, and exists only for convenient grouping of soils, mainly for educational purposes. Adding the modifiers to references forms soil types. The list of modifiers and the sequence of their use is not defined; thus, the system seems to be flexible and free from 'hidden hierarchy', and looks more like a reference base that is a structured collection of information used for comparing real objects with existing archetypes.

Anyway, the structure of 'incomplete' hierarchical classifications such as AFES (1998) and WRB (2006) should also have a mathematical structure. It may be that mathematical procedures used by Ibáñez and Ruiz-Ramos (2006) and Ibáñez et al (2006) cannot be used for these latter types of classifications, yet it is possible to analyse hierarchical classifications with regard to diagnostic horizons, properties and materials. If so, then both 'full' and 'incomplete' hierarchical classifications might be rigorously compared using mathematical protocols such as Monte Carlo simulations and entropy analyses.

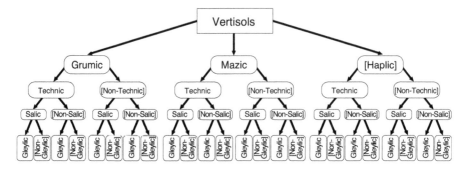

Figure 3.7 *Reconstructed 'hidden' hierarchy of the Vertisols reference group in the WRB (a fragment)*

References

AFES (1998) *A Sound Reference Base for Soils (The 'Referentiel pedologique': text in English)*, INRA, Paris, 322pp

Avery, B. W. (1987) *Soil Survey Methods: A Review*, Technical Monograph no 18, Soil Survey & Land Resource Centre, Silsoe, 86pp

Beckett, P. H. T. and Bie, S. W. (1978) *Use of Soil and Land-System Maps to Provide Soil Information in Australia*, CSIRO Aust. Div. Soils, Technical Paper no 33, 76pp

Burlando, B. (1990) 'The fractal dimension of taxonomic systems', *Journal of Theoretical Biology*, vol 146, no 2, pp99–114

Burlando, B. (1993) 'The fractal geometry of evolution', *Journal of Theoretical Biology*, vol 163, no 3, pp161–172

Clayton, W. D. (1972) 'Some aspects of the genus concept', *Kew Bulletin*, vol 27, no 4, pp281–287

Dent, B. D. (1985) *Principles of Thematic Map Design*, Addison-Wesley Longman Publishing Co, Reading, Mass., 413pp

Dent, D. and Young, A. (1981) *Soil Survey and Land Evaluation*, George Allen & Unwin, London, 278pp

Dial, K. P. and Marluff, J. M. (1989) 'Nonrandom diversification within taxonomic assemblages', *Systematic Zoology*, vol 38, no 1, pp26–37

Forbes, T. R., Rossiter, D. and Van Wambeke, A. (1982) *Guidelines for Evaluating the Adequacy of Soil Resource Inventories*, SMSS Technical Monograph no 4, Cornell University Department of Agronomy, Ithaca, NY, 51pp

Guo, Y., Amundson, R., Gong, P. and Ahrens, R. (2003) 'Taxonomic structure, distribution, and abundance of soils in the USA', *Soil Science Society of America Journal*, vol 67, no 11, pp1507–1516

Ibáñez, J. J. and Ruiz-Ramos, M. (2006) 'Biological and pedological classifications: A mathematical comparison', *Eurasian Soil Science*, vol 39, no 7, pp712-719

Ibáñez, J. J., Jiménez-Ballesta, R. and García-Álvarez, A. (1990) 'Soil landscapes and drainage basins in Mediterranean mountain areas', *Catena*, vol 17, no 4, pp573–583

Ibáñez, J. J., De-Alba, S., Bermúdez, F. F. and García-Álvarez. A. (1995) 'Pedodiversity: Concepts and measures', *Catena*, vol 24, no 2, pp215–232

Ibáñez, J. J., De-Alba, S., Lobo, A. and Zucarello, V. (1998) 'Pedodiversity and global soil patterns at coarser scales (with Discussion)', *Geoderma*, vol 83, no 2, pp171–192

Ibáñez, J. J., Ruiz-Ramos, M., Zinck, J. A. and Brú, A. (2005a) 'Classical pedology questioned and defended', *Eurasian Soil Science*, vol 38, supp 1, pp75–80

Ibáñez, J. J., Caniego, J., San-José, F. and Carrera, C. (2005b) 'Pedodiversity–area relationships for islands', *Ecological Modeling*, vol 182, no 3, pp257–269

Ibáñez, J. J., Ruiz-Ramos, M. and Tarquis, A. (2006) 'The mathematical structures of biological and pedological taxonomies', *Geoderma*, vol 134, no 3, pp360–372

Ibáñez, J. J., Arnold, R. W. and Ahrens, R. J. (2009) 'The fractal mind of pedologists (soil taxonomists and soil surveyors)', *Ecological Complexity* (in press)

IUSS Working Group WRB (2006) *World Reference Base for Soil Resources*, 2nd edition, World Soil Resources Reports no 103, UN Food and Agriculture Organization, Rome, 128pp

Jaynes, E. T. (1957) 'Information theory and statistical mechanics', *Physical Review*, vol 106, no 11, pp620–630

Mayr, E. (1969) *Principles of Systematic Zoology*, McGraw-Hill, New York, 428pp

Minelli, A., Fusco, G. and Sartori, S. (1991) 'Self-similarity in biological classifications', *BioSystematics*, vol 26, no 1, pp89–97

Minelli, A. (1993) *Biological Systematics. The State of the Art*, Chapman & Hall, London, 387pp

O'Donnell, A. G., Goodfellow, M. and Hawksworth, D. L. (1995) 'Theoretical and practical aspects of the quantification of biodiversity among microorganisms', in D.L. Hawksworth (ed.) *Biodiversity: Measurement and Estimation*. Chapman and Hall in association with the Royal Society, London, pp65-73

Pastor-Satorras, R. and Wagensberg, J. (1998) 'The maximum entropy principle and the nature of fractals', *Physica*, vol A251, no 3, pp291–302

Siddiqi, M. R. (2000) *Tylenchida, Parasites of Plants and Insects*, 2nd edition, CAB International, Wallingford, UK, 833pp

Soil Survey Division Staff (1993) *Soil Survey Manual*, United States Department of Agriculture Handbook 18, Washington DC, 437pp

Soil Survey Staff (1999) *Soil Taxonomy: A Basic System of Soil Classification for Making and Interpreting Soil Surveys*, USDA, Handbook 436, 2nd edition, United States Government Printing Office, Washington DC, 696pp

Strahler, A. N. (1957) 'Quantitative analysis of watershed geomorphology', *Transactions of American Geophysical Union*, vol 38, no 7, pp913–920

Swanson, D. K. (1999) 'Remaking Soil Taxonomy', *Soil Survey Horizons*, vol 40, no 3, pp81–88

Walters, S. M. (1986) 'The name of the rose: A review of ideas on the European bias in angiosperm classification', *New Phytology*, vol 104, no 4, pp527–546

Willis, J. C. and Yule, G. U. (1922) 'Some statistics of evolution and geographical distribution in plants and animals, and their significance', *Nature*, vol 109, no 2, pp177–179

Part 2
Soil Classifications and their Correlations

P. Krasilnikov and R. Arnold

4
World Reference Base for Soil Resources – A Tool for International Soil Correlation

Objectives and scope

The World Reference Base for Soil Resources (WRB) (IUSS Working Group WRB, 2006) is not a soil classification sensu stricto. It was created as a common project of two international groups, working on improving the legend for the FAO-UNESCO World Soil Map, and an International Reference Base. The main objective of the WRB is to serve as a tool for the evaluation of global soil resources. Another important task is to help correlate numerous national classifications. It was not intended to replace national classifications, rather to permit working out a common language for the specialists, a kind of 'soil Esperanto'. Also it would help soil scientists communicate with specialists in close disciplines, who have been dissatisfied with the absence of a uniform simple soil classification. The last, but not the least function of the WRB is to serve as a legend for international soil maps. Though the precursor of the WRB, the legend for the World Soil Map (FAO-UNESCO, 1974), was made for a map of the scale 1:5,000,000, and reflected only the most general traits of soils; the latest versions were much more detailed. It permitted its use for medium-scale soil mapping in several countries, for example in Mexico, which used the FAO-UNESCO legend for soil resources inventory and then adopted the WRB as a basis for soil survey (Cruz Gaistardo et al, 2006). A good example of successful use of the WRB for soil cartography is the recently published Soil Atlas of Europe (Jones et al, 2006).

The geographical extension of the WRB is worldwide, as it originated from the World Soil Map legend. The scope of the system is also rather broad as it includes most superficial terrestrial bodies (Nachtergaele, 2005) (Table 4.1).

Table 4.1 *The scope of the World Reference Base*

Superficial bodies	Representation in the system
Natural soils	Worldwide coverage
Urban soils	Included in the reference level (Technosols)
Transported materials	Included in the reference level (Technosols)
Bare rock	Included in the 2nd level (Nudilithic Leptosols)
Subaquatic soils	Included in the 2nd level in several reference groups
Soils strongly transformed by agricultural activities	Separated as a special reference group (Anthrosols)

Theoretical background

Though FAO-UNESCO (1974) from the very beginning planned the legend of the World Soil Map to be a seed for universal soil classification, it never declared its basic theoretical principles. The legend was mainly a compilation of historically recognized soil archetypes. To a great extent the decision on including or not including was a matter of consensus between developed national schools of pedology. Though there was an agreement on most of the groups, the wish of some schools to maintain their 'favourite' archetypes resulted in certain curiosities of the legend. For example, Soviet schools disagreed with including so-called 'podzolic soils' into a broader group of Luvisols (soils with a base-saturated horizon of clay illuviation, but not necessarily with a bleached superficial horizon), and a special group of Podzoluvisols (then renamed as Glossisols, and later as Albeluvisols) was included in the legend. Podzoluvisols (Albeluvisols) must have an albic (bleached) horizon with bleached material tongues penetrating into an underlying argic horizon. Since Luvisols also may have an optional albic horizon, the difference between the two groups is subtle and has a regional significance: the penetration of albic material into a deeper horizon depends to a great extent on the periglacial history of the area (Driessen et al, 2001). Moreover, in some cultivated Albeluvisols the tongues of light-coloured material are absent because of deep ploughing in places accompanied with sheet erosion, and these soils convert to Luvisols. However, this group persists until now in the classification. Some soil groups were included due to a continuous personal effort of some experts. For example, the Stagnosols were included as a reference group mainly thanks to the energy of German specialists, and particularly of Prof Hans-Peter Blume. Every classification results from the will and ideas of its authors; in the case of the WRB it was a collective will of a group of experts.

The first edition of the WRB (Spaargaren, 1994; FAO-ISRIC-ISSS, 1998) already declared some theoretical principles, though in a simplified form. It gave a definition of soils classified in the system, declared the principles of diagnostics and tried to explain the structure of the system. The second edition (IUSS Working Group WRB, 2006) gave slightly extended explanations of the principles and the structure of the system – with little insight, however, about their theoretical roots. It was difficult to elaborate a common background for this classification, because it was a result of common effort of many experts with sometimes contradictory ideas, rather than a product of a comprehensive, well documented concept. To a great extent the WRB remains a kind of *Kollektivunbewusste* of the soil science community.

Development

The history of the WRB started with the establishment of a group working on the legend of the Soil Map of the World in the 1960s. However, officially work on an International Reference Base of Soil Resources started in 1980–81 after meetings in Sofia as a follow-up to the Soil Map of the World (IUSS Working Group WRB, 2006). In 1992 the International Reference Base was renamed as the World Reference Base, and it was decided that the World Soil Map legend should serve as a basis for the system whereas initially it had been developed independently. In 1994 the first draft of the WRB was presented (Spaargaren, 1994), and in 1998 the first official edition was published (FAO-ISRIC-ISSS, 1998). The second edition appeared in 2006 (IUSS Working Group WRB, 2006); since then minor corrections have been made, but no official publications are available. Current progress in the development of the WRB can be monitored at the web sites and one can find more details on the history of the WRB in the texts of both editions (FAO-ISRIC-ISSS, 1998; IUSS Working Group WRB, 2006) and in a number of relevant publications (Nachtergaele et al, 2000; Deckers et al, 2005).

The transformation of the system since the first version of the legend of the Soil Map of the World to the latest published edition of the WRB is illustrated in Table 4.2. It presents soil groups with second-level taxa in the case of the World Soil Map legend, and with modifiers in the case of the WRB; in the latest version of the WRB the modifiers are divided into prefixes and suffixes, and are arranged in the same order as in the classification key. Since this order is different from that being used in the previous WRB edition (FAO-ISRIC-ISSS, 1998), the latter represents a new version that may serve as a basis for classification.

It should be mentioned that Table 4.2 does not provide a conversion tool for older databases or maps into the latest version of classification. For most groups the diagnostic criteria have been changed or modified. The WRB uses much more complex criteria than the legend of the Soil Map of the World. Even between the first (FAO-ISRIC-ISSS, 1998) and the second (IUSS Working Group WRB, 2006) editions of the WRB there are significant differences in the definitions of diagnostic horizons and reference groups, which may be responsible for inclusion or exclusion of soil objects from corresponding taxa. Schad (2008) gives some examples of such inconsistency, and recommends reclassifying soils when soil map conversions are needed.

The structure is a weak point of the WRB because its hierarchical levels and the criteria for their establishment are not clearly defined. On the one hand it is declared to be only a reference base; on the other hand, it has a pronounced

Table 4.2 *The development of international soil classification/correlation system*

World Soil Map Legend (1974)	World Soil Map Legend (1988)	World Reference Base (1994)	World Reference Base (1998)	World Reference Base (2006) Prefixes	Suffixes
Acrisols	**Acrisols**	**Acrisols**	**Acrisols**	**Acrisols**	
			Vetic	Vetic	
			Lamellic	Lamellic	
				Cutanic	
				Technic	
			Leptic	Leptic	
				Fractiplinthic	
				Petroplinthic	
				Pisoplinthic	
Plinthic	Plinthic	Plinthic	Plinthic	Plinthic	
Gleyic	Gleyic	Gleyic	Gleyic	Gleyic	
			Vitric	Vitric	
			Andic	Andic	
				Nitic	
Humic	Humic		Stagnic	Stagnic	
			Umbric	Umbric	
Orthic	Haplic	Haplic	Haplic	Haplic	
					Antric
		Albic	Albic		Albic
					Fragic
					Sombric
					Manganiferric
Ferric	Ferric	Ferric	Ferric		Ferric
			Abruptic		Abruptic
					Ruptic
			Alumic		Alumic
		Humic	Humic		Humic
			Hyperdystric		Hyperdystric
					Epieutric
					Oxiaquic
					Greyic
			Profondic		Profondic
			Hyperochric		Hyperochric
					Nudiargic
					Densic
			Skeletic		Skeletic
		Arenic	Arenic		Arenic
					Siltic
					Clayic
			Rhodic		Rhodic
			Chromic		Chromic
					Transportic
					Novic
			Geric		
Podzoluvisols	**Podzoluvisols** =	**Glossosols** =	**Albeluvisols**	**Albeluvisols**	
		Fragic	Fragic	Fragic	
				Cutanic	
				Folic	
			Histic	Histic	
				Technic	
Gleyic	Gleyic	Gleyic	Gleyic	Gleyic	
	Stagnic	Stagnic	Stagnic	Stagnic	

World Soil Map Legend (1974)	World Soil Map Legend (1988)	World Reference Base (1994)	World Reference Base (1998)	World Reference Base (2006)	
				Prefixes	*Suffixes*
		Umbric	Umbric	Umbric	
				Cambic	
		Haplic	Haplic	Haplic	
			Abruptic		
			Ferric		
					Anthric
					Manganiferric
					Ferric
					Abruptic
					Ruptic
			Alumic		Alumic
Dystric	Dystric				Dystric
Euthric	Euthric	Euthric	Endoeutric		Eutric
	Gelic	Gelic	Gelic		Gelic
					Oxiaquic
					Greyic
					Densic
			Arenic		Arenic
			Siltic		Siltic
					Clayic
					Drenic
					Transportic
					Novic
			Alic		
	Alisols	**Alisols**	**Alisols**	**Alisols**	
				Hiperalic	
			Lamellic	Lamellic	
		Luvic		Cutanic	
			Albic	Albic	
				Technic	
				Leptic	
		Vertic	Vertic	Vertic	
				Fractiplinthic	
				Petroplinthic	
				Pisoplinthic	
	Plinthic	Plinthic	Plinthic	Plinthic	
	Gleyic	Gleyic	Gleyic	Gleyic	
				Vitric	
	Ferric	Ferric	Andic	Andic	
			Nitic	Nitic	
	Stagnic		Stagnic	Stagnic	
			Umbric	Umbric	
	Haplic	Haplic	Haplic	Haplic	
					Antric
					Fragic
					Manganiferric
			Ferric		Ferric
			Abruptic		Abruptic
					Ruptic
					Alumic
	Humic	Humic	Humic		Humic
			Hyperdystric		Hiperdystric
					Epieutric
					Turbic
					Gelic
					Oxiaquic
					Greyic
			Profondic		Profundic
					Hyperochric
					Nudiargic

Table 4.2 *(Continued)*

World Soil Map Legend (1974)	World Soil Map Legend (1988)	World Reference Base (1994)	World Reference Base (1998)	World Reference Base (2006)	
				Prefixes	Suffixes
					Densic
			Skeletic		Skeletic
			Arenic		Arenic
					Siltic
					Clayic
			Rhodic		Rhodic
		Chromic	Chromic		Chromic
					Transportic
					Novic
Andosols	**Andosols**	**Andosols**	**Andosols**	**Andosols**	
Vitric	Vitric	Vitric	Vitric	Vitric	
		Alic			
				Aluandic	
			Eutrisilic	Eutrosilic	
		Silic	Silic	Silandic	
			Melanic	Melanic	
			Fulvic	Fulvic	
		Hydric	Hydric	Hydric	
				Folic	
			Histic	Histic	
Mollic	Mollic			Technic	
			Leptic	Leptic	
	Gleyic		Gleyic	Gleyic	
			Mollic	Mollic	
				Gypsic	
Humic =				Petroduric	
			Duric	Duric	
			Vetic	Calcic	
	Umbric		Umbric	Umbric	
	Haplic		Haplic	Haplic	
					Anthric
					Fragic
			Calcaric		Calcaric
					Coluvic
			Acroxic		Acroxic
Ochric			Sodic		Sodic
			Dystric		Dystric
	Gelic	Eutric	Eutric		Eutric
					Turbic
					Gelic
					Oxiaquic
			Placic		Placic
					Greyic
					Thixotropic
			Skeletic		Skeletic
			Arenic		Arenic
					Siltic
					Clayic
					Drenic
					Transportic
					Novic
		Pachic	Pachic		
		Hydralic			
		Pachalic			
			Luvic		
			Thaptic		

World Soil Map Legend (1974)	World Soil Map Legend (1988)	World Reference Base (1994)	World Reference Base (1998)	World Reference Base (2006)	
				Prefixes	Suffixes
	Anthrosols	**Anthrosols**	**Anthrosols**	**Anthrosols**	
	Aric	Hydragric	Hydragric	Hydragric	
		Irragric	Irragric	Irragric	
	Cumulic	Cumulic	Terric	Terric	
			Plaggic	Plaggic	
		Hortic	Hortic	Hortic	
				Escalic	
				Technic	
				Fluvic	
				Salic	
			Gleyic	Gleyic	
			Spodic	Spodic	
			Ferralic	Ferralic	
			Stagnic	Stagnic	
			Regic	Regic	
					Sodic
					Alcalic
					Dystric
					Eutric
					Oxiaquic
					Densic
			Arenic		Arenic
					Siltic
					Clayic
					Novic
			Luvic		
	Fimic			→Technosols	
	Urbic			→Technosols	
Arenosols	**Arenosols**	**Arenosols**	**Arenosols**	**Arenosols**	
			Lamellic	Lamellic	
Luvic	Luvic	Luvic	Hypoluvic	Hypoluvic	
				Hyperalbic	
Albic	Albic	Albic	Albic	Albic	
			Rubic	Rubic	
Cambic	Cambic	Cambic		Brunic	
				Hydrophobic	
		Protic	Protic	Protic	
				Folic	
				Technic	
				Endosalic	
				Endogleyic	
				Fractiplinthic	
				Petroplinthic	
				Pisoplinthic	
			Plinthic	Plinthic	
Ferralic	Ferralic	Ferralic	Ferralic	Ferralic	
				Endostagnic	
	Haplic	Haplic	Haplic	Haplic	
					Ornithic
		Gypsiric	Gypsiric		Gypsiric
	Calcaric	Calcaric	Calcaric		Calcaric
			Tephric		Tephric
			Hyposalic		Hyposalic
			Dystric		Dystric
			Eutric		Eutric
	Gleyic		Gleyic		Petrogleyic
					Turbic
			Gelic		Gelic

Table 4.2 *(Continued)*

World Soil Map Legend (1974)	World Soil Map Legend (1988)	World Reference Base (1994)	World Reference Base (1998)	World Reference Base (2006) Prefixes	Suffixes
					Greyic
					Placic
					Hyperochric
			Yermic		Yermic
			Aridic		Aridic
					Transportic
					Novic
		Leptic			
			Hypoduric		
			Fragic		
	Calcisols	**Calcisols**	**Calcisols**	**Calcisols**	
	Petric	Petric	Petric	Petric	
			Hypercalcic	Hypercalcic	
			Hypocalcic	Hypocalcic	
				Technic	
				Hyperskeletic	
			Leptic	Leptic	
			Vertic	Vertic	
			Endosalic	Endosalic	
			Gleyic	Endogleyic	
				Gypsic	
	Luvic	Luvic	Luvic	Luvic	
				Lixic	
	Haplic	Haplic	Haplic	Haplic	
					Ruptic
		Sodic	Sodic		Sodic
			Takyric		Takyric
			Yermic		Yermic
			Aridic		Aridic
			Hyperochric		Hyperochric
					Densic
			Skeletic		Skeletic
					Arenic
					Siltic
					Clayic
					Chromic
					Transportic
					Novic
		Cambic			
Cambisols	**Cambisols**	**Cambisols**	**Cambisols**	**Cambisols**	
				Folic	
				Antraquic	
				Hortic	
		Fluvic	Fluvic	Irragric	
				Plaggic	
				Terric	
				Technic	
			Leptic	Leptic	
Vertic	Vertic	Vertic	Vertic	Vertic	
				Thionic	
				Fluvic	
			Endosalic	Endosalic	
Gleyic	Gleyic	Gleyic	Gleyic	Endogleyic	

World Soil Map Legend (1974)	World Soil Map Legend (1988)	World Reference Base (1994)	World Reference Base (1998)	World Reference Base (2006)	
				Prefixes	Suffixes
			Vitric	Vitric	
			Andic	Andic	
				Fractiplinthic	
				Petroplinthic	
				Pisoplinthic	
			Plinthic	Plinthic	
Ferralic	Ferralic	Ferralic	Ferralic	Ferralic	
				Fragic	
			Gelistagnic	Gelistagnic	
			Stagnic	Stagnic	
			Haplic	Haplic	
					Manganiferric
					Ferric
					Ornithic
					Coluvic
			Gypsiric		Gypsiric
	Calcaric	Calcaric	Calcaric		Calcaric
					Tephric
					Alumic
			Sodic		Sodic
					Alcalic
Humic	Humic				Humic
Dystric	Dystric	Dystric	Dystric		Dystric
Euthric	Euthric	Euthric	Eutric		Eutric
					Laxic
					Turbic
Gelic	Gelic	Gelic	Gelic		Gelic
					Oxiaquic
					Greyic
					Ruptic
Calcic					Pisocalcic
			Hyperochric		Hyperochric
			Takyric		Takyric
			Yermic		Yermic
			Aridic		Aridic
					Densic
			Skeletic		Skeletic
					Siltic
					Clayic
			Rhodic		Rhodic
Chromic	Chromic	Chromic	Chromic		Chromic
					Escalic
					Transportic
					Novic
		Mollic	Mollic	→Mollic Umbrisols	
Chernozems	**Chernozems**	**Chernozems**	**Chernozems**	**Chernozems**	
			Chernic	Voronic	
			Vermic	Vermic	
				Technic	
				Leptic	
		Vertic	Vertic	Vertic	
				Endofluvic	
				Endosalic	
	Gleyic	Gleyic	Gleyic	Gleyic	
				Vitric	
				Andic	
				Stagnic	
				Petrogypsic	

Table 4.2 *(Continued)*

World Soil Map Legend (1974)	World Soil Map Legend (1988)	World Reference Base (1994)	World Reference Base (1998)	World Reference Base (2006)	
				Prefixes	Suffixes
				Gypsic	
				Petroduric	
				Duric	
				Petrocalcic	
Calcic	Calcic	Calcic	Calcic	Calcic	
Luvic	Luvic	Luvic	Luvic	Luvic	
Haplic	Haplic	Haplic	Haplic	Haplic	
					Anthric
Glossic	Glossic	Glossic	Glossic		Glossic
					Tephric
					Sodic
					Pachic
					Oxiaquic
					Greyic
					Densic
					Skeletic
					Arenic
			Siltic		Siltic
					Clayic
		Cryosols	Cryosols	Cryosols	
			Glacic	Glacic	
			Turbic	Turbic	
				Folic	
		Histic	Histic	Histic	
				Technic	
				Hyperskeletic	
			Leptic	Leptic	
			Natric	Natric	
			Salic	Salic	
				Vitric	
				Spodic	
			Mollic	Mollic	
			Calcic	Calcic	
			Umbric	Umbric	
				Cambic	
			Thionic	Haplic	
					Gypsiric
			Stagnic		Calcaric
			Haplic		Ornithic
					Dystric
					Eutric
					Reductaquic
			Oxyaquic		Oxyaquic
		Thixotropic			Thixotropic
			Aridic		Aridic
					Skeletic
					Arenic
					Siltic
					Clayic
					Drenic
					Transportic
					Novic
			Lithic		
			Andic		
			Gleyic		

World Soil Map Legend (1974)	World Soil Map Legend (1988)	World Reference Base (1994)	World Reference Base (1998)	World Reference Base (2006)	
				Prefixes	Suffixes
			Gypsic		
			Yermic		
			Durisols	**Durisols**	
			Petric	Petric	
				Fractipetric	
				Technic	
			Leptic	Leptic	
			Vertic	Vertic	
				Endogleyic	
			Gypsic	Gypsic	
				Petrocalcic	
			Calcic	Calcic	
			Luvic	Luvic	
				Lixic	
			Haplic	Haplic	
					Ruptic
					Sodic
			Takyric		Takyric
			Yermic		Yermic
			Aridic		Aridic
			Hyperochric		Hyperochric
			Arenic		Arenic
					Siltic
					Clayic
			Chromic		Chromic
					Transportic
					Novic
Ferralsols	**Ferralsols**	**Ferralsols**	**Ferralsols**		**Ferralsols**
		Gibbsic	Gibbsic	Gibbsic	
			Posic	Posic	
	Geric	Geric	Geric	Geric	
			Vetic	Vetic	
				Folic	
				Technic	
			Andic	Andic	
				Fractiplinthic	
				Petroplinthic	
				Pisoplinthic	
Plinthic	Plinthic	Plinthic	Plinthic	Plinthic	
			Mollic	Mollic	
Acric			Acric	Acric	
		Lixic	Lixic	Lixic	
			Umbric	Umbric	
Orthic	Haplic	Haplic	Haplic	Haplic	
					Sombric
					Manganiferric
			Ferric	Ferric	
					Coluvic
Humic	Humic	Humic	Humic	Humic	
			Alumic	Alumic	
					Dystric
		Eutric		Eutric	
					Ruptic
					Oxiaquic
					Densic
			Arenic	Arenic	

Table 4.2 *(Continued)*

World Soil Map Legend (1974)	World Soil Map Legend (1988)	World Reference Base (1994)	World Reference Base (1998)	World Reference Base (2006)	
				Prefixes	Suffixes
					Siltic
					Clayic
Rhodic	Rhodic	Rhodic	Rhodic		Rhodic
Xantic	Xantic		Xanthic		Xanthic
					Transportic
					Novic
		Gleyic	Gleyic		
			Histic		
			Endostagnic		
			Hyperdystric		
			Hypereutric		
Fluvisols	**Fluvisols**	**Fluvisols**	**Fluvisols**		**Fluvisols**
				Subaquatic	
				Tidalic	
				Limnic	
				Folic	
			Histic	Histic	
				Technic	
	Salic	Salic	Salic	Salic	
			Gleyic	Gleyic	
			Stagnic	Stagnic	
	Mollic	Mollic	Mollic	Mollic	
				Gypsic	
				Calcic	
	Umbric	Umbric	Umbric	Umbric	
			Haplic	Haplic	
Thionic	Thionic	Thionic	Thionic		Thionic
					Anthric
			Gypsiric		Gypsiric
Calcaric	Calcaric	Calcaric	Calcaric		Calcaric
			Tephric		Tephric
					Petrogleyic
			Gelic		Gelic
					Oxiaquic
			Humic		Humic
			Sodic		Sodic
Dystric	Dystric	Dystric	Dystric		Dystric
Eutric	Eutric	Eutric	Eutric		Eutric
					Greyic
			Takyric		Takyric
			Yermic		Yermic
			Aridic		Aridic
					Densic
			Skeletic		Skeletic
			Arenic		Arenic
					Siltic
					Clayic
					Drenic
					Transportic
		Vertic			
Gleysols	**Gleysols**	**Gleysols**	**Gleysols**		**Gleysols**
				Folic	
		Histic	Histic		
		Antraquic	Antraquic		
				Technic	

World Soil Map Legend (1974)	World Soil Map Legend (1988)	World Reference Base (1994)	World Reference Base (1998)	World Reference Base (2006)	
				Prefixes	*Suffixes*
		Fluvic		Fluvic	
			Endosalic	Endosalic	
				Vitric	
			Andic	Andic	
				Spodic	
Plinthic	Plinthic	Plinthic	Plinthic	Plinthic	
Mollic	Mollic	Mollic	Mollic	Mollic	
			Gypsic	Gypsic	
Calcic	Calcic	Calcic	Calcic	Calcic	
				Alic	
				Acric	
				Luvic	
				Lixic	
	Umbric	Umbric	Umbric	Umbric	
		Haplic	Haplic	Haplic	
		Thionic	Thionic		Thionic
			Abruptic		Abruptic
			Calcaric		Calcaric
		Tephric	Tephric		Tephric
					Coluvic
Humic			Humic		Humic
			Sodic		Sodic
			Alcalic		Alcalic
			Alumic		Alumic
			Toxic		Toxic
Dystric	Dystric		Dystric		Dystric
Eutric	Eutric		Eutric		Eutric
					Petrogleyic
					Turbic
Gelic	Gelic		Gelic		Gelic
					Greyic
			Takyric		Takyric
		Arenic	Arenic		Arenic
					Siltic
					Clayic
					Drenic
					Novic
		Cryic	→Gleyic Cryosols		

Greyzems	**Greyzems**	→**Greyic Phaeozems**			
Gleyic	Gleyic				
Haplic	Haplic				

	Gypsisols	**Gypsisols**	**Gypsisols**	**Gypsisols**	
	Petric	Petric	Petric	Petric	
			Hypergypsic	Hypergypsic	
			Hypogypsic	Hypogypsic	
		Arzic	Arzic	Arzic	
				Technic	
				Hyperskeletic	
			Leptic	Leptic	
			Vertic	Vertic	
			Endosalic	Endosalic	
				Endogleyic	
				Petroduric	
			Duric	Duric	
				Petrocalcic	
	Calcic	Calcic	Calcic	Calcic	
	Luvic	Luvic	Luvic	Luvic	

Table 4.2 *(Continued)*

World Soil Map Legend (1974)	World Soil Map Legend (1988)	World Reference Base (1994)	World Reference Base (1998)	World Reference Base (2006)	
				Prefixes	Suffixes
	Haplic	Haplic	Haplic	Haplic	
					Ruptic
			Sodic		Sodic
			Hyperochric		Hyperochric
			Takyric		Takyric
			Yermic		Yermic
			Aridic		Aridic
			Skeletic		Skeletic
					Arenic
					Siltic
					Clayic
					Transportic
					Novic
		Cambic			
Histosols	**Histosols**	**Histosols**	**Histosols**	**Histosols**	
	Folic	Folic	Folic	Folic	
				Limnic	
				Lignic	
	Fibric	Fibric	Fibric	Fibric	
				Hemic	
	Terric		Sapric	Sapric	
				Flotic	
				Subaquatic	
			Glacic	Glacic	
			Ombric	Ombric	
			Rheic	Rheic	
				Technic	
			Cryic	Cryic	
				Hyperskeletic	
				Leptic	
				Vitric	
				Andic	
		Salic	Salic	Salic	
				Calcic	
	Thionic	Thionic	Thionic		Thionic
					Ornithic
					Calcaric
					Sodic
			Alcalic		Alcalic
			Toxic		Toxic
Dystric			Dystric		Dystric
Eutric			Eutric		Eutric
					Turbic
Gelic	Gelic	Gelic	Gelic		Gelic
					Petrogleyic
					Placic
					Skeletic
					Tidalic
					Drenic
					Transportic
					Novic
		Haplic			

World Soil Map Legend (1974)	World Soil Map Legend (1988)	World Reference Base (1994)	World Reference Base (1998)	World Reference Base (2006)	
				Prefixes	Suffixes
Kastanozems	**Kastanozems**	**Kastanozems**	**Kastanozems**	**Kastanozems**	
				Vermic	
				Technic	
				Leptic	
			Vertic	Vertic	
				Endosalic	
				Gleyic	
				Vitric	
				Andic	
				Stagnic	
				Petrogypsic	
	Gypsic	Gypsic	Gypsic	Gypsic	
				Petroduric	
				Duric	
				Petrocalcic	
Calcic	Calcic	Calcic	Calcic	Calcic	
Luvic	Luvic	Luvic	Luvic	Luvic	
Orthic	Haplic	Haplic	Haplic	Haplic	
			Anthric		Anthric
					Glossic
					Tephric
			Hyposodic		Sodic
					Oxiaquic
					Greyic
					Densic
					Skeletic
					Arenic
			Siltic		Siltic
					Clayic
			Chromic		Chromic
	Leptosols	**Leptosols**	**Leptosols**	**Leptosols**	
				Nudilithic	
Lithosols =	Lithic	Lithic	Lithic	Lithic	
			Hyperskeletic	Hyperskeletic	
Rendzinas =	Rendzic	Rendzic	Rendzic	Rendzic	
				Folic	
				Histic	
				Technic	
				Vertic	
				Salic	
			Gleyic	Gleyic	
				Vitric	
				Andic	
			Calcaric	Stagnic	
	Mollic	Mollic	Mollic	Mollic	
Rankers =	Umbric	Umbric	Umbric	Umbric	
				Cambic	
			Haplic	Haplic	
					Brunic
			Gypsiric		Gypsiric
					Calcaric
					Ornithic
					Tephric
					Protothionic
			Humic		Humic
					Sodic
	Dystric	Dystric	Dystric		Dystric
	Euthric	Euthric	Eutric		Eutric

Table 4.2 *(Continued)*

World Soil Map Legend (1974)	World Soil Map Legend (1988)	World Reference Base (1994)	World Reference Base (1998)	World Reference Base (2006)	
				Prefixes	Suffixes
	Gelic		Gelic		Oxiaquic
					Gelic
					Placic
					Greyic
			Yermic		Yermic
			Aridic		Aridic
		Skeletic			Skeletic
					Drenic
					Novic
		Cryic	→ Leptic Cryosols		
Lithosols – see Leptosols					
	Lixisols	**Lixisols**	**Lixisols**		**Lixisols**
			Vetic	Vetic	
			Lamellic	Lamellic	
				Cutanic	
				Technic	
			Leptic	Leptic	
	Gleyic	Gleyic	Gleyic	Gleyic	
			Vitric	Vitric	
			Andic	Andic	
				Fractiplinthic	
				Petroplinthic	
				Pisoplinthic	
	Plinthic	Plinthic	Plinthic	Plinthic	
				Nitic	
	Stagnic		Stagnic	Stagnic	
			Calcic	Calcic	
	Haplic	Haplic	Haplic	Haplic	
					Anthric
	Albic	Albic	Albic		Albic
					Fragic
					Manganiferric
	Ferric	Ferric	Ferric		Ferric
			Abruptic		Abruptic
					Ruptic
			Humic		Humic
					Epidystric
					Hypereutric
					Oxiaquic
					Greyic
			Profondic		Profondic
			Hyperochric		Hyperochric
					Nudiargic
					Densic
					Skeletic
		Arenic	Arenic		Arenic
					Siltic
					Clayic
			Rhodic		Rhodic
			Chromic		Chromic
					Transportic
					Novic
			Geric		

World Soil Map Legend (1974)	World Soil Map Legend (1988)	World Reference Base (1994)	World Reference Base (1998)	World Reference Base (2006)	
				Prefixes	Suffixes
Luvisols	**Luvisols**	**Luvisols**	**Luvisols**	**Luvisols**	
			Lamellic	Lamellic	
			Cutanic	Cutanic	
Albic	Albic	Albic	Albic	Albic	
				Escalic	
				Technic	
			Leptic	Leptic	
Vertic	Vertic	Vertic	Vertic	Vertic	
Gleyic	Gleyic	Gleyic	Gleyic	Gleyic	
			Vitric	Vitric	
			Andic	Andic	
				Nitic	
	Stagnic		Stagnic	Stagnic	
Calcic	Calcic	Calcic	Calcic	Calcic	
Orthic	Haplic	Haplic	Haplic	Haplic	
				Anthric	
				Fragic	
				Manganiferric	
Ferric	Ferric	Ferric	Ferric	Ferric	
				Abruptic	
				Ruptic	
				Humic	
			Hyposodic	Sodic	
		Dystric	Dystric	Epidystric	
				Hypereutric	
				Turbic	
				Gelic	
				Oxiaquic	
				Greyic	
			Profondic	Profondic	
			Hyperochric	Hyperochric	
				Nudiargic	
				Densic	
				Skeletic	
			Arenic	Arenic	
				Siltic	
				Clayic	
			Rhodic	Rhodic	
Chromic	Chromic	Chromic	Chromic	Chromic	
				Transportic	
				Novic	
Nitosols	**Nitosols**	**Nitisols**	**Nitisols**	**Nitisols**	
			Vetic	Vetic	
				Technic	
			Andic	Andic	
			Ferralic	Ferralic	
		Mollic	Mollic	Mollic	
		Alic	Alic	Alic	
				Acric	
				Luvic	
				Lixic	
		Umbric	Umbric	Umbric	
	Haplic		Haplic	Haplic	
Humic	Humic	Humic	Humic		Humic
			Alumic		Alumic
Dystric		Dystric	Dystric		Dystric
Euthric		Euthric	Eutric		Eutric
					Oxiaquic

Table 4.2 *(Continued)*

World Soil Map Legend (1974)	World Soil Map Legend (1988)	World Reference Base (1994)	World Reference Base (1998)	World Reference Base (2006) Prefixes	Suffixes
					Coluvic
					Densic
Rhodic	Rhodic	Rhodic	Rhodic		Rhodic
					Transportic
					Novic
Phaeozems	**Phaeozems**	**Phaeozems**	**Phaeozems**	**Phaeozems**	
			Vermic	Vermic	
		Greyic	Greyic	Greyic	
Gleyic	Gleyic	Gleyic	Gleyic	Technic	
				Rendzic	
			Leptic	Leptic	
		Vertic	Vertic	Vertic	
				Endosalic	
				Gleyic	
			Vitric	Vitric	
			Andic	Andic	
				Ferralic	
	Stagnic	Stagnic	Stagnic	Stagnic	
				Petrogypsic	
				Petroduric	
				Duric	
				Petrocalcic	
Calcic				Calcic	
Luvic	Luvic	Luvic	Luvic	Luvic	
Orthic	Haplic	Haplic	Haplic	Haplic	
					Anthric
			Albic		Albic
			Abruptic		Abruptic
		Glossic	Glossic		Glossic
	Calcaric		Calcaric		Calcaric
			Tephric		Tephric
			Sodic		Sodic
			Pachic		Pachic
					Oxiaquic
					Densic
			Skeletic		Skeletic
					Arenic
			Siltic		Siltic
					Clayic
			Chromic		Chromic
Planosols	**Planosols**	**Planosols**	**Planosols**	**Planosols**	
Solodic				Solodic	
				Folic	
		Histic	Histic	Histic	
				Technic	
		Vertic	Vertic	Vertic	
			Endosalic	Endosalic	
			Plinthic	Plinthic	
			Gleyic	Endogleyic	
Mollic	Mollic	Mollic	Mollic	Mollic	
			Gypsic	Gypsic	
				Petrocalcic	
			Calcic	Calcic	
			Alic	Alic	

World Soil Map Legend (1974)	World Soil Map Legend (1988)	World Reference Base (1994)	World Reference Base (1998)	World Reference Base (2006)	
				Prefixes	Suffixes
				Acric	
			Luvic	Luvic	
				Lixic	
	Umbric	Umbric	Umbric	Umbric	
			Haplic	Haplic	
			Thionic		Thionic
			Albic		Albic
					Manganiferric
			Ferric		Ferric
			Geric		Geric
					Ruptic
			Calcaric		Calcaric
			Sodic		Sodic
			Alcalic		Alcalic
			Alumic		Alumic
Dystric	Dystric	Dystric	Dystric		Dystric
Eutric	Eutric	Eutric	Eutric		Eutric
Gelic	Gelic	Gelic	Gelic		Gelic
					Greyic
			Arenic		Arenic
					Siltic
					Clayic
			Chromic		Chromic
					Drenic
					Transportic
Humic					
			Petroferric		
			Rhodic		
	Plinthisols	**Sesquisols**	**Plinthisols**	**Plinthisols**	
		Petric	Petric	Petric	
				Fractipetric	
				Pisolithic	
				Gibbsic	
				Posic	
			Geric	Geric	
			Vetic	Vetic	
				Folic	
				Histic	
				Technic	
		Stagnic	Stagnic	Stagnic	
			Acric	Acric	
				Lixic	
			Umbric	Umbric	
		Haplic	Haplic	Haplic	
	Albic	Albic	Albic		Albic
					Manganiferric
			Ferric		Ferric
			Endoduric		Endoduric
			Abruptic		Abruptic
					Coluvic
					Ruptic
			Alumic		Alumic
	Humic	Humic	Humic		Humic
	Dystric				Dystric
	Eutric	Eutric	Endoeutric		Eutric
					Oxiaquic
					Pachic
			Glossic		Umbriglossic

Table 4.2 *(Continued)*

World Soil Map Legend (1974)	World Soil Map Legend (1988)	World Reference Base (1994)	World Reference Base (1998)	World Reference Base (2006)	
				Prefixes	Suffixes
					Arenic
					Siltic
					Clayic
					Drenic
					Transportic
					Novic
		Aeric			
			Alic		
			Pachic		
Podzols	**Podzols**	**Podzols**	**Podzols**	**Podzols**	
Placic			Placic	Placic	
				Ortsteinic	
Humic =	Carbic =	Humic =	Carbic	Carbic	
Ferric	Ferric		Rustic	Rustic	
	Cambic	Cambic =	Entic	Entic	
				Albic	
				Folic	
			Histic	Histic	
				Technic	
				Hyperskeletic	
Leptic				Leptic	
Gleyic	Gleyic	Gleyic	Gleyic	Gleyic	
			Lamellic	Vitric	
				Andic	
		Stagnic	Stagnic	Stagnic	
		Umbric	Umbric	Umbric	
Orthic	Haplic	Haplic	Haplic	Haplic	
					Hortic
					Plaggic
					Terric
			Anthric		Anthric
					Ornithic
			Fragic		Fragic
					Ruptic
					Turbic
	Gelic	Gelic	Gelic		Gelic
					Oxiaquic
					Lamellic
			Densic		Densic
			Skeletic		Skeletic
					Drenic
					Transportic
					Novic
		Duric			

Podzoluvisols – see Albeluvisols

Rankers – see Leptosols

Regosols	**Regosols**	**Regosols**	**Regosols**	**Regosols**	
				Folic	
			Aric	Aric	
				Coluvic	
				Technic	
			Leptic	Leptic	
			Gleyic	Endogleyic	

World Soil Map Legend (1974)	World Soil Map Legend (1988)	World Reference Base (1994)	World Reference Base (1998)	World Reference Base (2006)	
				Prefixes	Suffixes
			Thaptovitric	Taptovitric	
			Thaptoandic	Taptoandic	
			Gelistagnic	Gelistagnic	
			Stagnic	Stagnic	
			Haplic	Haplic	
					Brunic
					Ornithic
	Gypsic =	Gypsiric	Gypsiric		Gypsiric
Calcic =	Calcaric	Calcaric	Calcaric		Calcaric
		Tephric	Tephric		Tephric
			Humic		Humic
			Hyposalic		Hyposalic
			Hyposodic		Sodic
Dystric	Dystric	Dystric	Dystric		Dystric
Euthric	Euthric	Euthric	Eutric		Eutric
					Turbic
Gelic	Gelic	Gelic	Gelic		Gelic
					Oxiaquic
			Vermic		Vermic
			Hyperochric		Hyperochric
			Takyric		Takyric
			Yermic		Yermic
			Aridic		Aridic
					Densic
			Skeletic		Skeletic
			Arenic		Arenic
					Siltic
					Clayic
					Escalic
					Transportic
		Anthropic	Anthropic	→ Technosols	
			Garbic	→ Garbic Technosols	
			Reductic	→ Technosols (Reductic)	
			Spolic	→ Spolic Technosols	
			Urbic	→ Urbic Technosols	
	Umbric		→ Umbrisols		

Rhendzinas – see Leptosols

Sesquisols – see Plinthisols

Solonchaks	**Solonchaks**	**Solonchaks**	**Solonchaks**	**Solonchaks**	
			Petrosalic	Petrosalic	
			Hypersalic	Hypersalic	
				Puffic	
				Folic	
			Histic	Histic	
				Technic	
			Vertic	Vertic	
Gleyic	Gleyic	Gleyic	Gleyic	Gleyico	
	Stagnic	Stagnic	Stagnic	Stagnic	
Mollic	Mollic	Mollic	Mollic	Mollic	
	Gypsic	Gypsic	Gypsic	Gypsic	
			Duric	Duric	
	Calcic	Calcic	Calcic	Calcic	

Table 4.2 *(Continued)*

World Soil Map Legend (1974)	World Soil Map Legend (1988)	World Reference Base (1994)	World Reference Base (1998)	World Reference Base (2006)	
				Prefixes	Suffixes
Orthic	Haplic	Haplic	Haplic	Haplic	
		Sodic	Sodic		Sodic
			Aceric		Aceric
			Chloridic		Chloridic
			Sulphatic		Sulphatic
			Carbonatic		Carbonatic
	Gelic		Gelic		Gelic
					Oxiaquic
Takyric			Takyric		Takyric
			Yermic		Yermic
			Aridic		Aridic
					Densic
					Arenic
					Siltic
					Clayic
					Drenic
					Transportic
					Novic
			Ochric		
Solonetz	**Solonetz**	**Solonetz**	**Solonetz**	**Solonetz**	
				Technic	
			Vertic	Vertic	
Gleyic	Gleyic	Gleyic	Gleyic	Gleyic	
		Salic	Salic	Salic	
	Stagnic	Stagnic	Stagnic	Stagnic	
Mollic	Mollic	Mollic	Mollic	Mollic	
	Gypsic	Gypsic	Gypsic	Gypsic	
			Duric	Duric	
				Petrocalcic	
	Calcic	Calcic	Calcic	Calcic	
Orthic	Haplic	Haplic	Haplic	Haplic	
					Glossalbic
		Albic	Albic		Albic
					Abruptic
					Coluvic
					Ruptic
			Magnesic		Magnesic
			Humic		Humic
					Oxiaquic
			Takyric		Takyric
			Yermic		Yermic
			Aridic		Aridic
					Arenic
					Siltic
					Clayic
					Transportic
					Novic
		Stagnosols	–	**Stagnosols**	
				Folic	
		Histic	(Histic Stagnic Cambisols)	Histic	
				Technic	
		Vertic	(Vertic Stagnic Cambisols)	Vertic	

World Soil Map Legend (1974)	World Soil Map Legend (1988)	World Reference Base (1994)	World Reference Base (1998)	World Reference Base (2006)	
				Prefixes	Suffixes
				Endosalic	
				Plinthic	
		Gleyic	(Gleyic Stagnic Cambisols)	Endogleyic	
		Mollic	(Stagnic Phaeo-zems)	Mollic	
				Gypsic	
				Petrocalcic	
				Calcic	
				Alic	
				Acric	
		Luvic	(Stagnic Luvisols)	Luvic	
				Lixic	
				Umbric	
		Haplic	(Stagnic Cambi-sols)	Haplic	
					Thionic
		Albic	(Albic Planosols)		Albic
					Manganiferric
					Ferric
					Ruptic
					Geric
					Calcaric
					Ornithic
					Sodic
					Alcalic
					Alumic
					Dystric
					Eutric
		Gelic			Gelic
					Greyic
					Placic
					Arenic
					Siltic
					Clayic
					Rhodic
					Chromic
					Drenic

Technosols

Ekranic
Linic
Urbic
Spolic
Garbic
Folic
Histic
Cryic
Leptic
Fluvic
Gleyico
Vitric
Stagnic
Mollic
Alic
Acric
Luvic
Lixic
Umbric

Table 4.2 *(Continued)*

World Soil Map Legend (1974)	World Soil Map Legend (1988)	World Reference Base (1994)	World Reference Base (1998)	World Reference Base (2006)	
				Prefixes	*Suffixes*
					Calcaric
					Toxic
					Reductic
					Humic
					Oxiaquic
					Densic
					Skeletic
					Arenic
					Siltic
					Clayic
					Drenic
					Novic
		Umbrisols	**Umbrisols**		**Umbrisols**
				Folic	
				Histic	
				Technic	
			Leptic	Leptic	
				Fluvic	
			Gleyic	Endogleyic	
				Vitric	
				Andic	
			Ferralic	Ferralic	
			Stagnic	Stagnic	
				Mollic	
		Cambic		Cambic	
		Haplic	Haplic	Haplic	
			Anthric		Anthric
		Albic	Albic		Albic
					Brunic
					Ornithic
					Thionic
					Glossic
			Humic		Humic
					Alumic
					Hyperdystric
					Endoeutric
					Pachic
					Turbic
		Gelic	Gelic		Gelic
					Oxiaquic
					Greyic
					Laxic
					Placic
					Densic
		Skeletic	Skeletic		Skeletic
		Arenic	Arenic		Arenic
					Siltic
					Clayic
					Chromic
					Drenic
					Novic
Vertisols	**Vertisols**	**Vertisols**	**Vertisols**		**Vertisols**
			Grumic	Grumic	

World Soil Map Legend (1974)	World Soil Map Legend (1988)	World Reference Base (1994)	World Reference Base (1998)	World Reference Base (2006)	
				Prefixes	Suffixes
			Mazic	Mazic	
			Natric	Technic	
				Endoleptic	
		Salic	Salic	Salic	
				Gleyico	
			Alic	Sodic	
				Stagnic	
				Mollic	
	Gypsic	Gypsic	Gypsic	Gypsic	
			Duric	Duric	
	Calcic	Calcic	Calcic	Calcic	
		Haplic	Haplic	Haplic	
		Thionic	Thionic		Thionic
					Albic
					Manganiferric
					Ferric
			Gypsiric		Gypsiric
					Calcaric
					Humic
					Hyposalic
		Sodic	Hyposodic		Hyposodic
		Dystric	Dystric	Mesotrophic	Mesotrophic
	Euthric		Eutric		Hypereutric
			Pellic		Pellic
Chromic		Chromic	Chromic		Chromic
					Novic
Umbric					
Xerosols					
Orthic	→ (Aridic Regosols)				
Calcaric	→ (Aridic Calcisols)				
Gypsic	→ (Aridic Gypsisols)				
Luvic	→ (Aridic Luvisols)				
Takyric	→ (Takyric Regosols)				
Yermosols					
Orthic	→ (Yermic Regosols)				
Calcaric	→ (Yermic Calcisols)				

Table 4.3 *The structure of the World Reference Base for Soil Resources*

Level	Taxon name	Taxon characteristics	Borders between classes	Diagnostics	Terminology
0	Soils	Kingdom			
1	References	Generic	Formal	Chemico-morphological	Mixed
2	Prefixes	Specific 1	Formal	Chemico-morphological	Artificial
3	Suffixes	Specific 2	Formal	Chemico-morphological	Artificial

hierarchical structure. The reference groups are defined as follows: 'At the higher categorical level, classes are differentiated mainly according to the primary pedogenetic process that has produced the characteristic soil features, except where special soil parent materials are of overriding importance' (IUSS Working Group WRB, 2006). This definition is similar to ones used for generic levels in the Russian classification (Shishov et al, 2004) and in the US Soil Taxonomy (Soil Survey Staff, 1999). However, the declaration is not fulfilled: many reference groups lack a unique pedogenetic process and even a set of process-related properties (like Cambisols, Regosols or Leptosols), others have similar properties resulting from several different processes (Acrisols, Luvisols, etc.), and other groups have the same pedogenetic process, but are separated because of historical or practical reasons (e.g. Chernozems and Kastanozems). This appears to be a result of an 'inverse' construction; instead of developing the theory and then separating soils according to this theory, it first accepts traditional archetypes, and then tries to make a retrospective justification of their existence.

The modifiers are divided actually into two groups: prefixes, which are specific for a reference group or represent a transition to another group, and suffixes, which indicate non-specific properties, like texture, colour or stoniness. The prefixes are similar in this sense to subtypes in Russian classification (Shishov et al, 2004) or subgroups in the US Soil Taxonomy (Soil Survey Staff, 1999); the suffixes correspond to genus level in Russia and to families in the US. Practically these two groups of modifiers represent two different hierarchical levels in the the WRB system, and their use is even recommended for soil mapping at different scales; however, it is not recognized openly in the text of the WRB (IUSS Working Group WRB, 2006). Although the logic of separation of modifiers into two groups follows the common logic of most soil classifications, the realization of the principle was not very successful. It is not clear why some similar modifiers are regarded differently in the system: for example, the modifier Endosalic is regarded as a prefix, and Endoduric as a suffix, though both of them represent a transition to other groups. The modifier Albic appears as a prefix in some groups, and as a suffix in the others. These contradictions seem to represent mainly mistakes rather than a defect of the system, and hopefully will be corrected in future editions.

Diagnostics

The diagnostics of soil groups and lower level units is done using formal quantitative criteria. The authors stress that, though pedogenesis is basic for separating soil taxa, no subjective concepts are used directly for classifying soils. The object of classification is a soil profile, and the placement is made by using diagnostic horizons, properties and materials with strict definitions. Information on the factors of soil formation and water and thermal regimes of soils is not used.

The legend of the World Soil Map (FAO-UNESCO, 1974) initially declared that the diagnostics should consist mainly of field criteria. The main reason was that in developing countries pedologists had no access to advanced laboratory equipment, and could not make costly amd complicated analyses for classifying soils during soil surveys. In contrast, the WRB now requires more chemical analyses than any national soil classification. Some excessively complex procedures, such as the determination of total reserve of bases in clay fraction for diagnostics of Alisols, were excluded from classification (IUSS Working Group WRB, 2006). However, a lot of complex procedures remain in the WRB, making diagnostic determinations difficult in some countries of the developing world. For example, in Mexico many Nitisols are under question now, because soil surveys have limited access to laboratories where selective extraction of Fe necessary for determining a nitic horizon can be made (Cruz Gaistardo et al, 2006).

In the second edition (IUSS Working Group WRB, 2006) the authors made certain efforts towards harmonization of quantitative diagnostic criteria of the WRB with those used in the other worldwide system, the US Soil Taxonomy (Soil Survey Staff, 1999). However, most diagnostic criteria do not coincide, making it difficult to obtain correct correlations of terms of the two classifications.

Terminology

The terminology used for reference groups in the WRB is compiled from many sources, as was the preceeding World Soil Map legend (FAO-UNESCO, 1974). The latter one from the beginning was a product of consensus of various national soil science schools. Some soil names were adopted from Russian (Chernozems, Solonchaks, Solonetz, Podzols), the others were artificial terms proposed for the new soil taxonomy of the US (Histosols, Vertisols); some were modified terms of national soil science schools (gley → Gleysols, kashtanovye pochvy → Kastanozems, ando → Andosols), and the others were constructed specially for the legend on the basis of Latin roots (Leptosols, Fluvisols, Ferralsols, etc.). In some cases it led to certain misunderstanding because the term podzol of the Russian school was not the same as Podzol in the World Map legend, and Histosols in the US classification have different diagnostic criteria and absolutely different lower level groups than the Histosols of the Soil Map of the World. To some extent the WRB inherited these disparities from the World Soil Map legend. The reference group names adopted in the post-legend epoch are completely artificial and derived from Latin roots. The name Umbrisols may seem an exception, since it repeats the term Umbrisoluri of Romanian classification; however, to our surprise, it was constructed independently, directly from Latin. The terminology used for modifiers is completely artificial and uses almost exclusively Latin and Greek roots (Table 4.2).

References

Cruz Gaistardo, C. O., García Calderón, N. E. and Krasilnikov, P. (2006) 'Avances en la cartografía de México con WRB', Memorias del X Congreso nacional y II Internacional de la Ciencia del Suelo, Suelo, Seguridad Alimentaria y Pobreza, 6–10 November 2006, Lima, Perú, pp193–196

Deckers, J., Spaargaren, O., Nachtergaele, F., Berding, F., Ahrens, R., Michéli, E. and Schad, P. (2005) 'Rationale for the key and the qualifiers of the WRB 2006', *Eurasian Soil Science*, vol 38, suppl 1, pp6–12

Driessen, P., Deckers, J., Spaargaren, O. and Nachtergaele, F. (2001) *Lecture Notes on the Major Soils of the World*, Soil Resources Report 96, UN Food and Agriculture Organization, Rome, 334pp

FAO-ISRIC-ISSS (1998) *World Reference Base for Soil Resources*, Soil Resources Report no 84, UN Food and Agriculture Organization, Rome

FAO-UNESCO (1974) *Legend of the Soil Map of the World*, UN Food and Agriculture Organization, Rome

IUSS Working Group WRB (2006) *World Reference Base for Soil Resources*, 2nd edition, World Soil Resources Reports no 103, UN Food and Agriculture Organization, Rome, 128pp

Jones, A., Montanarella, L., Jones, R. (eds) (2006) *Soil Atlas of Europe*, Joint Research Centre, Ispra, 128pp

Nachtergaele, F. (2005) 'The "soils" to be classified in the World Reference Base for Soil Resources', *Eurasian Soil Science*, vol 38, suppl 1, pp13–19

Nachtergaele, F. O., Spaargaren, O., Deckers, J. A. and Ahrens, R. A. (2000) 'New developments in soil classification World Reference Base for Soil Resources', *Geoderma*, vol 96, no 4, pp345–357

Schad, P. (2008) 'New wine in old wineskins: Why soil maps cannot simply be "translated" from WRB 1998 into WRB 2006', in W. H. Blum, M. H. Gerzabek and M. Vodrazka (eds) *Eurosoil 2008, Book of Abstracts*, BOKU, Vienna, Austria, 4 August 2008, p120

Shishov, L. L., Tonkonogov, V. D., Lebedeva, I. I. and Gerasimova, M. I. (2004) *Classification and Diagnostics of Soils of Russia*, Oykumena, Smolensk, 342pp (in Russian)

Soil Survey Staff (1999) *Soil Taxonomy: A Basic System of Soil Classification for Making and Interpreting Soil Surveys*, USDA, Handbook 436, 2nd edition, United States Government Printing Office, Washington DC, 696pp

Spaargaren, O. (ed) (1994) *World Reference Base for Soil Resources (Draft)*, World Soil Resources Reports no 84, UN Food and Agricultural Organization, Rome, 88pp

5
The United States Soil Taxonomy

Objectives and scope

The main problems that motivated American soil scientists to work on a new classification were the following:

- The previous classification did not give objective criteria for distinguishing the taxa of the higher levels, the borders between most classes were fuzzy, the diagnostic soil properties were generally subjective, and global relationships among soils were not clear.
- Ambiguous terminology led to misunderstanding: as the scale of soil survey increased, different soils were often described under the same name.
- The absence of quantitative criteria for soil diagnostics led to difficulties in the work of practical soil surveyors.
- The old classification did not readily incorporate soil series.

These problems were addressed by creating a number of approximations of the US Soil Taxonomy. Though numerous criticisms were aroused by the publication of this classification, it appeared to be an important stage in the development of a world soil classification. Soil Taxonomy is used as an official classification not only in the US, but also in dozens of other countries. Many national and international classifications, created later (in Canada, China, World Soil Map legend FAO-UNESCO and WRB, and many others), borrowed basic ideas from this system. Even classifications based on different principles (such as in Russia or Australia) have some elements in structure, diagnostics and terminology, similar to Soil Taxonomy. For the moment Soil Taxonomy is the most precisely developed classification in the world and in practice has an international status.

Theoretical background

Several basic principles support the American soil classification. First, the object of classification is the profile or a small representative volume, not processes

Table 5.1 *The scope of the US Soil Taxonomy*

Superficial bodies	Representation in the system
Natural soils	Worldwide coverage
Urban soils	Classified as if they are natural soils
Man-transported materials	Classified as if they are natural soils
Bare rock	Not recognized as soils
Subaquatic soils	Not recognized as soils
Soils deeply transformed by agricultural activities	Mostly classified as if they are natural soils; in the order of Inceptisols recognized at great group level
Small spatial entities	Not recognized

or factors of soil formation. Second, all the levels should be separated on the basis of quantitative diagnostic soil properties. Whittaker (1975) notes that in a hidden form Soil Taxonomy implies the idea of continuity of soil cover, which is artificially separated into classes by formal criteria.

One of the main concerns of the authors of Soil Taxonomy about earlier classifications was that they paid too much attention to the factors of soil formation. It was stated that natural objects should be classified according to their internal properties. It is interesting to note, however, that Soil Taxonomy appears to be more 'climatic' than any other. It uses surrogate atmospheric climatic data to estimate and extend measurable internal temperature and moisture parameter trends in soils. Most suborders and two orders (*Aridisols* and *Gelisols*) are distinguished by soil climatic conditions that limit current soil-forming processes.

Traditionally Soil Taxonomy was erroneously contrasted with 'genetic' classifications, as it was considered to be more soil survey-oriented. The diagnostic horizons in Soil Taxonomy are distinguished not only on the basis of their practical significance for agriculture, but also because they are characteristic of many conditions affecting soil-forming processes. For some land management there is little difference if there is a *spodic* horizon in a soil or not. This horizon is thought to relate to the process of migration of aluminium, silica, humus and/or iron in a soil profile. The interpretation of the genetic nature of these diagnostic horizons is illustrated by the following change of methodology. For a long time a spodic horizon was recognized by the ratio of Al + 1/2 Fe, extracted by pyrophosphate, to the same elements extracted by dithionite-citrate-bicarbonate because it was considered that podzol formation was connected mainly with the migration of aluminium and iron-organic complexes which were believed to be extracted by pyrophosphate. Currently podzol formation is believed to be mainly the result of accumulation of X-ray amorphous alumosilicates in the *spodic* horizon. These criteria were changed. An acid oxalate buffered solution is used for iron and aluminium extractions which are believed to dissolve amorphous allophanes and imogolite minerals.

Development

Before the year 1949 two independent systems of soil classification existed in the US. One system, used for larger-scale mapping, was based on empirical separation of soils into so-called soil series and soil types: 'groups of soils developed from the same kind of parent material by the same genetic combination of processes, and whose horizons are quite similar in their arrangement and general characteristics' (Brady, 1974). This system was used mainly for agricultural soil survey. The names of series, like in geology, were given according to soil texture (type), and the name of the place of their first description (series): for example, 'Manitoba clay loam', 'Malibu blue clay' and so on. For scientific purposes a genetic soil classification by Kellogg, Thorp and Baldwin (Buol et al, 1977) was used. This system was similar to those proposed by early works of the Russian school with Great Groups placed into Suborders under Azonal, Intrazonal and Zonal categories. These two systems did not fit well with each other. In 1949 work on a new taxonomic classification began. The idea was to create taxonomy with soil series as the lowest level. An outstanding soil scientist, Guy D. Smith, headed the work. In 1960 the first complete version, known as the 7th Approximation, was published. In 1975 the US Department of Agriculture recommended the first final edition for routine use. For following years additions and modification recommended by specific international committees were published as a series of Keys to Soil Taxonomy, and in the year 1999 the second edition of *Soil Taxonomy* (significantly enlarged and modified) was published (Soil Survey Staff, 1999).

Structure

In the US Soil Taxonomy there are the following levels: orders, suborders, great groups, subgroups, families and series. Soil type was dropped and considered to be a surface texture phase of a soil series. Phases are outside the formal classification system and commonly use defined properties or conditions that are relevant to the use and management of soils, such as stoniness, rockiness, slope degree and complexity, salty surface, protected from flooding, irrigated where not common and degree and kind of erosion. Phases may be designed for use with map units named at any categorical level, for example, Gently undulating Mollisols, or Steep, stony Udepts. Commonly they are carefully defined and controlled for use in soil survey activities.

Because the system was designed to assist in making and interpreting soil surveys, the diagnostic properties and features selected to satisfy category definitions include both dynamic and static properties of soils. The margin of temporal and spatial attributes has often been misunderstood as inconsistency in applying diagnostics; however, the groupings of soils have permitted many pragmatic interpretations as well as identifying bodies of soils in the pedosphere and suggesting some aspects of their order in nature.

Table 5.2 indicates the general structure of Soil Taxonomy with its six categories – Orders to Series. The Order may be defined as soils having

Table 5.2 *The structure of the US Soil Taxonomy*

Level	Taxon name	Taxon characteristics	Borders between classes	Diagnostics	Terminology
0	Soils	Kingdom			
1	Order	Collective	Formal	Chemico-morphological and regimes	Artificial
2	Suborder	Collective	Formal	Regimes and morphological	Artificial
3	Great group	Generic 1	Formal	Chemico-morphological	Artificial
4	Subgroup	Specific Varietal	Formal	Chemico-morphological	Artificial
5	Family	Specific Varietal	Formal	Chemico-mineralogical	Mixed
6	Series	Generic 2	Formal	Chemico-morphological	Traditional

properties or conditions resulting from, or reflecting, major soil-forming processes that are sufficiently stable in a pedologic sense (Arnold and Eswaran, 2003). Insofar as highly organic natural accumulations, those of volcanic debris, those of highly weathered and resistant minerals, and those of high shrink–swell clays can be recognized and identified as soils, classes of Histosols, Andisols, Oxisols and Vertisols are separated as Order classes. Soils that have cold temperatures and reflect freezing and thawing, and other soils with relict features or current regimes of aridity are separated as Gelisol and Aridisol Order classes, respectively. Thus diagnostic soil features are selected to specify the details for consistent recognition and placement of soil entities into the designated classes.

The Suborder category may be defined as soils within an Order class having additional properties or conditions that are major controls, or reflect such controls, on the current set of soil-forming processes. At this level more dynamic features are selected as evidence of influences on pedogenesis. Some are relict properties such as fragipans, but many are dynamic temporal properties such as moisture regimes and/or temperature regimes if not previously used at a higher level in the system for that group of soils. Thus Alfisols may be separated into Aqualf, Cryalf, Udalf, Ustalf and Xeralf classes at the suborder level. In other Orders the major controls may be materials such as the sandy Arents, or the salty Salids. A quick examination of the Keys to Soil Taxonomy reveals the judgements about priorities for process controls in different environmental settings.

The Great Groups are soils within a Suborder having additional properties that constitute subordinate or additional controls, or reflect such controls on the current set of soil-forming processes. Classes of Suborders consequently provide additional information useful for interpreting soil behaviour in various landscape settings. Priorities are given in the keys to guide consistent placement of soil entities. Because this approach subdivides the pedosphere into more homogeneous groups there is commonly one class that is the residual from the Suborder class being considered. It is designated as the Haplo- class and likely includes soils that may be separated in the future.

The Subgroups are more complicated as they are soils within a Great Group having additional properties resulting from a blending or overlapping of sets of processes in space and time that cause one kind of soil to develop from, or towards, another kind of soil. These classes are intergrades with linkages to other Great Groups, Suborders, or Orders. Also included are other soils called extragrades, having sets of processes or conditions that had not been recognized as diagnostic for any class at a higher level, including non-soil features. A bedrock contact (lithic) at a shallow depth would be such an extragrade.

Families are soils within a Subgroup having properties that often are indicative of the potential for further pedogenic development. Such properties are often characteristic of chemical and physical capacity to change. Included are soil textures including coarse fragments at specified locations in a profile, additional soil temperature variations, mineralogy and activity of clays. Most diagnostics at the family level are relevant to use and management of soils. Details of soil series classification are not shown in the primary structure of Soil Taxonomy or its Keys as they pertain to many properties not applied in higher level classes, such as horizon thickness, colours, structural units, in-place biological features, and other information about the parent materials present in the family class. Most laboratory and other support data are provided in electronic data bases (NASIS) for designated soil series.

Diagnostics

The placement into most taxa is made on the basis of the presence of certain diagnostic horizons, materials and properties in a soil profile. The diagnostic horizons, materials and properties are defined quantitatively. For most definitions special tests are needed; these tests require laboratory equipment, and only a few criteria are possible to determine in the field. In some cases not only simple chemical analyses are required, but also mineralogical composition and microstructure are investigated to confirm the presence of a certain horizon. These analytical requirements limit the possibilities for field soil diagnostics. However, the second edition of *Soil Taxonomy* has less diagnostic criteria requiring expensive and time-consuming laboratory analyses; more attention is given to field diagnostics.

A peculiarity of the US Soil Taxonomy is that it requires measured or estimated information on water and temperature regimes of soils. It is theoretically impossible to classify any soil without climatic information; however, thousands of 30-year or more climate records exist around the world and current models permit estimations and extrapolations of such information. No other soil classification in the world is so climate-oriented. This was due, in large part, because the US contains various climatic conditions, from arctic to tropical ones. From a practical point of view it was important to indicate favourable and limiting conditions for agriculture, grazing and forestry.

The use of quantitative criteria allows even a non-specialist to name a soil, if necessary information on the profile is available.

The diagnostics in this classification can be characterized as quantitative climatic-chemico-morphological.

Terminology

In the 1960s the situation of using traditional terminology in soil science was becoming critical. On the one hand, the same terms were used in different meanings by different authors, and on the other hand, some soils had dozens of synonyms. The American specialists proposed a radical departure. No traditional names are used in Soil Taxonomy to minimize ambiguity and avoid unnecessary associations. Especially for this classification a new system of terms based on Greek and Latin roots was devised. In cases when it was difficult to find roots fitting the archetype, meaningless syllables were used (e.g. for the names of the orders *Alfisols* and *Entisols*). American soil scientists paid great attention to the construction of these terms. For practically all the taxa several variants were proposed; many variants were rejected as difficult for pronunciation or as causing misunderstanding (Heller, 1963).

An interesting idea was to have mnemonic nomenclature involved, including the roots of the names of higher taxa in the names of lower ones. As a result it is possible to understand from a single name not only specific properties of a taxon but also its place in the classification. Table 5.3 illustrates the nomenclature structure of Soil Taxonomy.

For example, the subgroup of Humic Vitrixerands has a specific property of high humus content in the surface horizon; also it is possible to understand that it is included in the great group of Vitrixerands, which are characterized with low water-holding capacity due to high volcanic glass content in respect to volcanic ash and tuffs; also it is clear that this soil belongs to the suborder of Xerands, soils with a dry water regime; finally, the ending of the term points to the fact that this soil is included in the order of Andisols, formed on pyroclastic sediments. This system is similar to an 'analytic language' of John Wilkins, described in an essay by J. L. Borges by the same name (1942):

Table 5.3 *Example of nomenclature structure used in Soil Taxonomy*

Andisols		ends with –<u>sols</u>; root is vowel and consonants –<u>and</u>; Andi comes from Ando ('dark soil' in Japanese for soils from volcanic materials)
Xerand	two syllables	<u>Xer</u>- refers to xeric moisture regime
Vitrixerand	three/four syllables	<u>Vitri</u>- from vitreous glassy materials
Humic Vitrixerand	separate modifier(s) ending with –ic	<u>humic</u>- organic enriched surface layer
eutric, isothermic Humic Vitrixerand		high base saturation, isothermic temperature regime

> *He divided the universe into forty categories or genera, these being further subdivided into differences, which were subdivided into species. He assigned to each genus a monosyllable of two letters; to each difference, a consonant; to each species, a vowel. For example: de, which means an element; deb, the first of the elements, fire; deba, a part of the element fire, a flame.*

Though the terminology of the American classification is strictly regulated, and new terms are included only officially under auspices of the US Department of Agriculture, the use of Soil Taxonomy in other countries is difficult to control. For example, in Papua New Guinea the 7th Approximation was modified (Haantjens and Bleeker, 1975), and distorted names of soil orders were used (Alfosols, Vertosols, Oxosols, Alfosols, Ultosols, Cambosols, Humosols Entosols and Histisols). Some changes in Soil Taxonomy have not been as rigorously tested as hoped for, thus some discrepancies have occurred (Beinroth and Eswaran, 2003). Such examples serve as a caution to work with acceptable global guidelines and protocols to maintain a universal system.

Correlation

Here we propose a correlation of the terminology used in the second edition of *Soil Taxonomy* (Soil Survey Staff, 1999) with the *World Reference Base for Soil Resources* (WRB) (IUSS Working Group WRB, 2006). Unfortunately, the lack of space in this edition does not allow us to present correlation before the level of subgroups. We should stress that it is only correlation, an approximate correspondence: even in the cases when the WRB borrowed some diagnostic horizons or soil taxa from Soil Taxonomy, the quantitative criteria are in most cases different. For example, Histosols are distinguished in Soil Taxonomy when the organic material depth is more than 40cm (>60cm if it is *fibric* material), and in the WRB this depth should be more than 50cm. Also the WRB is lacking the climatic criteria used for distinguishing suborders in most orders of Soil Taxonomy: that is why most of great groups in different suborders are correlated in the same WRB taxa.

Alfisols – soil order. ≈ Lixisols / Luvisols / Solonetz / Albeluvisols / Planosols / Stagnosols
Aqualfs – soil suborder. ≈ Planosols / Stagnic Solonetz / Stagnosols / Luvisols / Albeluvisols
The following great groups are distinguished within the suborder:

- *Albaqualfs* ≈ Planosols (Albic)
- *Cryaqualfs* ≈ Gelic Planosols / Gelic Stagnosols
- *Duraqualfs* ≈ Planosols (Petroduric) / Stagnosols (Petroduric)
- *Endoaqualfs* ≈ Gleyic Luvisols
- *Epiaqualfs* ≈ Haplic Stagnosols
- *Fragaqualfs* ≈ Planosols (Fragic) / Stagnosols (Fragic)

- *Glossaqualfs* ≈ Stagnic Albeluvisol
- *Kandiaqualfs* ≈ Planosols / Stagnosols / Stagnic Albeluvisols
- *Natraqualfs* ≈ Stagnic Solonetz
- *Plintaqualfs* ≈ Plinthic Planosols / Plinthic Stagnosols
- *Vermaqualfs* ≈ Planosols (Vermic) / Stagnosols (Vermic)

Cryalfs – soil suborder. ≈ Albeluvisols (Gelic) / Luvisols (Gelic)
The following great groups are distinguished within the suborder:

- *Glossocryalfs* ≈ Albeluvisols (Gelic) / Solonetz (*Gelic*)
- *Haplocryalfs* ≈ Luvisols (Gelic) / Lixisols (Gelic) / Solonetz (*Gelic*)
- *Palecryalfs* ≈ Albeluvisols (Gelic) / Solonetz (*Gelic*)

Udalfs – soil suborder. ≈ Luvisols / Albeluvisols / Lixisols / Solonetz
The following great groups are distinguished within the suborder:

- *Ferrudalfs* ≈ Ferric Albeluvisols
- *Fragiudalfs* ≈ Luvisols (Fragic) / Lixisols (Fragic)
- *Fraglossudalfs* ≈ Albeluvisols (*Fragic*)
- *Glossudalfs* ≈ Albeluvisols
- *Hapludalfs* ≈ Luvisols
- *Kandiudalfs* ≈ Lixisols (Profondic)
- *Kanhapludalfs* ≈ Lixisols
- *Natrudalfs* ≈ Solonetz
- *Paleudalfs* ≈ Luvisols (Profondic)
- *Rhodudalfs* ≈ Luvisols (Rhodic)

Ustalfs – soil suborder. ≈ Luvisols / Lixisols / Luvic Durisols / Solonetz / Plinthisols
The following great groups are distinguished within the suborder:

- *Durustalfs* ≈ Luvic Petric Durisols
- *Haplustalfs* ≈ Luvisols
- *Kandiustalfs* ≈ Lixisols (Profondic)
- *Kanhaplustalfs* ≈ Lixisols
- *Natrustalfs* ≈ Solonetz
- *Paleustalfs* ≈ Luvisols (Profondic)
- *Plinthustalfs* ≈ Lixic Plinthisols / Petroplinthic Lixisols / Pisoplinthic Lixisols / Plinthic Lixisols
- *Rhodustalfs* ≈ Rhodic Luvisols

Xeralfs – soil suborder. ≈ Luvisols / Luvic Durisols / Solonetz
The following great groups are distinguished within the suborder:

- *Durixeralfs* ≈ Petric Luvic Durisols
- *Fragixeralfs* ≈ Luvisols (Fragic) / Lixisols (Fragic)

- *Haploxeralfs* ≈ Haplic Luvisols / Haplic Lixisols
- *Natrixeralfs* ≈ Solonetz
- *Palexeralfs* ≈ *Petrocalcic* Luvisols / *Petrocalcic* Lixisols
- *Plinthoxeralfs* ≈ Lixic Plinthisols / Petroplinthic Lixisols / Pisoplinthic Lixisols / Plinthic Lixisols
- *Rhodoxeralfs* ≈ Rhodic Luvisols / Rhodic Lixisols

Andisols – soil order. ≈ Andosols
Aquands – soil suborder. ≈ Histic Andosols / Gleyic Andosols
The following great groups are distinguished within the suborder:

- *Cryaquands* ≈ Histic Andosols (Gelic) / Gleyic Andosols (Gelic)
- *Duraquands* ≈ Petroduric Histic Andosols / Petroduric Gleyic Andosols
- *Endoaquands* ≈ Gleyic Andosols
- *Epiaquands* ≈ *Stagnic* Andosols
- *Melanaquands* ≈ Gleyic Melanic Andosols
- *Placaquands* ≈ Gleyic Andosols (Placic)
- *Vitraquands* ≈ Gleyic Vitric Andosols

Cryands – soil suborder. ≈ Andosols (Gelic)
The following great groups are distinguished within the suborder:

- *Duricryands* ≈ Petroduric Andosols (Gelic)
- *Fulvicryands* ≈ Fulvic Andosols (Gelic)
- *Haplocryands* ≈ Andosols (Gelic)
- *Hydrocryands* ≈ Hydric Andosols (Gelic)
- *Melanocryands* ≈ Melanic Andosols (Gelic)
- *Vitricryands* ≈ Vitric Andosols (Gelic)

Torrands – soil suborder. ≈ Andosols
The following great groups are distinguished within the suborder:

- *Duritorrands* ≈ Petroduric Andosols / Petrocalcic Andosols
- *Haplotorrands* ≈ Andosols
- *Vitritorrands* ≈ Vitric Andosols

Udands – soil suborder. ≈ Andosols
The following great groups are distinguished within the suborder:

- *Durudands* ≈ Petroduric Andosols
- *Fulvudands* ≈ Fulvic Andosols
- *Hapludands* ≈ Andosols
- *Hydrudands* ≈ Hydric Andosols
- *Melanudands* ≈ Melanic Andosols
- *Placudands* ≈ Andosols (Placic)

Ustands – <u>soil suborder</u>. ≈ Andosols
The following <u>great groups</u> are distinguished within the suborder:

- *Durustands* ≈ Petroduric Andosols
- *Haplustands* ≈ Andosols

Vitrands – <u>soil suborder</u>. ≈ Vitric Andosols
The following <u>great groups</u> are distinguished within the suborder:

- *Udivitrands* ≈ Vitric Andosols
- *Ustivitrands* ≈ Vitric Andosols

Xerands – <u>soil suborder</u>. ≈ Andosols
The following <u>great groups</u> are distinguished within the suborder:

- *Haploxerands* ≈ Andosols
- *Melanoxerands* ≈ Melanic Andosols
- *Vitrixerands* ≈ Vitric Andosols

<u>Aridisols</u> – <u>soil order</u>. ≈ Calcisols / Gypsisols / Durisols / Solonchaks / Solonetz / Calcic Luvisols
Argids – <u>soil suborder</u>. ≈ Luvisols / Solonetz
The following <u>great groups</u> are distinguished within the suborder:

- *Calciargids* ≈ Calcic Luvisols / Luvisols (Bathycalcic) / Luvic Calcisols
- *Gypsiargids* ≈ Luvisols (*Gypsic*) / Luvic Gypsisols
- *Haplargids* ≈ Luvisols
- *Natriargids* ≈ Solonetz
- *Paleargids* ≈ Profondic Luvisols
- *Petroargids* ≈ Luvisols (*Bathypetrocalcic*) / Luvisols (*Bathypetroduric*) / Luvisols (*Bathypetrogypsic*)

Calcids – <u>soil suborder</u>. ≈ Calcisols
The following <u>great groups</u> are distinguished within the suborder:

- *Haplocalcids* ≈ Calcisols
- *Petrocalcids* ≈ Petric Calcisols

Cambids – <u>soil suborder</u>. ≈ Cambisols / Irragric Anthrosols
The following <u>great groups</u> are distinguished within the suborder:

- *Anthracambids* ≈ Irragric Cambisols / Irragric Anthrosols
- *Aquicambids* ≈ Gleyic Cambisols / Irragric Cambisols
- *Haplocambids* ≈ Cambisols
- *Petrocambids* ≈ Cambisols (*Bathypetrocalcic*) / Cambisols (*Bathypetroduric*) / Cambisols (*Bathypetrogypsic*)

Cryids – <u>soil suborder</u>. ≈ Calcisols (Gelic) / Gypsisols (Gelic) / Cambisols (Gelic) / Luvisols (Gelic) / Solonetz (Gelic) / Solonchaks (Gelic) / Durisols (Gelic)
The following <u>great groups</u> are distinguished within the suborder:

- *Argicryids* ≈ Luvisols (Gelic) / Solonetz (*Gelic*)
- *Calcicryids* ≈ Calcisols (Gelic)
- *Gypsicryids* ≈ Gypsisols (Gelic)
- *Haplocryids* ≈ Cambisols (Gelic)
- *Petrocryids* ≈ Petric Calcisols (Gelic) / Petric Gypsisols (Gelic) / Petric Durisols (Gelic)
- *Salicryids* ≈ Solonchaks (Gelic)

Durids – <u>soil suborder</u>. ≈ Durisols / Petroduric Solonetz
The following <u>great groups</u> are distinguished within the suborder:

- *Argidurids* ≈ Luvic Petric Durisols
- *Haplodurids* ≈ Petric Durisols
- *Natridurids* ≈ Petroduric Solonetz

Gypsids – <u>soil suborder</u>. ≈ Gypsisols / Gypsic Solonetz
The following <u>great groups</u> are distinguished within the suborder:

- *Argigypsids* ≈ Luvic Gypsisols
- *Calcigypsids* ≈ Calcic Gypsisols
- *Haplogypsids* ≈ Gypsisols
- *Natrigypsids* ≈ Gypsic Solonetz
- *Petrogypsids* ≈ Petric Gypsisols

Salids – <u>soil suborder</u>. ≈ Solonchaks
The following <u>great groups</u> are distinguished within the suborder:

- *Aquisalids* ≈ Gleyic Solonchaks
- *Haplosalids* ≈ Solonchaks

Entisols – <u>soil order</u>. ≈ Arenosols / Regosols / Gleysols / Stagnosols / Fluvisols
Aquents – <u>soil suborder</u>. ≈ Gleysols / Gleyic Fluvisols
The following <u>great groups</u> are distinguished within the suborder:

- *Cryaquents* ≈ Gleysols (Gelic) / Gleyic Fluvisols (Gelic)
- *Endoaquents* ≈ Gleysols
- *Epiaquents* ≈ Stagnosols
- *Fluvaquents* ≈ Gleyic Fluvisols
- *Hydraquents* ≈ Gleysols / Fluvisols
- *Psammaquents* ≈ Gleysols (Arenic) / Gleyic Fluvisols (Arenic)
- *Sulfaquents* ≈ Gleysols (Thionic) / Gleyic Fluvisols (Thionic)

Arents – <u>soil suborder</u> (these soils are characterized by a presence of fragments of diagnostic horizons in profile). ≈ Anthrosols / Technosols
The following <u>great groups</u> are distinguished within the suborder:

- *Torriarents* ≈ Anthrosols / Technosols
- *Udarents* ≈ Anthrosols / Technosols
- *Ustarents* ≈ Anthrosols / Technosols
- *Xerarents* ≈ Anthrosols / Technosols

Fluvents – <u>soil suborder</u>. ≈ Fluvisols
The following <u>great groups</u> are distinguished within the suborder:

- *Cryofluvents* ≈ Fluvisols (Gelic)
- *Torrifluvents* ≈ Fluvisols (Eutric)
- *Udifluvents* ≈ Fluvisols
- *Ustifluvents* ≈ Fluvisols
- *Xerofluvents* ≈ Haplic Fluvisols

Orthents – <u>soil suborder</u>. ≈ Regosols
The following <u>great groups</u> are distinguished within the suborder:

- *Cryorthents* ≈ Regosols (Gelic)
- *Torriorthents* ≈ Regosols (Eutric)
- *Udorthents* ≈ Regosols
- *Ustorthents* ≈ Regosols
- *Xerorthents* ≈ Regosols

Psamments – <u>soil suborder</u>. ≈ Arenosols
The following <u>great groups</u> are distinguished within the suborder:

- *Cryopsamments* ≈ Arenosols (Gelic)
- *Quartzipsamments* ≈ Arenosols
- *Torripsamments* ≈ Arenosols (Eutric)
- *Udipsamments* ≈ Arenosols
- *Ustipsamments* ≈ Arenosols
- *Xeropsamments* ≈ Arenosols

<u>Gelisols</u> – <u>soil order</u>. ≈ Cryosols / Cryic Histosols
Histels – <u>soil suborder</u>. ≈ Cryic Histosols
The following <u>great groups</u> are distinguished within the suborder:

- *Fibrihistels* ≈ Cryic Fibric Histosols
- *Folistels* ≈ Cryic Folic Histosols
- *Glacistels* ≈ Cryic Glacic Histosols
- *Hemistels* ≈ Cryic Hemic Histosols
- *Sapristels* ≈ Cryic Sapric Histosols

Orthels – <u>soil suborder</u>. ≈ Cryosols
The following <u>great groups</u> are distinguished within the suborder:

- *Anhyorthels* ≈ Cryosols (Aridic)
- *Aquorthels* ≈ Cryosols (Reductaquic)
- *Argiorthels* ≈ (*Luvic*) Cryosols
- *Haplorthels* ≈ Cryosols
- *Historthels* ≈ Histic Cryosols
- *Mollorthels* ≈ Mollic Cryosols
- *Psammorthels* ≈ Cryosols (Arenic)
- *Umbrorthels* ≈ Umbric Cryosols

Turbels – <u>soil suborder</u>. ≈ Turbic Cryosols
The following <u>great groups</u> are distinguished within the suborder:

- *Anhyturbels* ≈ Turbic Cryosols (Aridic)
- *Aquiturbels* ≈ Turbic Cryosols (Reductaquic)
- *Haploturbels* ≈ Turbic Cryosols
- *Histoturbels* ≈ Histic Turbic Cryosols
- *Molliturbels* ≈ Mollic Turbic Cryosols
- *Psammoturbels* ≈ Turbic Cryosols (Arenic)
- *Umbriturbels* ≈ Umbric Turbic Cryosols

Histosols – <u>soil order</u>. ≈ Histosols
Fibrists – <u>soil suborder</u>. ≈ Fibric Histosols
The following <u>great groups</u> are distinguished within the suborder:

- *Cryofibrists* ≈ Fibric Histosols (Gelic)
- *Haplofibrists* ≈ Fibric Histosols
- *Sphagnofibrists* ≈ Ombric Fibric Histosols

Folists – <u>soil suborder</u>. ≈ Folic Histosols
The following <u>great groups</u> are distinguished within the suborder:

- *Cryofolists* ≈ Folic Histosols (Gelic)
- *Torrifolists* ≈ Folic Histosols
- *Udifolists* ≈ Folic Histosols
- *Ustifolists* ≈ Folic Histosols

Hemists – <u>soil suborder</u>. ≈ Histosols
The following <u>great groups</u> are distinguished within the suborder:

- *Cryohemists* ≈ Hemic Histosols (Gelic)
- *Haplohemists* ≈ Hemic Histosols
- *Sulfihemists* ≈ Hemic Histosols (Protothionic)
- *Sulfohemists* ≈ Hemic Histosols (Orthothionic)
- *Tropohemists* ≈ Hemic Histosols

Saprists – <u>soil suborder</u>. ≈ Sapric Histosols
The following <u>great groups</u> are distinguished within the suborder:

- *Cryosaprists* ≈ Sapric Histosols (Gelic)
- *Haplosaprists* ≈ Sapric Histosols
- *Sulfisaprists* ≈ Sapric Histosols (Protothionic)
- *Sulfosaprists* ≈ Sapric Histosols (Orthothionic)

<u>Inceptisols</u> – <u>soil order</u>. ≈ Cambisols / Regosols / Durisols / Plinthisols / Gleysols / Anthrosols / Umbrisols / Gleyic Solonetz
Anthrepts – <u>soil suborder</u>. ≈ Umbrisols (Anthric) / Plaggic Anthrosols
The following <u>great groups</u> are distinguished within the suborder:

- *Haplanthrepts* ≈ Umbrisols (Anthric)
- *Plagganthrepts* ≈ Plaggic Anthrosols

Aquepts – <u>soil suborder</u>. ≈ Gleysols / Gleyic Solonetz / Plinthisols
The following <u>great groups</u> are distinguished within the suborder:

- *Cryaquepts* ≈ Gleysols (Gelic)
- *Endoaquepts* ≈ Gleysols
- *Epiaquepts* ≈ Stagnosols
- *Fragaquepts* ≈ Fragic Gleysols
- *Halaquepts* ≈ Sodic Gleysols / Gleyic Solonchaks
- *Humaquepts* ≈ Histic Gleysols / Umbric Gleysols / Mollic Gleysols
- *Petraquepts* ≈ Petrolinthic Gleysols / Petrogypsic Gleysols / Petrocalcic Gleysols
- *Sulfaquepts* ≈ Thionic Gleysols
- *Vermaquepts* ≈ Gleysols (*Vermic*)

Cryepts – <u>soil suborder</u>. ≈ Gelic Cambisols
The following <u>great groups</u> are distinguished within the suborder:

- *Dystrocryepts* ≈ Cambisols (Dystric, Gelic)
- *Eutrocryepts* ≈ Cambisols (Eutric, Gelic)

Udepts – <u>soil suborder</u>. ≈ Cambisols / Petric Durisols
The following <u>great groups</u> are distinguished within the suborder:

- *Durudepts* ≈ Petric Durisols
- *Dystrudepts* ≈ Cambisols (Dystric)
- *Eutrudepts* ≈ Cambisols (Eutric)
- *Fragiudepts* ≈ Cambisols (*Fragic*)
- *Sulfudepts* ≈ Cambisols (*Thionic*)

Ustepts – <u>soil suborder</u>. ≈ Cambisols / Petric Durisols / Calcisols

The following great groups are distinguished within the suborder:

- *Calciustepts* ≈ Calcisols
- *Duriustepts* ≈ Petric Durisols
- *Dystrustepts* ≈ Cambisols (Dystric)
- *Haplustepts* ≈ Cambisols

Xerepts – soil suborder. ≈ Cambisols / Petric Durisols / Calcisols
The following great groups are distinguished within the suborder:

- *Calcixerepts* ≈ Calcisols
- *Durixerepts* ≈ Petric Durisols
- *Dystroxerepts* ≈ Cambisols (Dystric)
- *Fragixerepts* ≈ Cambisols (*Fragic*)
- *Haploxerepts* ≈ Cambisols

Mollisols – soil order. ≈ Chernozems / Phaeozems / Kastanozems / Mollic Solonetz
Albolls – soil suborder. ≈ Luvic Phaeozems (Albic, *Ferric*) / Mollic Solonetz (Albic, *Ferric*)
The following great groups are distinguished within the suborder:

- *Argialbolls* ≈ Luvic Phaeozems (Albic, *Ferric*)
- *Natrialbolls* ≈ Mollic Solonetz (Albic, *Ferric*)

Aquolls – soil suborder. ≈ Gleyic Chernozems / Gleyic Phaeozems / Mollic Gleyic Solonetz
The following great groups are distinguished within the suborder:

- *Argiaquolls* ≈ Luvic Gleyic Chernozems / Luvic Gleyic Phaeozems / Luvic Gleyic Kastanozems
- *Calcaquolls* ≈ Gleyic Chernozems / Gleyic Kastanozems
- *Cryaquolls* ≈ Gleyic Chernozems (*Gelic*) / Gleyic Phaeozems (*Gelic*) / Gleyic Kastanozems (*Gelic*)
- *Duraquolls* ≈ Petroduric Gleyic Chernozems / Petroduric Gleyic Phaeozems / Petroduric Gleyic Kastanozems
- *Endoaquolls* ≈ Gleyic Phaeozems
- *Epiaquolls* ≈ Stagnic Phaeozems
- *Natraquolls* ≈ Mollic Gleyic Solonetz

Cryolls – soil suborder. ≈ Chernozems / Phaeozems / Mollic Solonetz
The following great groups are distinguished within the suborder:

- *Argicryolls* ≈ Luvic Chernozems (Gelic) / Luvic Phaeozems (Gelic) / Luvic Kastanozems (Gelic)
- *Calcicryolls* ≈ Calcic Chernozems (Gelic) / Calcic Kastanozems (Gelic)

- *Duricryolls* ≈ Petroduric Chernozems (Gelic) / Petroduric Phaeozems (Gelic) / Petroduric Kastanozems (Gelic)
- *Haplocryolls* ≈ Chernozems (Gelic) / Phaeozems (Gelic) / Kastanozems (Gelic)
- *Natricryolls* ≈ Mollic Solonetz (*Gelic*)
- *Palecryolls* ≈ Luvic Chernozems (*Gelic*) / Luvic Phaeozems (*Gelic*) / Luvic Kastanozems (Gelic)

Rendolls – soil suborder. ≈ Rendzic Leptosols / Rendzic Phaeozems
The following great groups are distinguished within the suborder:
- *Cryrendolls* ≈ Rendzic Leptosols (*Gelic*) / Rendzic Phaeozems (*Gelic*)
- *Haprendolls* ≈ Rendzic Leptosols / Rendzic Phaeozems

Udolls – soil suborder. ≈ Chernozems / Phaeozems / Kastanozems / Mollic Solonetz
The following great groups are distinguished within the suborder:

- *Argiudolls* ≈ Luvic Chernozems / Luvic Phaeozems
- *Calciudolls* ≈ Calcic Chernozems / Petrocalcic Chernozems / Petrocalcic Phaeozems
- *Hapludolls* ≈ Chernozems / Phaeozems
- *Natrudolls* ≈ Mollic Solonetz
- *Paleudolls* ≈ Luvic Chernozems (*Profondic*) / Luvic Phaeozems (*Profondic*) / Chernozems (Bathypetrocalcic) / Phaeozems (Bathypetrocalcic)
- *Vermiudolls* ≈ Vermic Chernozems / Vermic Phaeozems

Ustolls – soil suborder. ≈ Chernozems / Kastanozems / Phaeozems / Mollic Solonetz
The following great groups are distinguished within the suborder:

- *Argiustolls* ≈ Luvic Chernozems / Luvic Kastanozems
- *Calciustolls* ≈ Calcic Chernozems / Calcic Kastanozems / Petrocalcic Chernozems / Petrocalcic Kastanozems / Petrocalcic Phaeozems / Gypsic Chernozems / Gypsic Kastanozems
- *Durustolls* ≈ Petroduric Chernozems / Petroduric Kastanozems / Petroduric Phaeozems
- *Haplustolls* ≈ Chernozems / Kastanozems / Phaeozems
- *Natrustolls* ≈ Mollic Solonetz
- *Paleustolls* ≈ Luvic Chernozems (*Profondic*) / Luvic Kastanozems (*Profondic*) / Luvic Phaeozems (*Profondic*) / Chernozems (Bathypetrocalcic) / Kastanozems (Bathypetrocalcic) / Phaeozems (Bathypetrocalcic)
- *Vermiustolls* ≈ Vermic Chernozems / Vermic Kastanozems / Vermic Phaeozems

Xerolls – soil suborder. ≈ Kastanozems / Chernozems / Mollic Solonetz
The following great groups are distinguished within the suborder:

- *Argixerolls* ≈ Luvic Kastanozems / Luvic Chernozems
- *Calcixerolls* ≈ Calcic Kastanozems / Calcic Chernozems / Gypsic Kastanozems / Gypsic Chernozems
- *Durixerolls* ≈ Duric Kastanozems / Duric Chernozems
- *Haploxerolls* ≈ Kastanozems / Chernozems
- *Natrixerolls* ≈ Mollic Solonetz
- *Palexerolls* ≈ Luvic Kastanozems (*Profondic*) / Luvic Chernozems (*Profondic*) / Petrocalcic Kastanozems / Petrocalcic Chernozems

Oxisols – soil order. ≈ Ferralsols
Aquoxes – soil suborder. ≈ Gleyic Ferralsols / Plinthosols
The following great groups are distinguished within the suborder:

- *Acraquoxes* ≈ Gleyic Geric Ferralsols
- *Eutraquoxes* ≈ Gleyic Ferralsols (Eutric)
- *Haplaquoxes* ≈ Gleyic Ferralsols
- *Plinthaquoxes* ≈ Plinthosols / Ferralsols (Bathyplinthic)

Peroxes – soil suborder. ≈ Ferralsols
The following great groups are distinguished within the suborder:

- *Acroperoxes* ≈ Geric Ferralsols
- *Eutriperoxes* ≈ Ferralsols (Eutric)
- *Haploperoxes* ≈ Ferralsols
- *Kandiperoxes* ≈ Acric Ferralsols / Lixic Ferralsols
- *Sombriperoxes* ≈ Ferralsols (Sombric)

Torroxes – soil suborder. ≈ Ferralsols
The following great groups are distinguished within the suborder:

- *Acrotorroxes* ≈ Geric Ferralsols
- *Eutrotorroxes* ≈ Ferralsols (Eutric)
- *Haplotorroxes* ≈ Ferralsols

Udoxes – soil suborder. ≈ Ferralsols
The following great groups are distinguished within the suborder:

- *Acrudoxes* ≈ Geric Ferralsols
- *Eutrudoxes* ≈ Ferralsols (Eutric)
- *Hapludoxes* ≈ Ferralsols
- *Kandiudoxes* ≈ Acric Ferralsols / Lixic Ferralsols
- *Sombriudoxes* ≈ Ferralsols (Sombric)

Ustoxes – soil suborder. ≈ Ferralsols
The following great groups are distinguished within the suborder:

- *Acrustoxes* ≈ Geric Ferralsols
- *Eutrustoxes* ≈ Ferralsols (Eutric)
- *Haplustoxes* ≈ Ferralsols
- *Kandiustoxes* ≈ Acric Ferralsols / Lixic Ferralsols
- *Sombriustoxes* ≈ Ferralsols (Sombric)

Spodosols – soil order. ≈ Podzols
Aquods – soil suborder. ≈ Gleyic Podzols / Histic Podzols
The following great groups are distinguished within the suborder:

- *Alaquods* ≈ Gleyic Podzols
- *Cryaquods* ≈ Gleyic Podzols (Gelic)
- *Duraquods* ≈ Gleyic Podzols (*Densic*)
- *Endoaquods* ≈ Gleyic Podzols
- *Epiaquods* ≈ Stagnic Podzols
- *Fragaquods* ≈ Gleyic Podzols (Fragic)
- *Placaquods* ≈ Gleyic Placic Podzols

Cryods – soil suborder. ≈ Podzols / Gelic Podzols
The following great groups are distinguished within the suborder:

- *Duricryods* ≈ Densic Podzols (Gelic)
- *Haplocryods* ≈ Podzols (Gelic)
- *Humicryods* ≈ Carbic Podzols (Gelic)
- *Placocryods* ≈ Placic Podzols (Gelic)

Humods – soil suborder. ≈ Carbic Podzols
The following great groups are distinguished within the suborder:

- *Durihumods* ≈ Carbic Podzols (*Densic*)
- *Fragihumods* ≈ Carbic Podzols (Fragic)
- *Haplohumods* ≈ Carbic Podzols
- *Placohumods* ≈ Carbic Placic Podzols

Orthods – soil suborder. ≈ Podzols
The following great groups are distinguished within the suborder:

- *Alorthods* ≈ Haplic Podzols
- *Durorthods* ≈ Podzols (*Densic*)
- *Fragiorthods* ≈ Podzols (Fragic)
- *Haplorthods* ≈ Podzols
- *Placorthods* ≈ Placic Podzols

Ultisols – soil order. ≈ Alisols / Acrisols
Aquults – soil suborder. ≈ Gleyic Alisols / Gleyic Acrisols / Acric Plinthosols / Planosols

The following <u>great groups</u> are distinguished within the suborder:

- *Albaquults* ≈ Alic Planosols / Acric Planosols
- *Endoaquults* ≈ Gleyic Alisols / Gleyic Acrisols
- *Epiaquults* ≈ Stagnic Alisols / Stagnic Acrisols
- *Fragaquults* ≈ Gleyic Alisols (Fragic) / Gleyic Acrisols (Fragic)
- *Kandiaquults* ≈ Gleyic Acrisols (Profondic)
- *Kanhaplaquults* ≈ Gleyic Acrisols
- *Paleaquults* ≈ Gleyic Alisols (Profondic)
- *Plinthaquults* ≈ Acric Plinthosols / Gleyic Plinthic Alisols / Gleyic Plinthic Acrisols
- *Umbraquults* ≈ Umbric Gleyic Alisols / Umbric Gleyic Acrisols

Humults – <u>soil suborder</u>. ≈ Alisols (Humic) / Acrisols (Humic) / Plinthosols
The following <u>great groups</u> are distinguished within the suborder:

- *Haplohumults* ≈ Alisols (Humic) / Acrisols (Humic)
- *Kandihumults* ≈ Acrisols (Humic, Profondic)
- *Kanhaplohumults* ≈ Acrisols (Humic)
- *Palehumults* ≈ Alisols (Humic, Profondic)
- *Plinthohumults* ≈ Acric Plinthosols (Humic) / Plinthic Alisols (Humic) / Plinthic Acrisols (Humic)
- *Sombrihumults* ≈ Sombric Acrisols (Humic) / Sombric Alisols (Humic)

Udults – <u>soil suborder</u>. ≈ Alisols / Acrisols / Plinthosols
The following <u>great groups</u> are distinguished within the suborder:

- *Fragiudults* ≈ Alisols (Fragic) / Acrisols (Fragic)
- *Hapludults* ≈ Alisols / Acrisols
- *Kandiudults* ≈ Acrisols (Profondic)
- *Kanhapludults* ≈ Acrisols
- *Paleudults* ≈ Alisols (Profondic)
- *Plinthudults* ≈ Acric Plinthosols / Plinthic Alisols / Plinthic Acrisols
- *Rhodudults* ≈ Alisols (Rhodic) / Acrisols (Rhodic)

Ustults – <u>soil suborder</u>. ≈ Alisols / Acrisols
The following <u>great groups</u> are distinguished within the suborder:

- *Haplustults* ≈ Alisols / Acrisols
- *Kandiustults* ≈ Acrisols (Profondic)
- *Kanhaplustults* ≈ Acrisols
- *Paleustults* ≈ Alisols (Profondic)
- *Plinthustults* ≈ Acric Plinthosols / Plinthic Alisols / Plinthic Acrisols
- *Rhodustults* ≈ Alisols (Rhodic) / Acrisols (Rhodic)

Xerults – <u>soil suborder</u>. ≈ Alisols / Acrisols

The following <u>great groups</u> are distinguished within the suborder:

- *Haploxerults* ≈ Alisols / Acrisols
- *Palexerults* ≈ Alisols (Profondic) / Acrisols (Profondic)

<u>Vertisols</u> – <u>soil order</u>. ≈ Vertisols
Aquerts – <u>soil suborder</u>. ≈ Gleyic Vertisols
The following <u>great groups</u> are distinguished within the suborder:

- *Calcaquerts* ≈ Calcic Gleyic Vertisols
- *Duraquerts* ≈ Petroduric Gleyic Vertisols
- *Dystraquerts* ≈ Gleyic Vertisols (Dystric)
- *Endoaquerts* ≈ Gleyic Vertisols
- *Epiaquerts* ≈ Stagnic Vertisols
- *Natraquerts* ≈ Sodic Gleyic Vertisols
- *Salaquerts* ≈ Gleyic Salic Vertisols

Cryerts – <u>soil suborder</u>. ≈ Vertisols (*Gelic*)
The following <u>great soil groups</u> are distinguished within the suborder:

- *Haplocryerts* ≈ Vertisols (*Gelic*)
- *Humicryerts* ≈ Vertisols (Humi, *Gelic*)

Torrerts – <u>soil suborder</u>. ≈ Vertisols
The following <u>great groups</u> are distinguished within the suborder:

- *Calcitorrerts* ≈ Calcic Vertisols
- *Gypsitorrerts* ≈ Gypsic Vertisols
- *Haplotorrerts* ≈ Vertisols
- *Salitorrerts* ≈ Salic Vertisols

Uderts – <u>soil suborder</u>. ≈ Vertisols
The following <u>great groups</u> are distinguished within the suborder:

- *Dystruderts* ≈ Vertisols (Dystric)
- *Hapluderts* ≈ Haplic Vertisols

Usterts – <u>soil suborder</u>. ≈ Vertisols
The following <u>great groups</u> are distinguished within the suborder:

- *Calciusterts* ≈ Calcic Vertisols
- *Dystrusterts* ≈ Vertisols (Dystric)
- *Gypsusterts* ≈ Gypsic Vertisols
- *Haplusterts* ≈ Vertisols
- *Salusterts* ≈ Salic Vertisols

Xererts – <u>soil suborder</u>. ≈ Vertisols
The following <u>great groups</u> are distinguished within the suborder:

- *Calcixererts* ≈ Calcic Vertisols
- *Durixererts* ≈ Petroduric Vertisols
- *Haploxererts* ≈ Vertisols

References

Arnold, R. W. and Eswaran, H. (2003) 'Conceptual basis for soil classification: Lessons from the past', in H. Eswaran, T. Rice, R. Ahrens and B. A. Stewart (eds) *Soil Classification: A Global Desk Reference*, CRC Press, Boca Raton, FL, pp27–42

Beinroth, F. H. and Eswaran, H. (2003) 'Classification of soils in the tropics: A reassessment of Soil Taxonomy', in H. Eswaran, T. Rice, R. Ahrens and B. A. Stewart (eds) *Soil Classification: A Global Desk Reference*, CRC Press, Boca Raton, FL, pp231–244

Borges, J. L. (1942) 'El idioma analítico de John Wilkins', *La Nación*, 8 February 1942, English translation by Ruth L. C. Simms (1984) 'The Analytical Language of John Wilkins', *Other Inquisitions 1937–1952*, University of Texas Press, Austin, TX, pp34–39

Brady, N.C. (1974) *The Nature and Properties of Soils*, 8th edition, Macmillan Publishing Co, Inc, New York, NY, 639pp

Buol, S. W., McCracken, R. J. and Hole, F. D. (1977) *Soil Genesis and Classification*, 7th edition, Iowa State University Press, Ames, IA, 360pp

Haantjens, H. A. and Bleeker, P. (1975) 'Procedures for computer storage of soil capability and soil classification data', *Geoderma*, vol 13, no 2, pp115–128

Heller, J. L. (1963) 'The nomenclature of soils, or what's in a name?', *Soil Science Society of America Proceedings*, vol 27, no 2, pp216–220

IUSS Working Group WRB (2006) *World Reference Base for Soil Resources*, 2nd edition, World Soil Resources Reports no 103, UN Food and Agriculture Organization, Rome, 128pp

Soil Survey Staff (1999) *Soil Taxonomy: A Basic System of Soil Classification for Making and Interpreting Soil Surveys*, USDA, Handbook 436, 2nd edition, United States Government Printing Office, Washington DC, 696pp

Whittaker, R. H. (1975) *Communities and Ecosystems*, Macmillan, New York, 352pp

6
Soil Classification of Canada

Objectives and scope

The objectives of the Canadian soil classification are practically the same as those for creating the US Soil Taxonomy. Canadian soil science also uses a system of soil series. Poor development of higher levels of taxonomy did not allow successful grouping of these series, the number of which rapidly increased over the years. The concepts of the 7th Approximation, published in the 1960s, including diagnostic horizons, formal quantitative criteria for the taxa, and the structure of taxonomy, were adapted by the Canadians. We would like to introduce a term 'source system' – meaning a classification where the main concepts, structure and, partly, the terminology have been borrowed. The source system for the Canadian soil classification was the 7th Approximation. However, the Canadians did not copy Soil Taxonomy; its format was adapted to Canadian environments, and its terminology was almost completely reworked. No climatic criteria are included at the highest levels of the classification. The Canadian soil classification mostly uses traditional terms, and, some important traditional soil archetypes are also preserved. Soil Taxonomy practically eliminated some widely recognized soil archetypes, such as Gley and Solonetz; however, they exist in the Canadian classification. Alluvial soils are recognized only in the lower levels of classification (some *Cumulic Regosols* and *Organic soils*), illustrating that the importance of soil features is expert-dependent.

The Canadian soil classification is mainly used for soil survey within Canadian territory (Table 6.1). It does not recognize technogenic substrates, exposed rock and underwater sediments as soils. Agricultural and natural soils are not classified separately.

Theoretical background

The theoretical bases of the Canadian soil classification are stated clearly in the text. It is based on two levels of archetypes: soil orders and soil series. Soil order is a generic, synthetic archetype with unity of soils that are traditionally considered to be similar. The meaning of order is close to the concept of soil

Table 6.1 *The scope of the Canadian soil classification*

Superficial bodies	Representation in the system
Natural soils	National coverage
Urban soils	Classified as if they are natural soils
Man-transported materials	Classified as if they are natural soils
Bare rock	Not recognized as soils
Subaquatic soils	Not recognized as soils
Soils deeply transformed by agricultural activities	Classified as if they are natural soils

type in the early, Dokuchaev understanding. It is believed that soil orders represent the most general conceptual 'central images' of soils, and lower taxa are subdivisions of them. The authors recognize that soil taxa are subjective, and cannot be 'discovered': rather they are invented to arrange soils in the most convenient manner. Soil series is an analytical, individual archetype, which is regarded as a group of soils differing from the others. Diagnostic horizons and properties are used to separate orders in most cases.

Development

The classification of soils began in Canada in 1914 (Canada Soil Survey Committee, 1978). From the early stages it was strongly influenced by the US school of pedology. In the 1930s, soil classification was influenced, on the one hand, by the soil series concept having been proposed by Marbut (1951), and, on the other hand, soil genetic grouping at higher levels by Baldwin et al (1938). Later on, the progress in soil survey in Canada resulted in the development of classification of soil geographical units (regions, zones, catenas or associations, series) rather than soil taxonomic units *per se*. In the 1950s the first prototype of the present system appeared. The first official edition was published in the late 1970s (Canada Soil Survey Committee, 1978), and since that time two more editions have been published (Agriculture Canada Expert Committee on Soil Survey, 1987; Soil Classification Working Group, 1998). The Canadian classification is rather stable: only minor modifications have been made during the 30 years of its existence, except for the introduction of Vertisols in the third edition.

Structure

The structure of the Canadian soil classification is similar to that of Soil Taxonomy (Soil Survey Staff, 1999), but the meaning of the taxonomic levels is different (Table 6.2). The highest level is <u>soil order;</u> soil <u>great groups</u> and <u>subgroups</u> are distinguished on the basis of qualitative differences in soil profiles that reflect important controls of soil processes. The generic level is great group, the order level is collective. Within subgroups there are soil families, which are distinguished mainly according to quantitative criteria

Table 6.2 *The structure of the Canadian soil classification*

Level	Taxon name	Taxon characteristics	Borders between classes	Diagnostics	Terminology
0	Soils	Kingdom			
1	Order	Collective	Formal	Chemico-morphological	Mixed
2	Great group	Generic 1	Formal	Chemico-morphological	Artificial
3	Subgroup	Specific	Formal	Chemico-morphological	Artificial
4	Family	Specific / Varietal	Formal	Chemico-mineralogical	Mixed
5	Series	Generic 2	Formal	Chemico-morphological	Traditional

that are thought to reflect additional dynamics of soils: such as characterized temperature and water regimes, texture, and mineralogical composition. Similar to Soil Taxonomy, the name of a family consists of several descriptors (qualifiers) in a line. Soil series are distinguished independently, and later are referred to one or another soil family.

According to its structure the Canadian soil classification is a hierarchical taxonomy with formal borders between classes.

Diagnostics

The diagnostics used in the Canadian soil classification are very similar to those of Soil Taxonomy. Many require various laboratory analyses for a proper determination of diagnostic horizons and properties. The Canadian classification has less emphasis on climatic criteria. The only taxon of the highest level distinguished on the basis of temperature regime is *Cryosols*. The diagnostic soil properties in this classification can be called chemico-morphological.

Terminology

The Canadian system of soil classification preserved most of the traditional terms, mainly proposed by the Russian school: *chernozems, podzolic soils, solonetz, solod*. However, some terms were borrowed from Soil Taxonomy, or were constructed artificially from Greek, Latin and Russian roots (*Gleysols, Luvisols, Brunisols*). Since the works on construction of the Canadian system and the legend for the FAO-UNESCO Soil Map of the World proceeded almost simultaneously, these two classifications have a lot in common not only in structure, but also in terminology.

The names of great soil groups are formed in two ways: they may be separate names, differing from the name of the order, or repeat the name of the order with a modifier. The names of subgroups are constructed by adding modifiers to the names of great groups: *Grey Solodized Solonetz, Gleyed Sombric Humo-Ferric Podzol*, and so on.

Correlation

The Canadian soil classification has a close correspondence with World Reference Base for Soil Resources (WRB) soil groups. Almost all the orders and great groups can be correlated with one WRB taxon. The exceptions are great groups of the Regosolic and Gleysolic orders which are broader than the WRB taxa. It is important, however, to note that quantitative criteria for soil taxa are different in the two systems, and no direct conversion of Canadian terms into WRB soil names is possible. The names of the orders and of great groups within orders are listed alphabetically.

Brunisolic – soil order. ≈ Cambisols / Umbrisols
Four great groups are distinguished within the order:

- *Dystric Brunisol* ≈ Cambisols (Dystric)
- *Eutric Brunisol* ≈ Cambisols (Eutric)
- *Melanic Brunisol* ≈ Mollic Umbrisols
- *Sombric Brunisol* ≈ Umbrisols

Chernozemic – soil order. ≈ Kastanozems / Chernozems / Phaeozems
Four great groups are distinguished within the order:

- *Black Chernozem* ≈ Chernozems / Phaeozems
- *Brown Chernozem* ≈ Kastanozems / Phaeozems
- *Dark Brown Chernozem* ≈ Kastanozems / Phaeozems
- *Dark Grey Chernozem* ≈ Greyic Phaeozems / Chernozems (Greyic)

Cryosolic – soil order. ≈ Cryosols/ Histosols (Gelic)
Three great groups are distinguished within the order:

- *Organic Cryosol* ≈ Histic Cryosols / Histosols (Gelic)
- *Static Cryosol* ≈ Cryosols
- *Turbic Cryosol* ≈ Turbic Cryosols

Gleysolic – soil order. ≈ Gleysols / Planosols
Three great groups are distinguished within the order:
- *Gleysol* ≈ Gleysols
- *Humic Gleysol* ≈ Mollic Gleysols / Umbric Gleysols / Humic Gleysols
- *Luvic Gleysol* ≈ Luvic Gleysols / Planosols

Luvisolic – soil order. ≈ Luvisols / Albeluvisols
Two great groups are distinguished within the order:

- *Grey Brown Luvisol* ≈ Luvisols
- *Grey Luvisol* ≈ Albic Luvisols / Albeluvisols

Organic – soil order. ≈ Histosols

Four <u>great groups</u> are distinguished within the order:

- *Fibrisol* ≈ Fibric Histosols
- *Folisol* ≈ Folic Histosols
- *Humisol* ≈ Sapric Histosols
- *Mesisol* ≈ Hemic Histosols

Podzolic – <u>soil order</u>. ≈ Podzols
Three <u>great groups</u> are distinguished within the order:

- *Ferro-Humic Podzol* ≈ Podzols
- *Humic Podzol* ≈ Carbic Podzols
- *Humo-Ferric Podzol* ≈ Rustic Podzols / Haplic Podzols

Regosolic – <u>soil order</u>. ≈ Regosols / Fluvisols
Two <u>great groups</u> are distinguished within the order:

- *Humic Regosol* ≈ Humic Regosols / Mollic Fluvisols / Umbric Fluvisols / Humic Fluvisols
- *Regosol* ≈ Regosols / Fluvisols

Solonetzic – <u>soil order</u>. ≈ Solonetz / Solodic Planosols
Three <u>great groups</u> are distinguished within the order:

- *Solod* ≈ Solodic Planosols
- *Solodized Solonetz* ≈ Solonetz (Albic)
- *Solonetz* ≈ Solonetz
- *Vertic Solonetz* ≈ Vertic Solonetz / Sodic Vertisols

Vertisolic – <u>soil order</u>. ≈ Vertisols
Three <u>great groups</u> are distinguished within the order:

- *Vertisol* ≈ Vertisols
- *Humic Vertisol* ≈ Vertisols (Humic)

References

Agriculture Canada Expert Committee on Soil Survey (1987) *The Canadian System of Soil Classification*, 2nd edition, Agriculture and Agri-Food Canada Publication 1646, Supply and Services Canada, Ottawa, Ont., 164pp

Baldwin, M., Kellog, C. E. and Thorp, J. (1938) 'Soil classification', in United States Department of Agriculture, *Soils and Men: Yearbook of Agriculture 1938*, United States Government Printing Office, Washington DC, pp979–1001

Canada Soil Survey Committee (1978) *The Canadian System of Soil Classification*, Canadian Department of Agriculture Publication 1646, Supply and Services Canada, Ottawa, Ont., 164pp

Marbut, C. F. (1951) *Soils: Their genesis and classification. A memorial volume of lectures given in the Graduate School of the United States Department of Agriculture in 1928*, 2nd edition, Soil Science Society of America, Washington DC, 134pp

Soil Classification Working Group (1998) *The Canadian System of Soil Classification*, 3rd edition, Agriculture and Agri-Food Canada Publication 1646, Supply and Services Canada, Ottawa, Ont., 187pp

Soil Survey Staff (1999) *Soil Taxonomy: A Basic System of Soil Classification for Making and Interpreting Soil Surveys*, USDA, Handbook 436, 2nd edition, United States Government Printing Office, Washington DC, 696pp

7
French Soil Classification System

Objectives and scope

In a new version of soil classification (AFES, 1998) French soil scientists intended to create something more than another national soil classification. They tried to make a flexible reference base which could serve for constructing numerous practical classifications on its basis. The latest version of the French classification has worldwide coverage (Table 7.1). Soils transformed by agricultural or other human activities are included in the classification. Underwater soils are not represented; however, there is a special reference group HISTOSOLS FLOTTANTS for floating peat soils which represents soils not *under*, but *over* the water, corresponding to Floatic Histosols in the WRB.

Table 7.1 *The scope of the French soil classification*

Superficial bodies	Representation in the system
Natural soils	Worldwide coverage
Urban soils	Included in the reference group ANTHROPOSOLS ARTIFICIELS
Transported materials	Included in the reference group ANTHROPOSOLS ARTIFICIELS
Bare rock	Included in the reference group LITHOSOLS
Subaquatic soils	Not included in the classification
Soils deeply transformed by agricultural activities	Included in the reference group ANTHROPOSOLS TRANSFORMES

Theoretical background

The French soil classification is one of the most developed in methodological theory. All the concepts used are defined in detail, and the whole process of classifying soils is described.

The objects of soil investigation are always so-called soil mantles (*couverture Pédologique*), which are defined as real natural three-dimensional bodies. Soil section, *solum*, is a conceptual two-dimensional cut of soil mantle.

Soil mantle is not homogeneous, and requires profound investigation, which includes digging several sections, as well as research work using drill, auger and remote-sensing methods. The characteristics of soil are given on the basis of these complex data; the same as in biology, where the object of classification is not a single organism, but a population (Zarenkov, 1993). Soil profile is understood as a gradient of one or several properties in a vertical section of soil, for example, texture profile, salinity profile and so forth.

In the process of investigation a soil scientist divides the soil into horizons and, based on research in the field and laboratory, refers each real horizon to a reference one. The operation results in a transition from a real soil section to a conceptual solum having a sequence of reference horizons. The great achievement of the authors of the French classification is that they made clear the cognitive process of soil classification. In classifications based on a factor-genetic approach a specialist makes decisions using 'black box' principles based mainly on his non-verbal knowledge (Hudson, 1992), while in formalized classifications the role of a specialist to a great extent is reduced to fulfilling an algorithm. In the French system the process of recognizing soil horizons and their reference to the conceptual constructions is clearly described.

A conceptual solum is then referred to a certain soil reference (*reference Pédologique*), which is a sequence of reference horizons. The idea of a soil reference is the basic one for the French classification, and serves as an archetype of a certain soil.

Though reference horizons have quantitative diagnostic criteria, there is a basic difference in interpretation of these horizons in the French system and in the majority of other soil classifications, for example, in the US Soil Taxonomy (Soil Survey Staff, 1999). If a horizon does not fit quantitative criteria in Soil Taxonomy, it cannot be called diagnostic, and a soil is referred to another taxa. If a diagnostic horizon does not fit the definition given in the French classification for some properties, but is generally close to the 'central image' of such a horizon, it is still called that diagnostic. In fact quantitative criteria are given only to describe the 'central image' better, but around this there is a 'buffer zone', where similar soils exist, and they are also referred to the same class. Also in the French classification it is allowed to give mixed names for soils if they show transitional properties between two or more reference groups, for example, CALCOSOLS-PELOSOLS or ORGANOSOLS-ARENOSOLS-PEYROSOLS.

Development

For many years, the official system for soil classification in France and its overseas territories was the classification of the Commission on Pedology and Soil Cartography (CPCS, 1967). The system was also slightly modified and adapted for the use in tropical regions by ORSTOM (e.g. Segalen et al, 1979). The French classification was a hierarchical taxonomy with four main levels: classes, subclasses, groups and subgroups. The classes and subclasses were mainly collective levels. The classes partly reflected the stage of soil evolution, and partly general trends of soil-forming processes. The subclasses were grouped by various criteria, commonly by climatic conditions of soil formation. The level of groups was generic, though some groups were more specific particularly about soil structure. The diagnostic soil properties were mainly qualitative, but for some groups quantitative criteria were used, and required laboratory chemical analyses, for example, base saturation. As mentioned above, the classification at higher levels depended on climatic zones. The terminology was almost entirely traditional, with several artificial names (*Vertisols, Andosols*). Since this classification is still used for soil mapping in a number of French-speaking African countries, a correlation is provided for subclasses and groups. Since the classification follows the logical scheme of soil evolution, soil classes and lower taxa are presented in their original order.

Sols minéraux bruts – soil class, with three subclasses included.
Sols minéraux bruts non climatiques ≈ Arenosols / Fluvisols (Arenic) / Regosols (Skeletic, Arenic) / Technosols (Arenic)
There are six groups within the subclass:

- *Sols minéraux bruts d'érosion* ≈ Arenosols
- *Sols minéraux bruts d'apport alluvial* ≈ Fluvisols (Arenic)
- *Sols minéraux bruts d'apport colluvial* ≈ Regosols (Skeletic, Arenic)
- *Sols minéraux bruts d'apport éolien* ≈ Arenosols
- *Sols minéraux bruts d'apport volcanique* ≈ Arenosols (Tephric)
- *Sols minéraux bruts anthropiques* ≈ Technosols (Arenic)

Sols minéraux bruts climatiques des déserts froids ≈ Cryosols (Arenic)
There are three groups within the subclass:

- *Lithosols des déserts froids* ≈ Leptic Cryosols (Arenic)
- *Cryosols bruts inorganisés* ≈ Cryosols (Skeletic, Arenic)
- *Cryosols bruts organisés* ≈ Cryosols (Skeletic, Arenic)

Sols minéraux bruts des déserts chauds ≈ Leptosols (Aridic) / Arenosols (Aridic) / Regosols (Skeletic, Aridic)
There are five groups within the subclass:

- *Lithosols des déserts chauds* ≈ Leptosols (Aridic)
- *Sols minéraux bruts xériques inorganisés d'ablation* ≈ Arenosols (Aridic) / Regosols (Skeletic, Aridic)
- *Sols minéraux bruts xériques organisés d'ablation* ≈ Arenosols (Aridic) / Regosols (Skeletic, Aridic)
- *Sols minéraux bruts xériques inorganisés d'apport* ≈ Arenosols (Aridic) / Regosols (Skeletic, Aridic)
- *Sols minéraux xériques organisés d'apport* ≈ Arenosols (Aridic) / Regosols (Skeletic, Aridic)

<u>Sols peu évolués</u> – soil class, with four subclasses included.
Sols peu évolués à permagel ≈ Cryosols
There are four groups within the subclass:

- *Sols à forte ségrégation de glace non ordonnée* ≈ Glacic Cryosols
- *Sols à forte ségrégation de glace ordonnée en réseau* ≈ Glacic Cryosols
- *Sols sans ségrégation importante de glace à réseau organisé* ≈ Cryosols
- *Sols bruns arctiques* ≈ Cambic Cryosols

Sols peu évolués humifères ≈ Umbric Leptosols
There are three groups within the subclass:

- *Rankers* ≈ Umbric Leptosols
- *Sols humifères litho-calciques* ≈ Umbric Leptosols (Calcaric)
- *Sols peu évolués à allophane* ≈ Umbric Andic Leptosols

Sols peu évolués xériques ≈ Cambisols (Aridic) / Mollic Leptosols (Aridic)
There are two groups within the subclass:

- *Sols gris subdésertiques* ≈ Cambisols (Aridic)
- *Xerorankers* ≈ Mollic Leptosols (Aridic)

Sols peu évolués non climatiques ≈ Regosols / Fluvisols / Technosols
There are six groups within the subclass:

- *Sols d'érosion* ≈ Regosols
- *Sols d'apport alluvial* ≈ Fluvisols
- *Sols d'apport colluvial* ≈ Regosols
- *Sols d'apport éolien* ≈ Regosols
- *Sols d'apport volcanique friable* ≈ Regosols (Tephric)
- *Sols d'apport anthropique* ≈ Technosols

<u>Vertisols</u> – soil class, with two subclasses included.
Vertisols sans drainage externe ≈ Gleyic Vertisols
There are two groups within the subclass:

- *Vertisols sans drainage externe, à structure arrondie* ≈ Gleyic Vertisols
- *Vertisols sans drainage externe, à structure anguleuse* ≈ Gleyic Vertisols

Vertisols a drainage externe ≈ Vertisols
There are two groups within the subclass:

- *Vertisols à drainage externe, à structure arrondie* ≈ Vertisols
- *Vertisols à drainage externe, à structure anguleuse* ≈ Vertisols

<u>Andosols</u> – soil class, with two subclasses included.

Andosols des pays froids ≈ Andosols
There is one group within the subclass:

- *Andosols humifères désaturés* ≈ Andosols (Dystric)

Andosols des pays tropicaux ≈ Andosols
There are two groups within the subclass:

- *Andosols saturés* ≈ Andosols (Eutric)
- *Andosols désaturés* ≈ Andosols (Dystric)

Sols calcimagnésiques – soil class, with three subclasses included.
Sols carbonatés ≈ Rendzic Leptosols / Cambisols (Calcaric)
There are three groups within the subclass:

- *Rendzines* ≈ Rendzic Leptosols
- *Sols bruns calcaires* ≈ Cambisols (Calcaric)
- *Cryptorendzines* ≈ Leptic Cambisols (Calcaric)

Sols saturés ≈ Phaeozems / Calcisols / Cambisols (Calcaric, Humic)
There are three groups within the subclass:

- *Sols bruns calciques* ≈ *Cambic* Calcisols
- *Sols humiques carbonatés* ≈ Phaeozems (Calcaric) / Cambisols (Calcaric, Humic)
- *Sols calciques mélanisés* ≈ Petrocalcic Phaeozems

Sols gypseux ≈ Gypsisols / Leptosols (Gypsiric)
There are two groups within the subclass:

- *Sols gypseux rendziniformes* ≈ Leptosols (Gypsiric)
- *Sols bruns gypseux* ≈ Gypsisols

Sols isohumiques – soil class, with four subclasses included.
Sols isohumiques de climat relativement humide ≈ Phaeozems
There is one group within the subclass:

- *Brunizems* ≈ Phaeozems

Sols isohumiques à pédoclimat très froid ≈ Chernozems / Kastanozems / Cambisols (Humic)
There are three groups within the subclass:

- *Chernozems* ≈ Chernozems
- *Sols chatains* ≈ Kastanozems
- *Sols bruns isohumiques* ≈ Kastanozems / Cambisols (Humic)

Sols isohumiques à pédoclimat frais pendant la saison pluvieuse ≈ Kastanozems / Cambisols
There are two groups within the subclass:

- *Sols marrons* ≈ Kastanozems (Chromic)
- *Sierozems* ≈ Cambisols

Sols isohumiques à pédoclimat chaud pendant la saison pluvieuse ≈ Cambisols (Aridic)
There is one group within the subclass:

- *Sols bruns arides* ≈ Cambisols (Aridic)

Sols brunifiés – soil class, with four subclasses included.
Sols brunifiés des climats tempérés humides ≈ Cambisols / Luvisols / Phaeozems
There are two groups within the subclass:

- *Sols bruns* ≈ Cambisols (Humic) / *Cambic* Phaeozems
- *Sols lessivés* ≈ Luvisols (Humic) / Luvic Phaeozems

Sols brunifiés des climats tempérés continentaux ≈ Greyic Luvic Phaeozems / Luvisols / Albeluvisols
There are two groups within the subclass:

- *Sols gris forestiers* ≈ Greyic Luvic Phaeozems / Luvisols (Greyic)
- *Sols derno-podzoliques* ≈ Albic Luvisols / Albeluvisols

Sols brunifiés des climats boreaux ≈ Luvisols / Alisols
There is one group within the subclass:

- *Sols lessivés boreaux* ≈ Luvisols / Alisols

Sols brunifiés des climats tropicaux ≈ Phaeozems / Cambisols (Humic, Eutric)
There is one group within the subclass:

- *Sols bruns eutrophes tropicaux* ≈ Phaeozems / Cambisols (Humic, Eutric)

Sols podzolisés – soil class, with three subclasses included.
Sols podzolisés de climat tempéré ≈ Podzols
There are four groups within the subclass:

- *Podzols* ≈ Podzols
- *Sols podzoliques* ≈ Umbric Podzols
- *Sols ocrés podzoliques* ≈ Rustic Podzols
- *Sols cryptopodzoliques* ≈ Entic Podzols

Podzols des climats froids ≈ Arenosols
There are two groups within the subclass:

- *Podzols boreaux* ≈ Podzols
- *Podzols alpins* ≈ Leptic Podzols (Skeletic)

Sols podzolisés hydromorphes ≈ Gleyic Podzols / Stagnic Podzols
There are three groups within the subclass:

- *Podzols à gley* ≈ Gleyic Podzols
- *Molkens-podzols* ≈ Stagnic Podzols
- *Podzols à nappe tropicaux* ≈ Gleyic Podzols

Sols à sesquioxydes de fer – soil class, with two subclasses included.
Sols ferrugineux tropicaux ≈ Cambisols (Chromic) / Luvisols (Chromic)
There are three groups within the subclass:

- *Sols ferrugineux tropicaux peu lessivés* ≈ Cambisols (Chromic) / Luvisols (Chromic)
- *Sols ferrugineux tropicaux lessivés* ≈ Luvisols (Chromic)
- *Sols ferrugineux tropicaux appauvris* ≈ Cambisols (Chromic)

Sols fersiallitiques ≈ Cambisols / Luvisols / Alisols
There are two groups within the subclass:

- *Sols fersiallitiques à réserve calcique* ≈ Cambisols (Calcaric) / Luvisols (Calcaric)
- *Sols fersiallitiques lessivés (sans réserve calcique)* ≈ Alisols

Sols ferrallitiques – soil class, with three subclasses included.
Sols ferrallitiques faiblement désaturés en Bw ≈ Ferralsols (Eutric) / Nitisols (Eutric)
There are four groups within the subclass:

- *Sols ferrallitiques faiblement désaturés typiques* ≈ Ferralsols (Eutric) / Nitisols (Eutric)
- *Sols ferrallitiques faiblement désaturés appauvris* ≈ Ferralsols (Eutric) / Nitisols (Eutric)
- *Sols ferrallitiques faiblement désaturés remaniés* ≈ Ferralsols (Eutric) / Nitisols (Eutric)
- *Sols ferrallitiques faiblement désaturés rajeunis ou pénévolués* ≈ Ferralsols (Eutric, Novic) / Nitisols (Eutric, Novic)

Sols ferrallitiques moyennement désaturés en Bw ≈ Ferralsols (Dystric) / Nitisols (Dystric)
There are five groups within the subclass:

- *Sols ferrallitiques moyennement désaturés typiques* ≈ Ferralsols (Dystric) / Nitisols (Dystric)
- *Sols ferrallitiques moyennement désaturès humiferes* ≈ Ferralsols (Humic, Dystric) / Nitisols (Humic, Dystric)
- *Sols ferrallitiques moyennement désaturés appauvris* ≈ Ferralsols (Dystric) / Nitisols (Dystric)
- *Sols ferrallitiques moyennement désaturés remaniés* ≈ Ferralsols (Dystric) / Nitisols (Dystric)
- *Sols ferrallitiques moyennement désaturés rajeunis ou pénévolués* ≈ Ferralsols (Dystric, Novic) / Nitisols (Dystric, Novic)

Sols ferrallitiques fortement désaturés en Bw ≈ Ferralsols (Hyperdystric) / Nitisols (Hyperdystric) / Acrisols
There are six groups within the subclass:

- *Sols ferrallitiques fortement désaturés typiques* ≈ Ferralsols (Hyperdystric) / Nitisols (Hyperdystric)
- *Sols ferrallitiques fortement désaturés humifères* ≈ Ferralsols (Humic, Hyperdystric) / Nitisols (Humic, Hyperdystric)
- *Sols ferrallitiques fortement désaturés appauvris* ≈ Ferralsols (Hyperdystric) / Nitisols (Hyperdystric)
- *Sols ferrallitiques fortement désaturés remaniés* ≈ Ferralsols (Hyperdystric) / Nitisols (Hyperdystric)
- *Sols ferrallitiques fortement désaturés rajeunis ou pénévolués* ≈ Ferralsols (Hyperdystric, Novic) / Nitisols (Hyperdystric, Novic)
- *Sols ferrallitiques fortement désaturés lessivés* ≈ Acrisols

Sols hydromorphes – soil class, with three subclasses included.
Sols hydromorphes organiques ≈ Histosols
There are three groups within the subclass:

- *Sols de tourbe fibreuse* ≈ Fibric Histosols
- *Sols de tourbe semi-fibreuse* ≈ Hemic Histosols
- *Sols de tourbe altérée* ≈ Sapric Histosols

Sols hydromorphes moyennement organiques ≈ Gleysols / Stagnosols
There are two groups within the subclass:

- *Sols humiques à gley* ≈ Mollic Gleysols / Umbric Gleysols / Histic Gleysols / Gleysols (Humic)
- *Sols humiques à stagnogley* ≈ Mollic Stagnosols / Umbric Stagnosols / Histic Stagnosols / Stagnosols (Humic)

Sols hydromorphes peu humifères ≈ Gleysols / Stagnosols / Plinthosols
There are six groups within the subclass:

- *Sols à gley* ≈ Gleysols
- *Sols à pseudogley* ≈ Gleysols
- *Sols à stagnogley* ≈ Stagnosols
- *Sols à amphigley* ≈ Stagnosols
- *Sols à accumulation de fer en carapace ou cuirasse* ≈ Plinthosols / Gleysols (Petrogleyic)
- *Sols hydromorphes à redistribution de calcaire ou de gypse* ≈ Calcic Gleysols / Gypsic Gleysols

Sols sodiques – soil class, with two subclasses included.
Sols sodiques à structure non dégradée ≈ Solonchaks
There is one group within the subclass:

- *Sols salins (solonchak)* ≈ Solonchaks

Sols sodiques à structure dégradée ≈ Solonetz / Solodic Planosols
There are three groups within the subclass:

- *Sols salins à alcalins (solonchak-solonetz)* ≈ Salic Solonetz
- *Sols sodiques à horizon B (solonetz)* ≈ Solonetz
- *Sols sodiques à horizon blanchi (solodises)* ≈ Solodic Planosols (Albic)

Planosols – soil class, with two subclasses included.
Planosols pedomorphes ≈ Planosols
There are the no groups within the subclass.
Planosols lithomorphes ≈ Planosols (Ruptic)
There are the no groups within the subclass.

From 1986 onwards, the French Association for Soil Investigation (AFES) conducted works on creating a new soil classification. In 1990 the first version of the new classification was published under the name *Référentiel Pédologique*. In 1993 a reworked variant was printed and it was suitable for the classification of the soils of Europe. In 1995 the latest version was published, suitable for the classification of soils the world over; this variant was translated into English (AFES, 1998).

Structure

Référentiel Pédologique is declared not to be a hierarchical taxonomy (AFES, 1998). Instead it proposes the use of qualifiers which allows naming soils in a more precise manner. To a great extent the structure of the French classification is very similar to that of the World Reference Base for Soil Resources (WRB) (IUSS Working Group WRB, 2006). It is not surprising, because one of the main ideologists of the French classification, Alan Ruellan, led the WRB Working Group during the initial stage of its work (at that time IRB – International Reference Base). The qualifiers proposed can be purely descriptive ('at the bottom of a slope', 'on limestone', etc.), or can reflect a genetic interpretation of the observed properties ('podzolized', 'paleoluvic', etc.). To a great extent the problems existing in soil classifications were not solved by Référentiel Pédologique, but just hidden by moving them to the qualifiers level. For example, some of these qualifiers reflect not the properties of the solum itself, but opinions about soil genesis and factors of soil formation.

The use of such a system of qualifiers has both advantages and disadvantages. An advantage is that the problem of hierarchy of properties for classification does not exist: all the qualifiers are listed in a line. However, one of the most important functions of classification is lost: a function of curtailment of information. In fact Référentiel Pédologique makes a step back to a nominal system. The openness of the French classification can lead to uncontrolled growth of the quantity of qualifiers. The number of soil references also can grow significantly: the authors promised to increase them, mainly to account for tropical soils.

Though French soil classification is not hierarchical, it has at least two levels (Table 7.2). At the first level there are 'references', which are determined as conceptual sequences of reference horizons. Currently 102 references exist; in the future the authors plan to increase the quantity up to 150. The second level, soil types, are references with specified qualifiers. Currently there are 235 qualifiers. Conceptually the number of types is almost unlimited: if one uses four to six qualifiers for a single type, then the number of combinations is up to 10^{14}. Even if we take into account the restrictions for the use of qualifiers (they should not be repetitive or contradictive to the reference and other qualifiers), the number of theoretical soil types is very large.

The qualifiers reflect the following information on soils:

- soil texture, pH, base saturation, and some elements content;
- the origin of parent and underlying material, the type of organic profile;
- the presence of additional reference horizons;
- the source and intensity of additional water influx;
- the position of the solum in relief;
- the presence of paleosols and relict features;
- weakly developed additional soil-forming process;
- contradiction between the morphology of profile and actual soil regimes and processes;
- alteration of the normal sequence of horizons due to natural and artificial processes;
- additional information on landscape and soil-forming conditions.

References can be grouped into 'great groups of references' (*grand ensemble de References*), though the authors stress that they do not have a taxonomic significance, and are used only for the convenience of presentation of material in a text.

Thus, Référentiel Pédologique is a classification in the form of a reference base with fuzzy borders between the taxa, having some features of a nominal system.

Table 7.2 *The structure of the French soil classification*

Level	Taxon name	Taxon characteristics	Borders between classes	Diagnostics	Terminology
0	Soils	Kingdom			
1	Great group of References	Collective	Fuzzy	Chemico-morphological	Mostly artificial
2	Reference	Generic	Fuzzy	Chemico-morphological	Artificial
3	Type	Specific and variative	Formal and fuzzy	Chemico-landscape-morphological	Traditional

Diagnostics

The diagnostics of soil references are composed of properties of horizons themselves, and also landscape criteria. The diagnostics of soil horizons are made according to quantitative criteria, both morphological and those requiring laboratory determination (e.g. organic carbon content, the composition of exchangeable cations, aluminium and iron extracted by selective dissolution techniques, soil density, etc.). Qualifiers are diverse using both quantitative and qualitative criteria and include soil-forming conditions. The diagnostics can be called mainly quantitative landscape-morphologico-chemical.

Terminology

The terminology of the French soil classification is borrowed partially from artificial names of the legend of FAO-UNESCO Soil Map of the World and the US Soil Taxonomy, and is partially constructed by the authors. Some terms have Russian roots; the combination of these with ending *-sol* in places sounds strange (PODZOSOLS, CHERNOSOLS). The names of references consist of one or two words. To avoid mistakes all the references of Référentiel Pédologique should be listed only in French and only in capital letters. The names of qualifiers are mainly simple descriptions; they are listed in lower case letters, and may be translated into other languages.

Correlation

Many concepts and even qualitative characteristics of the French soil classification are close to those of the WRB (IUSS Working group WRB, 2006) due to the same source of the two classifications. However, it is important to remember that the WRB has strict limits on classes, whereas the limits of the references of the French classification are fuzzy. Thus, even if the concepts of soil groups are identical in the two systems, in reality the extent of the group is bigger in Référentiel Pédologique.

Here is a correlation of references of Référentiel Pédologique (AFES, 1998) with the terms of the WRB. The references are listed in alphabetical order, without noting great groups of the references.

ALOCRISOLS HUMIQUES ≈ Dystri-Humic Cambisols
ALOCRISOLS TYPIQUES ≈ Dystric Cambisols
ALUANDOSOLS HAPLIQUE ≈ Aluandic Andosols
ALUANDOSOLS HUMIQUES ≈ Melanic Aluandic Andosols / Fulvic Aluandic
     Andosols
ALUANDOSOLS PERHYDRIQUES ≈ Hydric Aluandic Andosols
ANTHROPOSOLS ARTIFICIELS ≈ Spolic Technosols
ANTHROPOSOLS RECONSTITUES ≈ Urbic Technosols
ANTHROPOSOLS TRANSFORMES ≈ Anthrosols
ARENOSOLS ≈ Arenosols
BRUNISOLS MESOSATURES ≈ Mollic Umbrisols
BRUNISOLS OLIGOSATURES ≈ Umbrisols
BRUNISOLS RESATURES ≈ Mollic Umbrisols (Anthric)
BRUNISOLS SATURES ≈ Mollic Umbrisols
CALCARISOLS ≈ Calcisols
CALCISOLS ≈ Hypereutric Cambisols
CALCOSOLS ≈ Calcaric Cambisols
CHERNOSOLS HAPLIQUE ≈ Calcic Chernozems
CHERNOSOLS MELANOLUVIQUES ≈ Luvic Chernozems
CHERNOSOLS TYPIQUES ≈ Haplic Chernozems
COLLUVIOSOLS ≈ Colluvic Regosols / Fluvisols
CRYOSOLS HISTIQUES ≈ Cryic Histosols
CRYOSOLS MINERAUX ≈ Cryosols
DOLOMITOSOLS ≈ Rendzic Leptosols (*Magnesic*) / Rendzic Phaeozems
     (*Magnesic*)
FERSIALSOLS CALCIQUES ≈ Cambisols (Hypereutric, Chromic)
FERSIALSOLS CARBONATIQUES ≈ Cambisols (Calcaric, Chromic)
FERSIALSOLS ELUVIQUES ≈ Albic Luvisols
FERSIALSOLS INSATURES ≈ Cambisols (Chromic)
FLUVISOLS BRUNIFIES ≈ Mollic Fluvisols / Umbric Fluvisols / Humic
     Fluvisols
FLUVISOLS BRUTS ≈ Arenic Fluvisols
FLUVISOLS TYPIQUES ≈ Fluvisols
GRISOLS DEGRADES ≈ Luvic Phaeozems (Albic) / Albic Luvisols
GRISOLS ELUVIQUES ≈ Luvic Phaeozems (Albic) / Albic Luvisols
GRISOLS HAPLIQUES ≈ Luvic Greyic Phaeozems / Luvisols
GYPSISOLS HAPLIQUES ≈ Gypsisols
GYPSISOLS PETROGYPSIQUES ≈ Petric Gypsisols
HISTOSOLS COMPOSITES ≈ Histosols (no equivalent modifiers in WRB)
HISTOSOLS FIBRIQUES ≈ Fibric Histosols
HISTOSOLS FLOTTANTS ≈ Floatic Histosols
HISTOSOLS LEPTIQUES ≈ Leptic Histosols
HISTOSOLS MESIQUES ≈ Hemic Histosols
HISTOSOLS RECOUVERTS ≈ Histosols (*Anthric*) / Terric Anthrosols
     (*Thaptohistic*)
HISTOSOLS SAPRIQUES ≈ Sapric Histosols

LEPTISMECTISOLS ≈ Vertic Leptosols
LITHOSOLS ≈ Lithic Leptosols
LITHOVERTISOLS ≈ Vertisols
LUVISOLS DEGRADES ≈ Albeluvisols
LUVISOLS DERNIQUES ≈ Umbric Albeluvisols / Albic Luvisols / Anthric
    Albeluvisols
LUVISOLS TRONQUES ≈ Anthric Luvisols
LUVISOLS TYPIQUES ≈ Abruptic Luvisols
MAGNESISOLS ≈ Cambisols (*Magnesic*)
NEOLUVISOLS ≈ Luvisols
ORGANOSOLS CALCAIRES ≈ Mollic Umbrisols (Calcaric) / Phaeozems
    (Calcaric)
ORGANOSOLS CALCIQUES ≈ Phaeozems
ORGANOSOLS INSATURES ≈ Umbrisols
ORGANOSOLS TANGELIQUES ≈ Mollic Folic Leptosols / Umbric Folic
    Leptosols
PARAVERTISOLS HAPLIQUES ≈ Vertisols / Vertic Cambisols
PARAVERTISOLS PLANOSOLIQUES ≈ Vertic Planosols (Albic)
PELOSOLS BRUNIFIES ≈ Cambisols (Chromic, Clayic)
PELOSOLS DIFFERENCIES ≈ Vertic Planosols
PELOSOLS TYPIQUES ≈ Vertic Cambisols
PEYROSOLS CAILLUOTIQUES ≈ Regosols (Skeletic) / Fluvisols (Skeletic) /
    Leptosols (Skeletic)
PEYROSOLS PIERRIQUES ≈ Hyperskeletic Leptosols
PHAEOSOLS HAPLIQUES ≈ Phaeozems
PHAEOSOLS MELANOLUVIQUES ≈ Luvic Phaeozems
PLANOSOLS DISTAUX ≈ Endostagnic Planosols
PLANOSOLS STRUCTURAUX ≈ Plinthic Planosols / Petroferric Planosols /
    Stagnic Plinthosols
PLANOSOLS TYPIQUES ≈ Planosols
PODZOSOLS DURIQUES ≈ Podzols (*Densic*)
PODZOSOLS ELUVIQUES ≈ Albic Arenosols / Leptosols (*Albic*)
PODZOSOLS HUMIQUES ≈ Umbric Podzols
PODZOSOLS HUMO-DURIQUES ≈ Umbric Podzols (*Fragic*)
PODZOSOLS MEUBLES ≈ Rustic Podzols / Haplic Podzols / Carbic Podzols
PODZOSOLS OCHRIQUES ≈ Entic Podzols
PODZOSOLS PLACIQUES ≈ Placic Podzols
POST-PODZOSOLS ≈ Anthric Podzols / Terric Podzols / Plaggic Podzols /
    Hortic Podzols
PSEUDO-LUVISOLS ≈ Ruptic Regosols
QUASI-LUVISOLS ≈ Ruptic Luvisols
RANKOSOLS ≈ Umbric Leptosols / Mollic Leptosols
REDOXISOLS ≈ Ruptic Planosols
REDUCTISOLS DUPLIQUES ≈ Gleysols (*Stagnic*) / Endogleyic Stagnosols
REDUCTISOLS STAGNIQUES ≈ Stagnosols
REDUCTISOLS TYPIQUES ≈ Gleysols

REGOSOLS ≈ Regosols
RENDISOLS ≈ Rendzic Leptosols
RENDOSOLS ≈ Leptosols (Calcaric)
SALISODISOLS ≈ Solonchaks (Sodic)
SALISOLS CARBONATES ≈ Solonchaks (Carbonatic)
SALISOLS CHLORIDO-SULFATES ≈ Solonchaks (Chloridic, Sulphatic)
SILANDOSOLS DYSTRIQUES ≈ Silandic Andosols
SILANDOSOLS EUTRIQUES ≈ Eutrisilic Andosols
SILANDOSOLS HUMIQUES ≈ Melanic Silandic Andosols / Fulvic Silandic
    Andosols
SILANDOSOLS PERHYDRIQUES ≈ Hydric Silandic Andosols
SODISALISOLS ≈ Salic Solonetz
SODISOLS INDIFFERENCIES ≈ Cambisols (Sodic)
SODISOLS SOLODISES ≈ Solodic Solonetz
SODISOLS SOLONETZIQUES ≈ Solonetz
SULFATOSOLS ≈ Fluvisols (Orthothionic)
THALASSOSOLS ≈ Tidalic Fluvisols
THIOSOLS ≈ Fluvisols (Thionic)
TOPOVERTISOLS ≈ Vertisols
VERACRISOLS ≈ Stagnic Acrisols (Vermic) / Stagnic Alisols (Vermic)
VITROSOLS ≈ Vitric Andosols

References

AFES (1998) *A Sound Reference Base for Soils* (The 'Référentiel Pédologique': text in English), INRA, Paris, 322pp

CPCS (1967) *Classification des sols*, Ecole nationale supérieure agronomique, Grignon, France, 87pp

Hudson, B. D. (1992) 'The soil survey as paradigm-based science', *Soil Science Society of America Journal*, vol 56, no 4, pp836–841

IUSS Working Group WRB (2006) *World Reference Base for Soil Resources*, 2nd edition, World Soil Resources Reports no 103, UN Food and Agriculture Organization, Rome, 128pp

Segalen, P., Fauck, R., Lamouroux, M, Perraud, A., Quantin, P., Roederer, P. and Vieillefon, J. (1979) *Projet de Classification des Sols*, ORSTOM, Paris, 301pp

Soil Survey Staff (1999) *Soil Taxonomy: A Basic System of Soil Classification for Making and Interpreting Soil Surveys*, USDA, Handbook 436, 2nd edition, United States Government Printing Office, Washington DC, 696pp

Zarenkov, N. A. (1993) 'Biological systematic as a particular problem of scientific theory of classification', in *The Theory and Methods of Biological Classifications*, Nauka Press, Moscow, pp29–45 (in Russian)

8
Soil Classification of the United Kingdom

Objectives and scope

The soil classification of the United Kingdom of Great Britain and Northern Ireland (UK) was made as a tool for soil inventory and survey at different scales. From the establishment of the Soil Survey of England and Wales in 1939, a system of soil series was used for mapping. Taxonomic soil classification was developed mainly for arranging and systematizing the existing soil series. The present soil classification was not intended for use in the overseas territories of the British Empire. It has only national coverage (Table 8.1), especially for England and Wales. Scotland and Northern Ireland used somewhat different systems of soil classification that had never been published. A similar version was used by the Soil Survey of the Republic of Ireland. Mainly soils used for agriculture were included in the classification. Bare rock and underwater soils are not considered as soils; however, soils completely transformed by human agricultural activity and transported substrates are regarded as man-made soils at the level of major groups or groups (Table 8.1).

Table 8.1 *The scope of soil classification of the UK*

Superficial bodies	Representation in the system
Natural soils	National coverage
Urban soils	Partly included in the major group of Man-made soils and the group of Man-made raw soils
Man-transported materials	Partly included in the major group of Man-made soils
Bare rock	Not considered as soils
Subaquatic soils	Not considered as soils
Soils deeply transformed by agricultural activities	Partly included in the groups of Man-made humus soils and Earthy peat soils

Theoretical background

Soil classification in the UK is based mainly on the ideas adopted from the North American pedological school. The British soil inventory is based on the use of a system of soil series that was introduced to pedology by the US Soil Survey (Simonson, 1989). The structure and terminology of the British classification are similar to those used in the older soil classification of the US (Baldwin et al, 1938). Soil classification of the UK also includes some elements from more recent concepts of the American school, such as the idea of diagnostic horizons (Soil Survey Staff, 1999).

Soil archetypes were not revised in the British classification. They are closer to the traditional European soil archetypes, or to the older US system (Baldwin et al, 1938), than to the artificial taxa of Soil Taxonomy (Soil Survey Staff, 1999). The general scheme of the classification involves the concept of soil 'evolution': the stages of soil development are included in the highest level of taxonomy that is well illustrated by alluvial soils, which are not brought together in one major group, but scattered among major groups depending on the stage of soil profile development.

Like the 1938 American classification the system of soil series was not always coincident with taxonomic classification. In practical soil mapping some series were divided into 'variants' that might belong to different taxonomic groups.

Development

The historical development of soil survey in Great Britain perhaps began in the 17th century when the Royal Society suggested collecting information on British soils (Clayden and Hollis, 1984). However, real soil mapping started in the UK in 1926, when the American method of soil series was introduced. The work on a taxonomic soil classification started in the late 1960s, after publication of the American and Dutch soil classifications. In a systematic form the classification was presented by B. V. Avery of Cranfield University (Avery, 1973). Minor corrections were introduced up to the 1980s, when the latest available version was published (Avery, 1980). Later on, the soil survey of the UK was closed, and there were no further updates of the system. The system is still widely used in the UK for education and scientific research (Avery, 1990).

Structure

The British soil classification has four hierarchical levels (Table 8.2). On the highest level there are ten major groups; each one includes several groups. The latter are divided into subgroups. The three highest levels are defined on the basis of the presence of certain horizons and properties at or within certain depths. Within subgroups there are series, defined as soil classes with identical sequences of horizons formed in similar conditions on the same parent material.

Table 8.2 *The structure of soil classification of the UK*

Level	Taxon name	Taxon characteristics	Borders between classes	Diagnostics	Terminology
0	Soils	Kingdom			
1	Major group	Collective	Mostly formal	Chemico-morphological	Traditional
2	Group	Generic 1	Mostly formal	Chemico-morphological	Traditional
3	Subgroup	Specific	Mostly formal	Chemico-morphological	Traditional
4	Type	Generic 2	Mostly	Fuzzy-morphological	Traditional

The structure of the classification is defined as a hierarchical taxonomy with partially formalized borders between classes; on the level of series it reduces to a nominal system.

Diagnostics

The object of diagnostics is the soil profile itself with no factors of soil formation. These factors may be taken into account at the series level, but they are not included directly as diagnostics. The definitions of horizons are less strict in British variants and fewer quantitative chemical criteria are used. In many cases a diagnostic is based only on the presence or absence of certain properties. The diagnostics could be called quantitative chemico-morphological.

Terminology

The classification of the UK uses traditional European soil terminology based mainly on the terms proposed by W. Kubiëna (1953). These terms are partly borrowed from the Russian soil science school (*podzol, gley, rendzina*), and partly constructed by Kubiëna himself (*stagnogley, cryptopodzol, pelosol*). This terminology was widely used both in the UK and in other European countries. In some cases the UK soil classification uses understandable English words (*raw soils, man-made soils*). Some artificial terms, borrowed from the US and other international classifications, are also used (*argillic, humic, ferric*, etc.).

Correlation

Correlation of the UK soil classification with the World Reference Base for Soil Resources (WRB) was relatively easy because its concepts and archetypes are clear and close to the concepts of the WRB. However, the limits of classes differ and one should be cautious with the correlation because certain overlapping of the groups occurs. Especially uncertain are the correlations of man-made soils; in British classification the concept is very general, while in the WRB these groups are strictly defined. The correlation is given down to the level of the groups. Major groups and groups within major groups are listed alphabetically.

Brown soils – soil <u>major group</u>. ≈ Cambisols / Luvisols / Arenosols
The following <u>groups</u> are distinguished within the major group:

- *Argillic brown earths* ≈ Luvisols
- *Brown alluvial soils* ≈ Fluvic Cambisols / Fluvisols
- *Brown calcareous alluvial soils* ≈ Fluvic Cambisols (Calcaric)
- *Brown calcareous earths* ≈ Cambisols (Calcaric)
- *Brown calcareous sands* ≈ Brunic Arenosols (Calcaric)
- *Brown earths (sensu stricto)* ≈ Cambisols
- *Brown sands* ≈ Brunic Arenosols
- *Paleo-argillic brown earths* ≈ Luvisols (Chromic)

Groundwater gley soils – soil <u>major group</u>. ≈ Gleysols / Gleyic Fluvisols / Gleyic Cambisols / Gleyic Luvisols
The following <u>groups</u> are distinguished within the major group:

- *Alluvial gley soils* ≈ Gleyic Fluvisols
- *Argillic gley soils* ≈ Gleyic Luvisols / Luvic Gleysols
- *Cambic gley soils* ≈ Gleyic Cambisols
- *Humic-alluvial gley soils* ≈ Mollic Gleyic Fluvisols / Umbric Gleyic Fluvisols / Gleyic Histic Fluvisols
- *Humic-gley soils (sensu stricto)* ≈ Mollic Gleysols / Umbric Gleysols / Histic Gleysols
- *Humic-sandy gley soils* ≈ Mollic Gleysols (Arenic) / Umbric Gleysols (Arenic) / Histic Gleysols (Arenic)
- *Sandy gley soils* ≈ Gleysols (Arenic)

Lithomorphic soils – soil <u>major group</u>. ≈ Leptosols / Phaeozems / Fluvisols
The following <u>groups</u> are distinguished within the major group:

- *Pararendzinas* ≈ Phaeozems (Calcaric)
- *Ranker-like alluvial soils* ≈ Leptic Mollic Fluvisols / Leptic Umbric Fluvisols / Leptic Histic Fluvisols
- *Rankers* ≈ Mollic Leptosols / Umbric Leptosols / Leptic Umbrisols
- *Rendzina-like alluvial soils* ≈ Mollic Fluvisols (Calcaric)
- *Rendzinas* ≈ Rendzic Leptosols
- *Sand-pararendzinas* ≈ Phaeozems (Calcaric, Arenic)
- *Sand-rankers* ≈ Mollic Leptosols (Arenic) / Umbric Leptosols (Arenic) / Umbrisols (Arenic)

Man-made soils – soil <u>major group</u>. ≈ Anthrosols / Technosols
The following <u>groups</u> are distinguished within the major group:

- *Disturbed soils* ≈ Technosols
- *Man-made humus soils* ≈ Terric Anthrosols

Peat soils – soil <u>major group</u>. ≈ Histosols / Terric Anthrosols (*Thaptohistic*)
The following <u>groups</u> are distinguished within the major group:

- *Earthy peat soils* ≈ Histosols (*Anthric*) / Terric Anthrosols (*Thaptohistic*)
- *Raw peat soils* ≈ Histosols

Pelosols – soil <u>major group</u>. ≈ Vertisols / Regosols / Cambisols / Luvisols
The following <u>groups</u> are distinguished within the major group:

- *Argillic pelosols* ≈ Luvisols (Clayic)
- *Calcareous pelosols* ≈ Cambisols (Calcaric, Clayic) / Regosols (Calcaric, Clayic) / Vertisols (Calcaric)
- *Non-calcareous pelosols* ≈ Cambisols (Clayic) / Regosols (Clayic) / Vertisols

Podzolic soils – soil <u>major group</u>. ≈ Podzols
The following <u>groups</u> are distinguished within the major group:

- *Brown podzolic soils* ≈ Umbric Podzols
- *Gley-podzols* ≈ Gleyic Podzols
- *Humic cryptopodzols* ≈ Entic Carbic Podzols
- *Podzols (sensu stricto)* ≈ Podzols
- *Stagnopodzols* ≈ Histi-Stagnic Podzols

Raw gley soils – soil <u>major group</u>. ≈ Gleysols / Gleyic Fluvisols / Gleyic Arenosols
The following <u>groups</u> are distinguished within the major group:

- *Raw sandy gley soils* ≈ Gleyic Fluvisols (Arenic) / Gleyic Arenosols
- *Unripened gley soils* ≈ Gleyic Fluvisols / Gleysols

Surface-water gley soils – soil <u>major group</u>. ≈ Planosols / Stagnosols
The following <u>groups</u> are distinguished within the major group:

- *Stagnogley soils (sensu stricto)* ≈ Stagnosols / Planosols
- *Stagnohumic gley soils* ≈ Histic Planosols / Mollic Planosols / Umbric Planosols / Histic Stagnosols / Mollic Stagnosols / Umbric Stagnosols

Terrestrial raw soils – soil <u>major group</u>. ≈ Regosols / Arenosols / Leptosols / Fluvisols
The following <u>groups</u> are distinguished within the major group:

- *Man-made raw soils* ≈ Technosols

- *Raw alluvial soils* ≈ Fluvisols (Arenic)
- *Raw earths* ≈ Regosols
- *Raw sands* ≈ Protic Arenosols
- *Raw skeletal soils* ≈ Hyperskeletic Leptosols

References

Avery, B. W. (1973) 'Soil classification in the soil survey of England and Wales', *Journal of Soil Science*, vol 24, no 3, pp324–338

Avery, B. W. (1980) *Soil Classification for England and Wales (Higher Categories)*, Soil Survey Technical Monograph no 14, Harpenden, 67pp

Avery, B. W. (1990) *Soils of the British Isles*, CAB International, Wallingford, 480pp

Baldwin, M., Kellog, C. E. and Thorp, J. (1938) 'Soil classification', in United States Department of Agriculture, *Soils and Men: Yearbook of Agriculture 1938*, United States Government Printing Office, Washington DC, pp979–1001

Clayden, B. and Hollis, J. M. (1984) *Criteria for Differentiating Soil Series*, Soil Survey Technical Monograph no 17, Harpenden, 159pp

Kubiëna, W. L. (1953) *Bestimmungsbuch und Systematik der Böden Europas*, Verlag Enke, Stuttgart, 392pp

Simonson, R. W. (1989) *Historical Highlights of Soil Survey and Soil Classification with Emphasis on the United States, 1899–1970*, International Soil Reference and Information Centre, Wageningen, Technical Paper no 18, 83pp

Soil Survey Staff (1999) *Soil Taxonomy: A Basic System of Soil Classification for Making and Interpreting Soil Surveys*, USDA, Handbook 436, 2nd edition, United States Government Printing Office, Washington DC, 696pp

9
German Soil Classification

written together with Peter Schad[1]

Objectives and scope

The German soil classification has a morpho-genetical approach. Its aim is to
order soils into natural groups for a better understanding of soil genesis and
geography, for education, and for soil mapping at various scales (Ad-hoc-AG
Boden, 2005). The current German classification was developed since the 1950s
and was strongly influenced by the classification of Kubiëna (1953). Kubiëna's
classification divided soils according to their water regime in a broad sense
(terrestrial, hydromorphous and subaquatic soils); the terrestrial soils were
arranged according to their stage of development. Traditionally, the German
classification includes shallow underwater soils as equal objects (Table 9.1).
Anthropogenic soils deeply transformed by cultivation are included as separate
taxa. Urban soils and technogenic or transported materials are not included
in the classification itself; however, the German soil science school has a rich
tradition of classifying urban and technogenic soils (e.g. Blume, 1989) which
are used apart from the main official classification. Additionally, the German
substrate classification offers a detailed classification of technogenic materials
(Ad-hoc-AG Boden, 2005).

Theoretical background

The main ideas of soil classification were outlined by Kubiëna (1953) in his
seminal book *Bestimmungsbuch und Systematik der Böden Europas*. Even
today German pedologists continue developing the theory of soil classification
(e.g. Albrecht et al, 2005). Basically, the authors of the classification clearly
understood the difference between an ideal natural grouping of entities

---

[1]Technische Universität München, Freising, Germany

Table 9.1 *The scope of the German soil classification*

Superficial bodies	Representation in the system
Natural soils	National coverage
Urban soils	Not included in the soil classification; separate classifications of urban soils exist; technogenic material classified by the substrate classification
Man-transported materials	Not included in the soil classification; separate classifications of urban soils exist; transported materials classified by the substrate classification
Bare rock	Not recognized as soils
Subaquatic soils	Included in the special branch of Semi-subaquatic and subaquatic soils (Semisubhydrische und Subhydrische Böden)
Soils deeply transformed by agricultural activities	Included in the classes of Anthropogenic terrestrial soils (Terrestrische Kultosole) and Anthropogenic peat soils (Erd- und Mulmmoore); a separate classification of cultivated peats exists (Kultivierte Moore)

('systematics') and a provisional grouping of soils in a real world ('classification'). The taxa are defined mostly by qualitative criteria: the German classification gives conceptual sequences of horizons rather than establishes strict quantitative limits between soil units. The archetypes remain almost the same as outlined by Kubiëna, which, in their turn, were derived from classical German studies at the beginning of the 20th century. The overall arrangement of the classification also follows the 'evolutionary' scheme of Kubiëna, from the simplest poorly developed soils to the most complex, polygenetic ones. Apart from a pure substantial characterization of soils, the classification partly includes a genetic interpretation of soils. Many deeply weathered soils are understood as paleosols, if their formation mainly occurred before the Holocene.

Development

A number of early soil classifications exist in Germany, even from the 19th century (e.g. Richthofen, 1886). In the 1920s and 1930s Hermann Stremme and his co-workers developed a system for soil mapping and published several soil maps of Germany (e.g. Stremme, 1936) and Europe. In the 1950s, the development of the current system started, which was done to some extent separately in West and East Germany. However, both systems were mainly based on the classification scheme of Kubiëna (1953). Eduard Mückenhausen elaborated the fundamental outline for the West German system (Mückenhausen, 1962, 1977). Afterwards, three editions of the classification system were published (AG Bodenkunde, 1965, 1971, 1982). The East German classification was issued by Ehwald et al (1966). Immediately after unifying the country in 1990, a uniform soil classification was established (Ad-hoc-AG Boden, 1994). In the meantime, the fifth edition (counting the three West German and the two common editions) of the system is in use (Ad-hoc-AG Boden, 2005).

Structure

The German soil classification has six hierarchical levels (Table 9.2). Soil branches (or orders) combine soils of certain levels of hydromorphism and of the nature of parent material (organic or mineral): terrestrial soils, semi-terrestrial soils, semi-subaquatic and subaquatic soils, and peat soils (fens and bogs). Soil classes (or suborders) are differentiated according to a similar stage of soil evolution and the dominant pedological processes. Soil types represent the basic unit of the classification. Each soil type is defined by a specific sequence of soil horizons. Subtypes are either modifications of the 'central image' or transitions between types (e.g. Braunerde-Podsol, Pseudogley-Fahlerde, etc.). Varieties and subvarieties reflect quantitative modifications of soil properties. Independent from the soil classification is the substrate classification. The combination of soil classification and substrate classification leads to soil forms (or series).

According to its structure the classification is a hierarchical taxonomy with fuzzy borders between taxa.

Table 9.2 *The structure of the German soil classification*

Level	Taxon name	Taxon characteristics	Borders between classes	Diagnostics	Terminology
0	Soil	Kingdom			
1	Branch	Collective 1	Formal	Chemico-morphological	Traditional
2	Class	Collective 2	Formal	Chemico-morphological	Traditional
3	Type	Generic	Formal	Chemico-morphological	Traditional
4	Subtype	Specific 1	Formal	Chemico-morphological	Traditional
5	Variety	Varietal 1	Formal		Traditional
6	Subvariety	Varietal 2	Formal		Traditional

Diagnostics

The object of classification in the German classification is the soil itself. Climatic conditions and other factors of soil formation are not taken into account. On higher levels of taxonomy the diagnostics are based on soil morphology; on lower levels some properties should be determined in a laboratory. The authors try to use mainly qualitative criteria or properties easily measured in the field (e.g. horizon depth, gravel content). The diagnostics in the German classification can be called semiquantitative chemico-morphological.

Terminology

Almost all the terminology of the German classification is borrowed from the soil classification (systematics) worked out by Kubiëna (1953), who partly borrowed it from the Russian school and folk soil names, and partly invented it himself. Since his classification affected greatly all the development of soil science in Europe, this terminology is well understood among European soil scientists.

Correlation

In the following, a correlation will be given between the soil terms of the German classification down to the type level and the corresponding World Reference Base for Soil Resources (WRB) terms (IUSS Working Group WRB, 2006). The soil names are listed in the same order as in the German classification. The translations of the terms of the German classification into English are derived from Wittmann (1997) with minor modifications. The German soil classification is difficult to correlate with the WRB, since the two systems use different criteria for soil taxa designation. Some soil units seem to have similar definitions in the German classification and in the WRB, but both their significance and the quantitative criteria are different. For example, for most shallow soils and for most surface horizons' depths, the German system uses the limit of 30cm, whereas the WRB applies various, case-dependent criteria. The criteria for *dystric* and *eutric* subtypes in German classification also differ from definitions of the WRB. First, base saturation is a relatively insignificant criterion, used at the lowest levels of taxonomy, while in the WRB it is one of the basic properties. Also, in Germany base saturation is measured in surface A horizon, while in the WRB the modifiers refer to base saturation at depths of more than 20cm.

The latest edition of the German classification (Ad-hoc-AG Boden, 2005) contains a rough correlation with the first edition of the WRB (FAO-ISRIC-ISSS, 1998). It is important to note that the correlations used in this book and in the text of the German classification are conceptually different: the text of the classification (Ad-hoc-AG Boden, 2005) tries to give all the intersections of the conceptual areas of the soil units in the two classifications. For example, the correlation of the type Ranker with the WRB terms includes: 'all *Leptosols*, except of Rendzic, Aridic, Gypsic and Calcaric Leptosols; when Ah < 1dm: *Lithic Leptosols*; when having a mollic horizon: *Mollic Leptosols*; when having an umbric horizon: *Umbric Leptosols*; when having an ochric horizon with a base saturation < 50 per cent: *Dystric Leptosols*; with a base saturation > 50 per cent: *Eutric Leptosols*, also *Gleyic, Hyperskeletic, Humic Leptosols*; when Ah > 2.5dm: *Leptic Phaeozem, Leptic Umbrisol, Leptic Cambisol*'. In the present book we included only the correspondence of the 'central images' of the type to certain WRB units. For example, Rankers are mainly shallow soils with humus-enriched topsoils that correspond to Mollic or Umbric Leptosols. Though Rankers may be shallower, slightly deeper, or poorer in organic matter than Mollic or Umbric Leptosols, we list only these two units in the correlation.

Terristrische Böden (Terrestrial soils) – soil branch.
The branch is divided into the following classes:
O/C-Böden (O/C soils) – soil class. ≈ Folic Histosols / Leptosols

The following soil types are distinguished within the class:

- *Felshumusboden (Rock-humus soil)* ≈ Leptic Folic Histosol / Lithic Leptosol
- *Skeletthumusboden (Skeletal humus soil)* ≈ Hyperskeletic Folic Histosol / Hyperskeletic Leptosol

Terristrische Rohböden (Terrestrial raw soils) – soil class. ≈ Leptosols / Regosols / Arenosols
The following soil types are distinguished within the class:

- *Syrosem* ≈ Lithic Leptosol
- *Lockersyrosem* ≈ Hyperskeletic Leptosol / Regosol / Arenosol

Ah/C-Böden (Ah/C soils) – soil class. ≈ Leptosols / Regosols / Arenosols / Phaeozems / Umbrisols
The following soil types are distinguished within the class:

- *Ranker* ≈ Mollic Leptosol / Umbric Leptosol
- *Regosol* ≈ Regosol / Arenosol / Umbrisol
- *Rendzina* ≈ Rendzic Leptosol / Rendzic Phaeozem
- *Pararendzina* ≈ Leptosol (Calcaric) / Regosol (Calcaric) / Arenosol (Calcaric) / Phaeozem (Calcaric)

Schwarzerden (Black earths) – soil class. ≈ Chernozems / Phaeozems / Kastanozems
The following soil types are distinguished within the class:

- *Tschernosem* ≈ Chernozem / Phaeozem
- *Kalktschernosem* ≈ Chernozem / Kastanozem

Pelosole (Soils developed from claystones) – soil class. ≈ Vertisols / Vertic Cambisols
Only one soil type is distinguished within the class:

- *Pelosol* ≈ Vertisol / Vertic Cambisol

Braunerden (Brown earths) – soil class. ≈ Cambisols / Cambic Umbrisols / Phaeozems / Brunic Arenosols / Aluandic Andosols / Silandic Andosols
Only one soil type is distinguished within the class:

- *Braunerde* ≈ Cambisol / Cambic Umbrisol / Phaeozem / Brunic Arenosol / Aluandic Andosol / Silandic Andosol

Lessivés (Soils with clay migration) – soil class. ≈ Luvisols / Alisols / Albeluvisols / Phaeozems

The following soil types are distinguished within the class:

- *Parabraunerde* ≈ Luvisol / Alisol / Luvic Phaeozem
- *Fahlerde* ≈ Alisol / Albeluvisol

Podsole (Soils with podzolization) – soil class. ≈ Podzols
Only one soil type is distinguished within the class:

- *Podsol* ≈ Podzol

Terrae calcis (Soils developed from intensive limestone weathering) – soil class.
≈ Cambisols (Clayic) / Luvisols (Clayic)
The following soil types are distinguished within the class:

- *Terra fusca* ≈ Cambisol (Clayic) / Luvisol (Clayic)
- *Terra rossa* ≈ Cambisol (Clayic, Chromic) / Luvisol (Clayic, Chromic)

Fersialitische und ferrallitische Paläoböden (Fersiallitic and ferralitic paleosols)
– soil class. ≈ Acrisols / Lixisols / Ferralsols
The following soil types are distinguished within the class:

- *... über Fersiallit* ≈ Acrisol / Lixisol
- *... über Ferallit* ≈ Ferralsol

Stauwasserböden (Stagnant water soils) – soil class. ≈ Stagnosols / Albeluvisols /
Planosols
The following soil types are distinguished within the class:

- *Pseudogley* ≈ Stagnosol / Albeluvisol / Planosol
- *Haftpseudogley* ≈ Stagnosol
- *Stagnogley* ≈ Stagnosol / Planosol

Reduktosole (Soils affected by geogenic or anthropogenic reducing gases) – soil
class. ≈ Technosols (Reductic) / Gleysols / Stagnosols
Only one soil type is distinguished within the class:

- *Reduktosol* ≈ Technosol (Reductic) / Gleysol / Stagnosol

Terristrische antropogene Böden (Terrestrial anthropogenic soils) – soil class. ≈
Anthrosols / Regsosols / Cambisols / Phaeozems / Umbrisols
The following soil types are distinguished within the class:

- *Kolluvisol* ≈ Coluvic Regosol / Phaeozem / Umbrisol
- *Plaggenesch* ≈ Plaggic Anthrosol / Plaggic Cambisol
- *Hortisol* ≈ Hortic Anthrosol / Hortic Cambisol
- *Rigosol* ≈ Aric Regosol
- *Treposols* ≈ Aric Regosol

Semiterrestrische Böden (Semi-terrestrial soils) – soil branch. No equivalents in WRB.

The branch is divided into the following classes:

Aucnböden (Floodplain soils) – soil class. ≈ Fluvisols / Hyperskeletic Leptosols / Cambisols / Pheeozems / Umbrisols

The following soil types are distinguished within the class:

- *Rambla* ≈ Fluvisol / Hyperskeletic Leptosol
- *Paternia* ≈ Fluvisol
- *Kalkpaternia* ≈ Fluvisol (Calcaric)
- *Tschernitza* ≈ Mollic Fluvisol / Phaeozem
- *Vega* ≈ Fluvic Cambisol / Phaeozem / Fluvic Umbrisol

Gleye (Groundwater soils) – soil class. ≈ Gleysols

The following soil types are distinguished within the class:

- *Gley* ≈ Gleysol
- *Naßgley* ≈ Gleysol
- *Anmoorgley* ≈ Umbric Gleysol / Mollic Gleysol
- *Moorgley* ≈ Histic Gleysol

Marschen (Soils developed from (peri)marine sediments) – soil class. ≈ Fluvisols / Gleysols / Stagnosols / Planosols / Histosols

The following soil types are distinguished within the class:

- *Rohmarsch* ≈ Gleyic Tidalic Fluvisol (Thionic)
- *Kalkmarsch* ≈ Gleyic Fluvisol (Calcaric) / Gleysol (Calcaric)
- *Kleimarsch* ≈ Gleyic Fluvisol / Gleysol
- *Haftnäßemarsch* ≈ Stagnic Endogleyic Fluvisol / Gleysol / Endogleyic Stagnosol
- *Dwogmarsch* ≈ Endogleyic Stagnosol (Novic)
- *Knickmarsch* ≈ Endogleyic Planosol (Ruptic) / Endogleyic Stagnosol (Ruptic)
- *Organomarsch* ≈ Histosol / Fluvisol / Gleysol

Strandböden (Soils of recent marine sediments) – soil class. ≈ Salic Tidalic Fluvisols (Arenic)

Only one soil type is distinguished within the class:

- *Strand* ≈ Salic Tidalic Fluvisol (Arenic)

Semisubhydrische und Subhydrische Böden (Semi-subaquatic and subaquatic soils) – soil branch.

Semisubhydrische Böden (Semi-subaquatic soils) – soil class. ≈ Tidalic Fluvisols

The following soil types are distinguished within the class:

- *Naßstrand* ≈ Salic Tidalic Fluvisol (Arenic)
- *Watt* ≈ Gleyic Tidalic Fluvisol (Thionic)

Subhydrische Böden (Subaquatic soils) – soil class. ≈ Limnic Subaquatic Fluvisols / Subaquatic Limnic Histosols
The following soil types are distinguished within the class:

- *Protopedon* ≈ Limnic Subaquatic Fluvisol
- *Gyttja* ≈ Limnic Subaquatic Fluvisol (Oxyaquic, Humic)
- *Sapropel* ≈ Subaquatic Limnic Histosol (Thionic) / Limnic Subaquatic Fluvisol (Thionic, Humic)
- *Dy* ≈ Subaquatic Limnic Histosol

Moore (Peat soils) – <u>soil branch</u>. ≈ Histosols
The branch is divided into the following <u>classes</u>:
Naturnahe Moore ((Semi-)Natural peat soils) – soil class. ≈ Histosols
The following soil types are distinguished within the class:

- *Niedermoor* ≈ Rheic Histosol
- *Hochmoor* ≈ Ombric Histosol

Erd- und Mulmmoore (Drained peat soils) – soil class. ≈ Histosols (Drainic)
Three soil types are distinguished within the class:

- *Erdniedermoor* ≈ Rheic Hemic Histosol (Drainic) / Rheic Sapric Histosol (Drainic)
- *Mulmniedermoor* ≈ Rheic Sapric Histosol (Drainic)
- *Erdhochmoor* ≈ Ombric Hemic Histosol (Drainic) / Ombric Sapric Histosol (Drainic)

Acknowledgement

The authors wish to thank H.-P. Blume (Kiel) for his many worthwhile comments.

References

Ad-hoc-AG Boden (1994) *Bodenkundliche Kartieranleitung*, herausgegeben von der Bundesanstalt für Geowissenschaften und Rohstoffe und den Geologischen Landesämtern in der Bundesrepublik Deutschland, Hannover, 4 Auflage, 392pp
Ad-hoc-AG Boden (2005) *Bodenkundliche Kartieranleitung*, herausgegeben von der Bundesanstalt für Geowissenschaften und Rohstoffe und den Geologischen Landesämtern in der Bundesrepublik Deutschland, Hannover, 5 Auflage, 438pp
AG Bodenkunde (1965) *Bodenkundliche Kartieranleitung*, herausgegeben von der Bundesanstalt für Geowissenschaften und Rohstoffe und den Geologischen Landesämtern in der Bundesrepublik Deutschland, Hannover, 1 Auflage, 169pp

AG Bodenkunde (1971) *Bodenkundliche Kartieranleitung*, herausgegeben von der Bundesanstalt für Geowissenschaften und Rohstoffe und den Geologischen Landesämtern in der Bundesrepublik Deutschland, Hannover, 2 Auflage, 169pp

AG Bodenkunde (1982) *Bodenkundliche Kartieranleitung*, herausgegeben von der Bundesanstalt für Geowissenschaften und Rohstoffe und den Geologischen Landesämtern in der Bundesrepublik Deutschland, Hannover, 3 Auflage, 331pp

Albrecht, C., Jahn, R. and Huwe, B. (2005) 'Bodensystematik und Bodenklassifikation. Teil I: Grundbegriffe', *Journal of Plant Nutrition and Soil Science*, vol 168, no 1, pp7–20

Blume, H.-P. (1989) 'Classification of soils in urban agglomerations', *Catena*, vol 16, no 3, pp269–275

Ehwald, E., Lieberoth, I. and Schwanecke, W. (1966) 'Zur Systematik der Böden der DDR besonders im Hinblick auf die Bodenkartierung', *Sitzungsberichte der Deutsche Akademie der Landwirtschaftswissenschaften zu Berlin*, Band IX, Heft 18

FAO-ISRIC-ISSS (1998) *World Reference Base for Soil Resources*, Soil Resources Report no 84, UN Food and Agriculture Organization, Rome, 88pp

IUSS Working Group WRB (2006) *World Reference Base for Soil Resources*, 2nd edition, World Soil Resources Reports, no 103, UN Food and Agriculture Organization, Rome, 128pp

Kubiëna, W. L. (1953) *Bestimmungsbuch und Systematik der Böden Europas*, Verlag Enke, Stuttgart, 392pp

Mückenhausen, E. (1962) *Entstehung, Eigenschaften und Systematik der Böden der Bundesrepublik Deutschland*, DLG-Verlag, Frankfurt, 1 Auflage, 148pp

Mückenhausen, E. (1977) *Entstehung, Eigenschaften und Systematik der Böden der Bundesrepublik Deutschland*, DLG-Verlag, Frankfurt, 2 Auflage, 300pp

Richthofen, F. F. von (1886) *Führer für Forschungsreisende. Anleitungen zu Beobachtungen über Gegenstände der physischen Geographie und Geologie*, Verlag Robert Oppenheim, Berlin

Stremme, H. E. (1936) *Die Böden des Deutschen Reiches und der Freien Stadt Danzig*, Erläuterungen zur Übersichtsbodenkarte 1 : 1 000 000 des Deutschen Reiches und der Freien Stadt Danzig. Ergänzungsheft Nr. 226 zu Petermanns Mitteilungen, Verlag Justus Perthes Gotha

Wittmann, O. (1997) 'Soil classification of the Federal Republic of Germany', *Mitteilungen der Deutschen Bodenkundlichen Gesellschaft*, Band 84, pp253–275

10
Soil Classification of Austria

Objectives and scope

The new Austrian soil classification (Nestroy et al, 2000) was prepared to replace the older system (Fink, 1969), which had been used for many years for soil survey and inventory in the country. Though the list of soils did not change much in the new version of the classification, the diagnostics and conceptual bases of the taxonomy were updated according to the actual demands, including the need for integration of national soil databases with the European soil information system and correlation and harmonization with the World Reference Base for Soil Resources (WRB) (Nestroy, 2001).

Geographically, the Austrian classification covers only the national territory (Table 10.1). Anthropogenically transformed soils, including transported substrates and urban soils, are represented in the classification, with a special class *Substratböden*. Soils deeply transformed by agriculture also have a special class *Kolluvien und Antrosolen*, comprising six soil types. Bare rock is not specifically mentioned in the classification, but primitive soils with a shallow raw-humus horizon or just forest floor are regarded as soils classified as *Fels-Auflagehumusboden* ('rock-raw-humus soils'). Underwater soils, following Kubiëna's tradition, are included in a separate class of the system.

Table 10.1 *The scope of the Austrian soil classification*

Superficial bodies	Representation in the system
Natural soils	National coverage
Urban soils	Included in the class *Substratböden*.
Man-transported materials	Included in the class *Substratböden*
Bare rock	Partly included in the type *Fels-Auflagehumusboden*
Subaquatic soils	Form a special class *Unterwasserböden*
Soils deeply transformed by agricultural activities	Form a special class *Kolluvien und Antrosolen*

Theoretical background

It is quite natural that the actual Austrian soil taxonomy has the classification of Walter Kubiëna (1953) as the source system. In Kubiëna's classification the lower levels of the taxonomy were not developed because he proposed a conceptual scheme, not a tool for large-scale mapping. That is why the classifications rooted in Kubiëna's system are often different. Though their basic ideas (like the evolutionary approach to soil grouping) and upper taxonomic levels are similar, at the lower levels of taxonomy the older and the new Austrian classifications differ. The same is true for the Austrian and German classifications. At first glance these two systems seem almost identical; however, they have a lot of minor differences which make difficult the transfer of cartographic data between the two countries.

Apart from the basic pedogenetic and evolutionary ideas of Kubiëna, the new Austrian classification partly accepts the novel ideas of quantitative diagnostics and the use of formal criteria for soil grouping.

Development

The first soil classification in Austria was proposed by Walter Kubiëna who had already in his famous book *Micropedology* (1938) schematically grouped European soils. Later he developed and finalized his classification (Kubiëna, 1953). As mentioned above, his classification was too general for soil inventory and large-scale mapping. The first Austrian working document that permitted soil survey in a uniform way was published in the late 1960s (Fink, 1969) about the same time as most countries published official classifications applicable for detailed soil mapping (USA, USSR, France, The Netherlands, etc.). This classification was successfully used in Austria for more than 30 years. The new soil classification (Nestroy et al, 2000) is not a revolutionary change of concepts; it is a continuation of the older version based on improved understanding of soils and soil cover.

Structure

The upper collective level of the Austrian classification includes two main orders: terrestrial and hydromorphic soils. Compared with the German classification (Ad-hoc-AG Boden, 2005) which developed from the same roots, the division on the highest level is made more general (two orders at the highest level vs three branches of the German classification). As in many classifications, the upper collective level serves mainly for showing the priorities of the authors (compare *sinlithogenic*, *postlithogenic* and *organic* trunks in Russian classification and branches and orders according to hydromorphism level in the German and Austrian classifications. Within the orders there are collective classes that group soils according to the stage of their profile development and general geochemistry. Soil types form a generic level of taxonomy. Subtypes specify particular features of soils, such as humus form and the presence of

Table 10.2 *The structure of the Austrian soil classification*

Level	Taxon name	Taxon characteristics	Borders between classes	Diagnostics	Terminology
0	Soils	Kingdom			
1	Order	Collective 1	Partly formalized	Conceptual (general water regime)	Traditional
2	Class	Collective 2	Formalized	Chemico-morphological	Traditional
3	Type	Generic	Formalized	Chemico-morphological	Traditional
4	Subtype	Specific	Formalized	Chemico-mineralogical	Traditional
5	Variety	Specific / Varietal	Formal	Chemico-mineralogical	Traditional

secondary carbonates. The <u>varieties</u> give additional characteristics of soils, both qualitative and quantitative. The structure of this classification is a hierarchical taxonomy with formalized borders between classes.

Diagnostics

The new Austrian soil classification includes a system of diagnostics based on quantitative evaluation of soil morphological and chemical parameters. The definitions of the horizons are less strict than in the WRB giving some flexibility to the soil surveyor. Only internal properties of soil profiles are recorded, no regime monitoring is required. The division of terrestrial and hydromorphic soils is made on the basis of soil morphology and observations about the position of the pedon in a landscape. The diagnostics, in general, are semiquantitative chemico-morphological.

Terminology

The terminology of the Austrian classification remains nearly the same as in Kubiëna's system. Some modifications in soil names were made, however most soil names are traditional terms which originated many years ago from folk soil terminology (*Podsol, Rendzina, Ranker*, etc.), and descriptive names based on common German words. Some artificial terminology is used for anthropogenically transformed soils that had not been recognized before: *Gartenboden, Rigolboden, Deponieboden* and so on.

Correlation

Though certain harmonization was made with the WRB (IUSS Working Group WRB, 2006), the archetypes of the Austrian classification differ from those in the WRB. Special work was done on the correlation of steppe soils in the Austrian classification and the WRB (Nestroy, 2001). The correspondence between the classes is in places very superficial, and may be misleading: for example, some *Tschernosem* soils in Austria correlate with *Kastanozems* in the WRB and these soils form in completely different pedoenvironments. Some eroded steppe soils (*Rump-Tschernosems*) are formally classified as *Calcisols* in the WRB system. We tried to avoid these analogies wherever possible.

For the correlation, the soils are listed in the same sequence as in the text of the classification so as not to disturb the logical evolutionary sequence proposed by the authors. The correlation is provided down to the level of soil types.

Terrestrische Böden – soil order. (No equivalent in WRB at this level of generalization)
Terrestrische Rohböden – soil class. ≈ Arenosols / Regosols / Hyperskeletic Leptosols
The following soil types are distinguished within the class:

- *Grobmaterial-Rohböden* ≈ Arenosols / Hyperskeletic Leptosols
- *Feinmaterial-Rohböden* ≈ Regosols

Auflagehumusböden und Entwickelte A-C-Böden – soil class. ≈ Chernozems / Phaeozems / Leptosols
The following soil types are distinguished within the class:

- *Fels-Auflagehumusboden* ≈ Lithic Leptosols
- *Rendzina* ≈ Rendzic Leptosols
- *Kalklehm-Rendzina* ≈ Rendzic Phaeozems
- *Pararendzina* ≈ Phaeozem (Calcaric)
- *Ranker* ≈ Mollic Leptosols / Umbric Leptosols / Leptic Folic Histosols
- *Tschernosem* ≈ Chernozems / Phaeozems
- *Rumpf-Tschernosem* ≈ (eroded Chernozems and Phaeozems; partially may be classified as Cambisols and Regosols)

Braunerden – soil class. ≈ Cambisols / Umbrisols / Luvisols
The following soil types are distinguished within the class:

- *Braunerde* ≈ Cambisols / Umbrisols
- *Parabraunerde* ≈ Luvisols

Podsole – soil class. ≈ Podzols
The following soil types are distinguished within the class:

- *Semipodsol* ≈ Entic Podzols
- *Podzol* ≈ Podzols
- *Staupodzol* ≈ Stagnic Podzols

Kalklehme – soil class. ≈ Cambisols (Calcaric) / Nitisols
The following soil types are distinguished within the class:

- *Kalkbraunlehm* ≈ Cambisols (Calcaric)
- *Kalkrotlehm* ≈ Cambisols (Calcaric, Chromic) / Nitisols (Chromic)

Substratböden – soil class. ≈ Technosols
The following soil types are distinguished within the class:

- *Farb-Substratboden* ≈ Technosols
- *Textur-Substratboden* ≈ Technosols (Ruptic)

Kolluvien und Anthrosole – soil class. ≈ Anthrosols / Chernozems / Phaeozems
The following soil types are distinguished within the class:

- *Kolluvisol* ≈ Chernozems (*Colluvic*) / Phaeozems (*Colluvic*)
- *Kultur-Rohboden* ≈ Regic Anthrosols
- *Gartenboden* ≈ Hortic Anthrosols
- *Rigolboden* ≈ Plaggic Anthrosols
- *Shüttungsboden* ≈ Anthrosols
- *Deponieboden* ≈ Terric Anthrosols

Hydromorphe Böden – soil order. (No equivalent in WRB at this level of generalization)
Pseudogleye – soil class. ≈ Gleysols / Stagnosols / Planosols / Gleyic Cambisols
The following soil types are distinguished within the class:

- *Typischer Pseudogley* ≈ Gleyic Cambisols
- *Stagnogley* ≈ Stagnosols
- *Hangpseudogley* ≈ Planosols
- *Haftnässe-Pseudogley* ≈ Stagnosols / Gleysols
- *Reliktpseudogley* ≈ Gleysols

Auböden – soil class. ≈ Fluvisols
The following soil types are distinguished within the class:

- *Auboden* ≈ Fluvisols
- *Augley* ≈ Gleyic Fluvisols
- *Schwemmboden* ≈ Mollic Gleyic Fluvisols
- *Rohauboden* ≈ Fluvisols (Arenic)

Gleye – soil class. ≈ Gleysols
The following soil types are distinguished within the class:

- *Gley* ≈ Gleysols
- *Naßgley* ≈ Gleysols (Oxiaquic)
- *Hanggley (Quellgley)* ≈ Gleysols

Salzböden – <u>soil class</u>. ≈ Solonetz / Solonchaks
The following soil <u>types</u> are distinguished within the class:

- *Solonstchak* ≈ Solonchaks
- *Solonetz* ≈ Solonetz
- *Solontschak-Solonetz* ≈ Salic Solonetz

Moore, Anmoore und Feuchtschwartzerden – <u>soil class</u>. ≈ Histosols
The following soil <u>types</u> are distinguished within the class:

- *Hochmoore* ≈ Ombric Histosols
- *Niedermoore* ≈ Rheic Histosols
- *Anmoore* ≈ Folic Histosols / Hemic Histosols
- *Feuchtschwartzerde* ≈ Sapric Histosols

Unterwasserböden – <u>soil class</u>. ≈ Aquic Histosols / Aquic Fluvisols
The following soil <u>types</u> are distinguished within the class:

- *Dy* ≈ Aquic Histosols
- *Gittya* ≈ Limnic Aquic Fluvisols (Humic)
- *Sapropel* ≈ Limnic Aquic Fluvisols (Clayic)

References

Ad-hoc-AG Boden (2005) *Bodenkundliche Kartieranleitung*, herausgegeben von der Bundesanstalt für Geowissenschaften und Rohstoffe, 5 verbesserte und erweiterte Auflage, Hannover, 438pp

Fink, J. (1969) 'Osterreichische Bodensystematik', *Mitteilungen der Osterreichischen Bodenkundlichen Gesellschaft*, vol 13, pp1–88

IUSS Working Group WRB (2006) *World Reference Base for Soil Resources*, 2nd edition, World Soil Resources Reports no 103, UN Food and Agriculture Organization, Rome, 128pp

Kubiëna, W. L. (1953) *Bestimmungsbuch und Systematik der Boden Europas*, Verlag Enke, Stuttgart, 392pp

Nestroy, O. (2001) 'Classification of Chernozems, Phaeozems and Calcisols in Austria according to the World Reference Base for Soil Resources (WRB)', *European Soil Bureau Research Report*, no 7, pp121–124

Nestroy, O., Dannenberg, O. H., English, M., Gessl, A., Herzenberger, E., Kilian, W., Nelhiebel, P., Pecina, E., Pehjamberger, A., Schneider, W. and Wagner, J. (2000) 'Systematische Gliederung der Boden Osterreichs (Osterreichische Bodensystematik 2000)', *Mitteilungen der Osterreichischen Bodenkundlichen Gesellschaft*, vol 60, pp1–104

11
Soil Classification of Switzerland

Objectives and scope

At the beginning of the 20th century Switzerland was a poor mountainous country, and now it is one of the most stable economies of the world. That is a result not only of successful financial policy, but also of the stable development of its agricultural sector. The development of agriculture required an inventory of soil resources. From the very beginning soil classification in Switzerland was mostly aimed at practical purposes. The recent version of classification (Arbeitsgruppe 'Bodenklassifikation und Nomenklatur', BGS, 2002) is suitable for soil mapping at any scale.

Though this classification was made for the use within the country, the scheme might be used for soils worldwide (Table 11.1). The authors tried to give a complete picture of soil classification and included also soil groups that were not present in Switzerland (like deep weathered tropical soils or alkaline soils). The soils of agricultural lands, which are of major importance for mapping, are classified as well as their natural analogies, with a special index at the subtype level. The substrates of urban and industrial areas are disregarded, and rocks and subaquatic sediments are also considered to be out of the scope of the classification. The soils deeply altered by agricultural practices are classified like natural soils.

Table 11.1 *The scope of soil classification of Switzerland*

Superficial bodies	Representation in the system
Natural soils	National coverage, but may be extended for wider areas
Urban soils	Not included in the classification
Man-transported materials	Not included in the classification
Bare rock	Not included in the classification
Subaquatic soils	Not included in the classification
Soils deeply transformed by agricultural activities	Classified as if natural soils

Theoretical background

Switzerland is unique. In its history, policy and culture it has always gone its own way, and the same is true for Swiss soil classification. At first glance, it is one of the many taxonomic classifications of Europe, with several collective levels; however, the classes, orders, branches and types of this classification, though they are called taxonomical levels, are in fact independent axes for characterizing soil. The classes in Swiss soil classification indicate the water regime of soil: percolative, periodically percolative, stagnative and so on. The classification of water regimes was borrowed from the Russian school (see Rode, 1961). The orders divide soils according to the substrate type: dominant primary minerals, dominant secondary (clay) minerals, a mixture of organic material with primary minerals, and so on. Though formally it is declared that classes have higher rank than the orders, in practice these are not taxonomic levels but two parallel systems of classifying soils. The next level – branches – adds another axis to soil characterization: mineralogical and chemical composition of soil substrates. The type level gives the dimension of soil-forming processes (substance migration in the profile). The combination of these four dimensions – hydrological, substrate, geochemical and pedological – provides a soil name. In theory, these four 'levels' may be arranged in any combination. In practice there are natural limits for the number of combinations because certain water regimes cannot be combined with some pedogenetic processes. Although the number of possible individual soils generated by combination of the four 'levels' of classification, with five to ten groups each, may exceed several hundred, the Swiss pedologists have named only 22 soils that have been detected in Switzerland.

Basically the classification reflects the concept of the authors, who see the world of soils as a transposition of certain pedogenic processes over existent substrates under the influence of water regimes. The concept corresponds to the mainstream paradigm of pedology and reflects the influence of Kubiëna and the Russian school. The two main axes of the classification – soil hydrology and the stage of the development of profile – are derived from Kubiëna's classification (1953). The classification of the hydrologic regimes and of soil-forming processes is strongly influenced by the concepts developed in Russia (Rode, 1961).

Development

The first soil classification of Switzerland was proposed by Pallman in 1947 (Bonnard, 1999). The work on the detailed soil classification of Switzerland started in the 1960s and 1970s (Frei and Juhasz, 1965; Frei, 1976). In the late 1970s, a special Working Group on Soil Classification was created (Arbeitsgruppe 'Bodenklassifikation', BGS, 1979, 1982), later renamed as a Working Group on Soil Classification and Nomenclature (Arbeitsgruppe 'Bodenklassifikation und Nomenklatur', BGS, 1996). They produced a number of approximations of the classification, the most recent one (30th version) in 2002 (Arbeitsgruppe 'Bodenklassifikation und Nomenklatur', BGS, 2002).

Structure

The structure of the Swiss soil classification has no analogies elsewhere in the world. First, it has no generic level. Soil names appear only as a combination of all the levels; however, soil archetypes are not noted at the lower level. At the lower level there are local forms which indicate the vegetation and other important landscape features. No soil name is included at that level. The Swiss classification is not a hierarchical taxonomy; the soils appear to be points or 'clouds' in a multidimensional space of soil characteristics, as in the French classification (AFES, 1998). The difference is that in the Swiss classification the axes for this multidimensional space are formalized and arranged according to their importance. It is possible to change the places of the classes, orders and types in the vertical sequence, with very few changes; however, the authors (Arbeitsgruppe 'Bodenklassifikation und Nomenklatur', BGS, 2002) regard their classification as a normal hierarchical taxonomy. This taxonomy arranges the categories and not real objects, like many other classifications. The ranking of these categories is shown in Table 11.2. The first category, classes, divides soils according to their hydrologic regime. The second category, orders, separates soils formed on different substrates. The third, branches, relate to the mineralogy and total chemistry of soils. These second and third categories seem to correspond to the family level in the US Soil Taxonomy (Soil Survey Staff, 1999); however, this level was split and put on the upper level of taxonomy. The fourth category, types, divides soils with different pedogenetic processes, mainly those connected with material transport in the profile. Apart from the four upper levels, which are called hierarchical ones, there are also lower 'non-hierarchical' levels. These levels are used to give additional characteristics of soil and the landscape. The subtype level gives additional information about the soil with a great variety of characteristics more or less corresponding to the modifiers in the World Reference Base for Soil Resources (WRB) (IUSS Working Group WRB, 2006). The level of soil forms completes soil characteristics with the texture, stoniness and ionic composition of the exchange complex, if necessary. Finally, the level of local form provides some characteristics of pedoenvironments: climate, relief and vegetation.

Diagnostics

The diagnostics in the Swiss soil classification are not very strict except in the lower levels where exact information on soil properties is needed. Apart from the profile itself, this classification requires certain information about soil-forming conditions and soil hydrological regimes. The water regime is required at the highest level of the classification; however, long monitoring is not required. Most diagnostics are based on soil morphology and landscape observations. Also the information on soil-forming conditions, namely climate, relief and vegetation, are noted in the description of local forms.

Table 11.2 *The structure of the Swiss soil classification*

Level	Taxon name	Taxon characteristics	Borders between classes	Diagnostics	Terminology
1	Class	Collective 1	Fuzzy	Water regime	Traditional
2	Order	Collective 2	Fuzzy	Substrate	Traditional
3	Branch	Collective 3	Fuzzy	Geochemistry	Traditional
4	Type	Collective 4	Fuzzy	Morphology	Traditional
5	Subtype	Specific 1	Fuzzy	Chemico-morphological	Traditional
6	Soil form	Varietal	Fuzzy	Chemico-morphological	Traditional
7	Local form	Specific 1	Fuzzy	Landscape	Traditional

Terminology

The terminology of the soil classification of Switzerland follows the traditions of Kubiëna's system (1953). Soil names are mainly the same as being used in all the German-language soil science literature. Also a number of terms proposed and/or distributed by the WRB (*Fluvisols, Regosols*, etc.) are included in the recent versions. An important feature of this classification is that every category has a numeral or letter code. It is not unusual that soils receive codes, for example, for naming soil polygons on maps, but in this particular case the final code of a soil is a sequence of codes of categories at every level. The traditional soil name is not very informative, but the code gives the information about all the soil features. For example, the soil code 1352 HD, E1 means that it is a soil with a percolative hydrological regime (class 1), composed of primary and secondary minerals with an admixture of organic matter (order 3), having iron oxides in a complex with clay minerals (branch 5), with neutral or slightly acid reaction, with calcium as the dominant exchangeable cation (type 2), with diffuse horizons (subtype HD) and a neutral reaction (subtype E1). A somewhat similar scheme was presented in the so-called 'Factual Key' in Australia (Northcote, 1971). To some extent, the system has something in common with the US Soil Taxonomy, where the names of soils include some fragments of the soil names from the upper levels, thus allowing reconstruction of the place of a soil in the system.

Correlation

The upper levels of the Swiss soil classification are difficult to correlate; however, the particular soils recognized in Switzerland are relatively easy to correlate as they correspond to well-known archetypes. It suggests that the classification was constructed *over* existing archetypes rather than deriving entities by formal separation of the soil universe.

Our correlation is for the soil groups without giving their complete characteristics at all the levels, but instead only the most general characteristics of soil groups associated with the table captions in Arbeitsgruppe

'Bodenklassifikation und Nomenklatur', BGS (2002, pp31–36). The codes of soil groups are not provided; for details we recommend consulting the original text of the classification.

The names of soils are presented as they are listed in the text of the classification.

PERCOLIERTE BÖDEN (Percolated soils)

Gesteinsböden (Initial soils)

- Kolluvialer Silikat-gesteinsboden ≈ Hyperskeletic Leptosols
- Sandiger Gesteinsfluvisols ≈ Fluvisols (Arenic) / Protic Arenosols

Humus-Gesteinsböden (Humus-initial soils)

- Rohhumoser Silikat-gesteinsboden ≈ Folic Regosols (Dystric) / Folic Arenosols (Dystric)
- Modrighumoser Karbonatlithosol ≈ Folic Leptosols (Calcaric)

A/C Böden

- Modrighumoser Silikatboden (Ranker) ≈ Umbric Leptosols / Leptic Umbrisols
- Teilweise entkarbonateten Fluvisol ≈ Fluvisols (Calcaric)
- Teilweise entkarbonateten Regosols ≈ Regosols (Calcaric)
- Mullhumose Rendzina ≈ Rendzic Leptosols

A/B/C Böden

- Schwach pseudogleyige Saure Braunerde ≈ Endogleyic Cambisols (Dystric)
- Neutrale Braunerde ≈ Cambisols (Eutric)
- Akkumulierte Kalkbraunerde ≈ Calcic Cambisols

Entwickelte Böden mit Bfe-Horizont

- Modrighumoser Braunpodsol ≈ Umbric Podzols

Entwickelte Böden mit E und I-Horizonten

- Ausgeprägte Parabraunerde ≈ Albic Luvisols
- Rohhumoser Podsol ≈ Podzols

SELTEN PERCOLIERTE BÖDEN (Periodically percolated soils)

- Teilweise entkarbonateten Regosols ≈ Regosols (Calcaric)

STAUWASSER-GEPRÄGTE BÖDEN (Soils affected by water stagnation)

- Schwach saurer Pseudogley ≈ Stagnosols (Dystric)

GRUND-ODER HANGWASSERGEPRÄGTE BÖDEN (Soils affected by groundwater)

<u>Mineralische Nassböden</u>

- Teilweise entkarbonateten Braunerde-Gley ≈ Endogleyic Cambisols (Calcaric)
- Verdichteter Buntgley ≈ Gleysols
- Anmooriger Fahlgley ≈ Histic Gleysols

Organische Nassböden

- Sapro-organisches, neutrales Halbmoor ≈ Rheic Sapric Histosols (Eutric)
- Tieftorfiges, zersetztes neutrales Moor ≈ Histosols (Eutric)

PERIODISCH ÜBERSCHWEMMTE BÖDEN (Periodically flooded soils)

- Humus-gesteins-aueboden ≈ Mollic Fluvisols

References

AFES (1998) *A Sound Reference Base for Soils* (The 'Référentiel pédologique': text in English), INRA, Paris, 322pp

Arbeitsgruppe 'Bodenklassifikation', BGS (1979) 'Bezeichnung der Horizonte der Bodenprofile', *BGS Bulletin*, no 3, pp84–85

Arbeitsgruppe 'Bodenklassifikation', BGS (1982) 'Verschlag für die Verwendung von Signaturen bei Profilskizzen', *BGS Bulletin*, no 6, pp177–182

Arbeitsgruppe 'Bodenklassifikation und Nomenklatur', BGS (1996) 'Schlüssel zur Klassifikation der Bodentypen der Schweiz', *BGS Bulletin*, no 20, pp24–85

Arbeitsgruppe 'Bodenklassifikation und Nomenklatur', BGS (2002) *Klassifikation der Böden de Schweiz, Version 30*, Eidgenössische Forschungsanstalt für Agrarökologie und Landbau, Zürich-Reckenholz, 96pp

Bonnard, L.-F. (1999) 'Soil Survey in Switzerland', in Bullock, P., Jones, R.J.A. and Montanarella, L. (eds) *Soil Resources of Europe*, European Soil Bureau Research Report no 6, pp153–158

Frei, E. (1976) 'Richtlinen für die Beschreibung und Klassifikation von Bodenprofilen', *Schweizerischen Landwirtschaft Forschung*, vol 15, pp339–347

Frei, E. and Juhasz, P. (1965) 'Beitrag zur Metodik der Bodenkartierung und der Auswertung von Bodenkarten unter schweizerischen Verhältnissen', *Schweizerischen Landwirtschaft Forschung*, vol 3, pp249–307

IUSS Working Group WRB (2006) *World Reference Base for Soil Resources*, 2nd edition, World Soil Resources Reports no 103, UN Food and Agriculture Organization, Rome, 128pp

Kubiëna, W. L. (1953) *Bestimmungsbuch und Systematic der Boden Europas*, Verlag Enke, Stuttgart, 392pp

Northcote, K. H. (1971) *A Factual Key for the Recognition of Australian Soils*, Rellim Technical Publication no VII, Kuratta Park, 122pp

Rode, A. A. (1961) *The Soil Forming Process and Soil Evolution*, Israel Program for Scientific Translations, Jerusalem, 269pp

Soil Survey Staff (1999) *Soil Taxonomy: A Basic System of Soil Classification for Making and Interpreting Soil Surveys*, USDA, Handbook 436, 2nd edition, United States Government Printing Office, Washington DC, 696pp

12
Soil Classification of The Netherlands

Objectives and scope

Soil classification of The Netherlands was one of the first soil taxonomies in Europe (De Bakker and Schelling, 1966). It reflects a very particular soil cover of a country where extensive areas are reclaimed marine sediments. Though practically not used nowadays, this classification was an important stage in the development of soil classification theory and application. Initially, the classification was designed for detailed soil survey of the country. It includes a lot of soils deeply transformed by agricultural management (Table 12.1) because these constitute a significant proportion of the soil cover of the country. However, technogenically transformed soils, as well as the bare rock and underwater sediments, are not included in the classification.

Table 12.1 *The scope of soil classification of The Netherlands*

Superficial bodies	Representation in the system
Natural soils	National coverage
Urban soils	Not included in the classification
Man-transported materials	Not included in the classification
Bare rock	Not considered as soils
Subaquatic soils	Not considered as soils
Soils deeply transformed by agricultural activities	Included at the level of subgroups to a number of groups; also a special suborder of cultivated organic soils (Earthy peat soils) exists

Theoretical background

To a great extent, the Dutch soil classification was an original product, both in approaches and in terminology (see below). It was also influenced by contemporary classifications, and like many other classifications of this period it borrowed some successful concepts of the early approximations of the US Soil Taxonomy, yet managed to preserve its originality. The classification was influenced by the ideas of Kubiëna (1953) including grouping soils according to the level of hydromorphism. There are no source systems for the classification of The Netherlands. A high degree of agricultural transformation and peculiarity of soil cover of The Netherlands resulted in the principles of dividing soil taxa differently from that in other countries. For example, peat soils are separated according to the stage of their agricultural transformation, presence of mineral layers on the surface, and evidences of transportation of peat particles in the profile – a process identical to *lessivage* in mineral soils. Some subgroups and suborders are recognized on the basis of the depth of anthropogenic disturbance of soils. In the Dutch classification there are no strict limits between organic and mineral soils because most sediments in The Netherlands are complex and polygenetic, and interlayering of mineral and organic deposits is common. The complexity of this composition is reinforced by many centuries of agricultural transformation of soils. For example, it is difficult to say if *'Dam' podzol soils* are mineral or organic: these soils consist of human-transported humus-rich sand overlying peat remaining after peat harvest which in turn lies over a podzol profile.

Development

The work on national classification was initiated after the Second World War. The final version was published in the 1960s (De Bakker and Schelling, 1966), almost simultaneously with the 7th Approximation of the US soil classification and the first official publication of the classification of the Soviet Union. After the first publication the classification was not modified and a well-known book on soils of The Netherlands (De Bakker, 1979) used the same version. Currently the classification is practically out of use as the soil survey finished its work in the country, and the classification is absent in most university courses. It is useful for understanding older soil maps and corresponding documents, and its terminology is interesting for people concerned with scientific methodology and the history of science (Siderius and de Bakker, 2003).

Structure

There are four levels in the taxonomy of Dutch soils: their names and, to some extent, meaning are similar to those in the US Soil Taxonomy (Table 12.2). The archetypes of this system are soil subgroups. They partially correspond to soil series in the US Soil Taxonomy (Soil Survey Staff, 1999) and of some other classifications, but are somewhat broader in scope. Also soil orders represent general archetypes, big entities, which are close to soil types in the initial, Dokuchaev meaning. They include soils with common arrangement of the profile. The presence of certain horizons is taken into account and they may be called diagnostic ones. The orders are divided into suborders according to the type of upper organic horizon, and to their water regime. Soil groups are separated according to the texture or some other specific features. The same principle is used for distinguishing <u>subgroups</u>, but they are more complex. Subgroups are also soil archetypes having features of the US soil series, and of other (non-generic) taxonomic levels. The structure of this classification is considered to be a hierarchical taxonomy with partially formalized limits between classes.

Table 12.2 *The structure of soil classification of The Netherlands*

Level	Taxon name	Taxon characteristics	Borders between classes	Diagnostics	Terminology
0	Soils	Kingdom			
1	Order	Generic 1	Mostly formal	Morphological	Traditional
2	Suborder	Specific 1	Mostly formal	Morphological	Traditional
3	Group	Specific 2	Mostly formal	Morphological	Traditional
4	Subgroup	Generic 2	Mostly fuzzy	Morphological	Traditional

Diagnostics

The diagnostics are based on the properties of the soil profile. If water regimes are included in the definitions, their diagnostics should be done by investigation of the morphology of soil profile, not by monitoring the dynamics of moisture in the soil. Only morphological features are used in diagnostics. Some of the criteria used are logical (presence/absence of property), some are quantitative (horizons' depth, colour according to Munsell soil colour charts), and some are semiquantitative ('weak development of B horizon'). The diagnostics of this classification can be called mainly morphological.

Terminology

The names of the highest levels of the Dutch soil taxonomy are partly traditional for European soil science (*Podzols, Peat soils*), some are constructed (*Xeropodzols, Hydropodzols*), and some are borrowed from Dutch folk classification. The latter names can raise some questions among specialists: for example, it is difficult to guess that 'Brick' is the name for soils with an *argillic* horizon, and 'Earth' is the name for soils with a humified well-structured surface horizon.

No analogies exist for soil terminology of the Dutch classification at the level of subgroups. H. De Bakker and J. Shelling (1966) explained its origin:

> On subgroup level names have been chosen consisting of frequent
> endings of (untranslated) Dutch toponims, combined with the
> name of the order. These soils can be found in the neighborhood
> of villages, hamlets etc., whose names end in this way. 'Aar' peat

soils can be found in the surroundings of Langeraar and Ter Aar, whereas most of the villages ending on 'koop' or 'kop' (Boskoop, Teckop) are situated on 'koop' peatsoils. Some of these names stem from medieval reclamation ... others are names of rivers, lakes, brooks, or just swampy or low lying areas ... some refer to a dominant soil use and some are just arbitrary.

Recently the origin of soil names in the Netherlands was presented in detail in a special paper (Siderius and de Bakker, 2003) which is highly recommended to those interested in soil terminology.

Correlation

The correlation of the terms of Dutch soil classification with the World Reference Base for Soil Resources (WRB) (IUSS Working Group WRB, 2006) is difficult mainly because of the complexity of soils and sediments in The Netherlands. Many soil profiles have no analogies in other parts of the world and are not included in the WRB system. To correlate some of the most complex profiles an excessive number of modifiers and specifiers had to be used.

The correlation of the terms is given for all the levels. The names of taxa are presented in English, except untranslatable names of subgroups. The order of presentation of taxa is not alphabetical, but the same as in the text of classification (De Bakker and Schelling, 1966). A monograph by H. De Bakker (1979) was used for correlation, because that book includes a lot of soils descriptions, each followed by correlation of Dutch soil names with the main world classifications, including the legend for the Soil Map of the World.

Peat soils – soil order. ≈ Histosols
Two suborders are distinguished within the order:
Earthy peat soils – soil suborder. ≈ Histosols (Eutric, Drainic, *Anthric*)
The suborder is divided into three soil groups:
Clayey earthy peat soils – soil group. ≈ Rheic Histosols (Eutric, Drainic, *Anthric*)
The group is divided into two soil subgroups:

- *'Aar' peat soils* ≈ Histosols (Eutric, Drainic, *Terric*)
- *'Koop' peat soils* ≈ Histosols (Eutric, *Anthric*)

Podzolic earthy peat soils – soil group. ≈ Histosols (Drainic, *Anthric, Luvic*)
Only one soil subgroup is included in this group:

- *'Bouwte' peat soils* ≈ Ombric Histosols (Dystric, Drainic, *Anthric, Luvic*)

Clay-poor earthy peat soils – soil group. ≈ Ombric Histosols (*Anthric*)
The group is divided into two soil subgroups:

- *'Bo' peat soils* ≈ Ombric Histosols (*Terric*)
- *'Made' peat soils* ≈ Ombric Histosols (*Anthric*)

<u>Raw peat soils</u> – <u>soil suborder</u>. ≈ Histic Gleysols / Histosols / Regosols (*Thaptohistic*) / Arenosols (*Thaptohistic*) / Anthrosols (*Thaptohistic*)
The suborder is divided into three soil groups:
Initial raw peat soils – soil group. ≈ Histic Gleysols
Only one soil subgroup is included in this group:

- *'Vliet' peat soils* ≈ Histic Gleysols

Podzolic raw peat soils – soil group. ≈ Histosols (*Luvic*) / Arenosols (*Thaptohistic*)
Only one soil subgroup is included in this group:

- *'Mond' peat soils* ≈ Ombric Histosols (*Luvic*) / Arenosols (*Thaptohistic*)

Ordinary raw peat soils – soil group. ≈ Histosols (*Dystric*) / Regosols (*Thaptohistic*) / Anthrosols (*Thaptohistic*) / Arenosols (*Thaptohistic*)
The group is divided into four soil subgroups:

- *'Weide' peat soils* ≈ Anthrosols (*Thaptohistic*)
- *'Waard' peat soils* ≈ Regosols (*Thaptohistic*)
- *'Meer' peat soils* ≈ Histosols (Dystric) / Arenosols (*Thaptohistic*)
- *'Vlier' peat soils* ≈ Histosols (Dystric)

<u>Podzol soils</u> – <u>soil order</u>. ≈ Umbrisols / Podzols
Three <u>suborders</u> are distinguished within the order:
<u>Moder podzol soils</u> – <u>soil suborder</u>. ≈ Umbrisols / Entic Podzols / Umbric Podzols
Only one soil group is distinguished:
Moder podzol soils – soil group. ≈ Umbrisols / Entic Podzols / Umbric Podzols
The group is divided into five soil subgroups:

- *'Holt' podzol soils with a sand cover* ≈ Umbric Podzols / Entic Podzols / Arenosols (*Thaptospodic*)
- *'Loo' podzol soils* ≈ Umbrisols (Anthric) / Entic Podzols (Anthric)
- *'Hoek' podzol soils* ≈ Umbrisols / Entic Podzols
- *'Horst' podzol soils* ≈ Umbric Podzols (Lamellic) / Entic Podzols (Lamellic)
- *'Holt' podzol soils* ≈ Umbric Podzols / Entic Podzols

<u>Hydropodzol soils</u> – <u>soil suborder</u>. ≈ Gleyic Podzols / Histosols
The suborder is divided into two soil groups:
Peaty podzol soils – soil group. ≈ Gleyic Histic Podzols

The group is divided into four soil subgroups:

- *'Moer' podzol soils with a clay cover* ≈ Gleyic Histic Carbic Podzols / Regosols (Thaptospodic, Thaptohistic)
- *'Moer' podzol soils with a sand cover* ≈ Gleyic Histic Carbic Podzols / (*Thaptospodic, Thaptohistic*) Arenosols
- *'Dam' podzol soils* ≈ Histosols (Novic, *Thaptospodic*)
- *'Moer' podzol soils* ≈ Gleyic Histic Carbic Podzols

Ordinary hydropodzol soils – soil group. ≈ Gleyic Podzols
The group is divided into four soil subgroups:

- *'Veld' podzol soils with a clay cover* ≈ Gleyic Carbic Podzols (Anthric) / Regosols (*Thaptospodic, Thaptohistic*)
- *'Veld' podzol soils with a sand cover* ≈ Gleyic Carbic Podzols (Anthric) / Regosols (*Thaptospodic, Thaptohistic*)
- *'Haar' podzol soils* ≈ Gleyic Placic Podzols
- *'Veld' podzol soils* ≈ Gleyic Carbic Podzols (Anthric)

<u>Xeropodzol soils</u> – <u>soil suborder</u>. ≈ Podzols / Albeluvisols
Only one soil group is distinguished:
Xeropodzol soils – soil group. ≈ Podzols / Albeluvisols
The group is divided into four soil subgroups:

- *'Haar' podzol soils with a sand cover* ≈ Placic Podzols / Arenosols (*Thaptospodic, Thaptohistic*)
- *'Kamp' podzol soils* ≈ Umbric Podzols
- *'Heuvel' podzol soils* ≈ Albeluvisols / Lamellic Podzols
- *'Haar' podzol soils* ≈ Placic Podzols

<u>Brick soils</u> – <u>soil order</u>. ≈ Planosols / Albeluvisols / Luvisols
Two <u>suborders</u> are distinguished within the order:
<u>Hydrobrick soils</u> – <u>soil suborder</u>. ≈ Endogleyic Planosols / Gleyic Albeluvisols
Only one soil group is distinguished:
Hydrobrick soils – soil group. ≈ Endogleyic Planosols / Gleyic Albeluvisols
The group is divided into two soil subgroups:

- *'Beemd' brick soils* ≈ Endogleyic Planosols / Gleyic Albeluvisols
- *'Kuil' brick soils* ≈ Endogleyic Planosols / Gleyic Albeluvisols

<u>Xerobrick soils</u> – <u>soil suborder</u>. ≈ Albeluvisols / Luvisols
Only one soil group is distinguished:
Xerobrick soils – soil group. ≈ Albeluvisols
The group is divided into five soil subgroups:

- *'Berg' brick soils* ≈ Luvisols (*Nudiargic*)
- *'Del' brick soils* ≈ Endogleyic Albic Luvisols (Abruptic)
- *'Rooi' brick soils* ≈ Albic Luvisols (Abruptic)
- *'Daal' brick soils* ≈ Endogleyic Albeluvisols / Endogleyic Luvisols
- *'Rade' brick soils* ≈ Albeluvisols / Albic Luvisols

Earth soils – soil order. ≈ Anthrosols / Gleysols / Gleyic Phaeozems / Gleyic Umbrisols / Technosols / Gleyic Fluvisols / Rendzic Leptosols
Three suborders are distinguished within the order:
Thick earth soils – soil suborder. ≈ Anthrosols / Technosols
The suborder is divided into two soil groups:
'Enk' earth soils – soil group. ≈ Anthrosols (Arenic)
The group is divided into two soil subgroups:

- *Brown 'enk' earth soils* ≈ Terric Anthrosols (Arenic)
- *Black 'enk' earth soils* ≈ Plaggic Anthrosols (Arenic)

'Tuin' earth soils – soil group ≈ Terric Anthrosols / Garbic Technosols
Only one soil subgroup is included in this group:

- *'Tuin' earth soils* ≈ Terric Anthrosols / Garbic Technosols

Hydroearth soils – soil suborder. ≈ Mollic Gleysols / Umbric Gleysols / Histic Gleysols / Gleyic Histic Fluvisols
The suborder is divided into three soil groups:
Peaty earth soils – soil group. ≈ Histic Gleysols
The group is divided into two soil subgroups:

- *'Plas' earth soils* ≈ Histic Gleysols (Thionic) / Gleyic Histic Fluvisols (Thionic)
- *'Broek' earth soils* ≈ Histic Gleysols (Arenic)

Sandy hydroearth soils – soil group. ≈ Mollic Gleysols (Arenic) / Umbric Gleysols (Arenic) / Gleyic Phaeozems (Arenic) / Gleyic Umbrisols (Arenic)
The group is divided into three soil subgroups:

- *Brown 'beek' earth soils* ≈ Gleysols (Humic, Arenic)
- *'Goor' earth soils* ≈ Gleyic Phaeozems (Arenic) / Gleyic Umbrisols (Arenic)
- *Black 'beek' earth soils* ≈ Mollic Gleysols (Arenic) / Umbric Gleysols (Arenic)

Clayey hydroearth soils – soil group. ≈ Mollic Gleysols / Umbric Gleysols
The group is divided into four soil subgroups:

- *'Lied' earth soils* ≈ Mollic Gleysols (*Thaptohistic*) / Umbric Gleysols (*Thaptohistic*)
- *'Toucht' earth soils* ≈ Umbric Gleysols (Thionic) / Umbric Gleyic Fluvisols (Thionic)
- *'Woud' earth soils* ≈ Mollic Gleysols
- *'Leek' earth soils* ≈ Mollic Gleysols

Xeroearth soils – soil suborder. ≈ Rendzic Leptosols / Anthrosols
The suborder is divided into three soil groups:
'Krijt' earth soils – soil group. ≈ Rendzic Leptosols
Only one soil subgroup is included in this group:

- *'Krijt' earth soils* ≈ Rendzic Leptosols

Sandy xeroearth soils – soil group. ≈ Anthrosols (Arenic)
The group is divided into two soil subgroups:

- *'Akker' earth soils* ≈ Spodic Anthrosols (Arenic)
- *'Kant' earth soils* ≈ Spodic Anthrosols (Arenic)

Clayey xeroearth soils – soil group. ≈ Regic Anthrosols
Only one soil subgroup is included in this group:

- *'Hof' earth soils* ≈ Regic Anthrosols

Vague soils – soil order. ≈ Fluvisols / Gleysols / Regosols / Arenosols / Cambisols
Two suborders are distinguished within the order:
Initial vague soils – soil suborder. ≈ Fluvisols
Only one soil group is distinguished:
Initial vague soils – soil group. ≈ Fluvisols
The group is divided into two soil subgroups:

- *'Gors' vague soils* ≈ Gleyic Salic Fluvisols
- *'Slik' vague soils* ≈ Gleyic Salic Tidalic Fluvisols

Hydrovague soils – soil suborder. ≈ Gleyic Fluvisols / Gleysols / Gleyic Arenosols
The suborder is divided into two soil groups:
Sandy hydrovague soils – soil group. ≈ Gleysols (Arenic) / Endogleyic Arenosols / Gleyic Fluvisols (Arenic)
Only one soil subgroup is included in this group:

- *'Vlak' vague soils* ≈ Endogleyic Arenosols (Calcaric) / Gleysols (Arenic) / Gleyic Fluvisols (Arenic)

Clayey hydrovague soils – soil group. ≈ Gleyic Fluvisols / Gleysols
The group is divided into three soil subgroups:

- *'Drecht' vague soils* ≈ Gleysols (*Thaptohistic*) / Gleyic Fluvisols (*Thaptohistic*)
- *'Nes' vague soils* ≈ Gleyic Fluvisols / Gleyic Salic Fluvisols
- *'Polder' vague soils* ≈ Gleyic Fluvisols / Gleyic Salic Fluvisols / Salic Gleysols

Xerovague soils – soil suborder. ≈ Arenosols / Fluvisols / Cambisols
The suborder is divided into two soil groups:
Sandy xerovague soils – soil group. ≈ Arenosols / Fluvisols (Arenic)
The group is divided into two soil subgroups:

- *'Duin' vague soils* ≈ Protic Arenosols
- *'Vorst' vague soils* ≈ Arenosols / Fluvisols (Arenic)

Clayey xerovague soils – soil group. ≈ Fluvisols / Cambisols
Only one soil subgroup is included in this group:

- *'Ooi' vague soils* ≈ Fluvisols (Eutric, Clayic) / Cambisols (Eutric, Clayic)

References

De Bakker, H. (1979) *Major Soils and Soil Regions in The Netherlands*, Junk B.V. Publishers and Pudoc, Wageningen, 203pp

De Bakker, H. and Schelling, J. (1966) *Systeem voor bodemklassifikatie voor Nederland*, De hogere niveaus, STIBOKA, Pudoc, Wageningen, 217pp

IUSS Working Group WRB (2006) *World Reference Base for Soil Resources*, 2nd edition, World Soil Resources Reports no 103, UN Food and Agriculture Organization, Rome, 128pp

Kubiëna, W. L. (1953) *Bestimmungsbuch und Systematic der Boden Europas*, Verlag Enke, Stuttgart, 392pp

Siderius, W. and de Bakker, H. (2003) 'Toponymy and soil nomenclature in the Netherlands', *Geoderma*, vol 111, nos 3–4, pp521–536

Soil Survey Staff (1999) *Soil Taxonomy: A Basic System of Soil Classification for Making and Interpreting Soil Surveys*, USDA, Handbook 436, 2nd edition, United States Government Printing Office, Washington DC, 696pp

13
Soil Classification of Poland

Objectives and scope

Poland has an old, developed school of soil science. Soil surveys started there in the early 1950s, and the country faced the need for a unified soil classification. The first classification (Anonymous, 1956), though based on a qualitative approach, was successfully used in Poland for mapping and classifying soils for more than 30 years. However, in the late 1980s it was decided that the classification should be updated according to new concepts in pedology, and that the diagnostics should be quantitative as in most classifications. The Polish classification is suited for soil mapping at any scale and for scientific soil research (Polish Society of Soil Science, 1989). This classification is designed for use only within the country (Table 13.1). Like most recent soil classifications, the soils of urban and industrial areas, and soils deeply transformed by agriculture, are included in the classification at high taxonomic levels. Bare rock and underwater sediments are not regarded as soils.

Table 13.1 *The scope of soil classification of Poland*

Superficial bodies	Representation in the system
Natural soils	National coverage
Urban soils	Included in a special order of industrial and urban soils
Man-transported materials	Included in a special order of industrial and urban soils
Bare rock	Not considered as soils
Subaquatic soils	Not considered as soils
Soils deeply transformed by agricultural activities	Included in a special order of cultivated soils (Kulturozems)

Theoretical background

Polish soil classification is based on pedogenetic understanding of soil processes and soil bodies, on traditional archetypes, and on quantitative diagnostics for limiting these archetypes. The classification represents a compromise between traditional grouping of soils and formalized methods of grouping (Charzyński et al, 2005). Many recently reworked classifications, especially in Eastern Europe, have followed the same way.

Development

Soil research has old traditions in Poland, even from the epoch of Dokuchaev and Glinka. Soil classification activities started there in the early 1950s, when soil mapping resulted in the need for a unified system of soil classification. The first complete text of soil classification was published in 1956 (Anonymous, 1956). For a long time this classification was applied to soil research in Poland. Eventually the classification was updated according to the evolving concepts of pedology. Polish soil science became more open to the influence of Western scientific schools, in part due to political reasons and partly to adjust the classification to current European approaches. In comparison with the older version, the new version (Polish Society of Soil Science, 1989) has few changes in the list of archetypes or in terminology. A new feature is the inclusion of human-modified soils and the structure and diagnostics of the classification are significantly improved. In the structure, two collective levels were included (Table 13.2), to help organize the information. The diagnostics are adjusted to the new quantitative concepts: the system of diagnostic horizons and properties used is similar to international standards for actual soil diagnostics.

Structure

There are six levels in the structure of the soil classification of Poland (Table 13.2). The highest level of taxonomy is soil <u>section</u>, which collects soils according to the main group of factors affecting soil formation (lithogenesis, hydromorphism, salinization, anthropogenic effect, etc.). Within the sections there are soil <u>orders</u>, grouping soils of the same leading pedogenetic process or the most general features of the profile. Soil <u>types</u> serve as a basic level of taxonomy, and are determined as soils with similar profiles and close chemical and physical properties. <u>Subtypes</u> are distinguished by additional soil-forming processes that result in the presence of additional horizons or properties or modifications of properties. <u>Genera</u> are distinguished according to the origin of parent material and the content of carbonates in this material. <u>Species</u> are similar to soil series in other classifications; they are defined as soil individuals as adopted by the Polish Society of Soil Science. It is important to mention that the system of soil series is not traditional for Poland, and species is not regarded as a generic level of taxonomy. In the countries with traditional use of soil series each soil is described and mapped, and then referred to a soil series, or a new series is proposed. Then, if necessary, this soil series is classified in terms of soil taxonomy. Where a soil series system was not used in a country, a soil is first classified in terms of soil taxonomy, and then, if the profile has some particular features, it may be named a special series. Though the difference seems to be just in the sequence of operations, in fact they are two different mental processes of synthesis and analysis.

Table 13.2 *The structure of soil classification of Poland*

Level	Taxon name	Taxon characteristics	Borders between classes	Diagnostics	Terminology
0	Soils	Kingdom			
1	Section	Collective 1	Fuzzy	Morphological / water regime	Traditional
2	Order	Collective 2	Fuzzy	Morphological	Traditional
3	Type	Generic 1	Formal	Chemico-morphological	Traditional
4	Subtype	Specific / Varietal	Formal	Chemico-morphological	Traditional
5	Genus	Specific	Formalized	Chemico-morphological	Traditional
6	Species	Generic 2 / Varietal	Formalized	Chemico-morphological	Traditional

The classification is a hierarchical taxonomy with partially formalized borders.

Diagnostics

The object of diagnostics in Polish classification is the soil profile. The definitions of taxa are quantitative, based on a system of diagnostic horizons and properties. The diagnostics for the more important levels of the taxonomy (types and lower taxa) are based mainly on morphological and chemical criteria. The collective levels are more genesis-oriented, based mainly on the general morphological features of soils and their genetic interpretation. The diagnostics are mainly quantitative chemico-morphological ones.

Terminology

The terminology of Polish soil classification is traditional and partially descriptive. Only a few soil taxa have additional artificial names listed in parentheses. For example, *undeveloped rock soils* have an international-style synonym *litosole*. The use of descriptive terminology makes the classification less compact, but more understandable. Some terms are traditional Polish soil names, which are easily understood by Polish pedologists, but may cause problems for an external reader. These terms caused some complications with the translation as discussed later.

Correlation

The correlation of Polish soil classification with the World Reference Base for Soil Resources (WRB) is not too difficult because the main concepts of the two classifications are similar. Recent studies compared diagnostic criteria, horizons and properties of the two classifications (Charzyński et al, 2005; Charzyński, 2006). The correlation of soil names is down to the type level.

The names of soil groups and types are listed in the same order as in the text of classification (Polish Society of Soil Science, 1989). Polish words are used for the most part in the classification and in most cases a direct translation of soil names into English can be made, but some names borrowed mainly from folk terminology seemed to be stable terms (like *Bielice* – Polish soil name for Podzol). In that case no translation was made, or is in parentheses in italics. The mixture of languages in soil names may appear awkward (like *Acid brunatne soils* or *Anthropogenic pararenzine*); however, we favoured not translating these terms.

The same situation exists with some Russian soil names. For example, *chernozem* would be translated as 'black earth', but nobody does as this term is well-known. Some Polish terms are less known and difficult to pronounce for foreigners, but we have left them untranslated.

Lithogenic soils – soil section. ≈ Leptosols / Regosols / Arenosols
The section is divided into two orders:
Mineral weakly developed soils without carbonates ≈ Leptosols / Regosols / Arenosols
The following soil types are distinguished within the order:

- Undeveloped rock soils (litosole) ≈ Lithic Leptosols
- Undeveloped loose soils (regosole) ≈ Protic Arenosols / Regosols
- Undeveloped clay soils (pelosole) ≈ Regosols (Clayic)
- Soils originating from massive rocks without carbonates (rankery) ≈ Leptosols
- Weakly developed soils from loose rocks (arenosole) ≈ Arenosols

Moderately developed carbonate soils ≈ Rendzic Leptosols / Rendzic Phaeozems / Umbrisols (Calcaric)
The following soil types are distinguished within the order:

- Redziny ≈ Rendzic Leptosols
- Pararedziny ≈ Rendzic Phaeozems / Umbrisols (Calcaric)

Autogenic soils – soil section. ≈ Chernozems / Cambisols / Luvisols / Podzols
The section is divided into three orders:
Czarnoziemic soils ≈ Chernozems
Only one soil type is distinguished within the order:

- Czarnoziemy ≈ Chernozems

Brown soils ≈ Cambisols / Luvisols
The following soil types are distinguished within the order:

- Typical brunatne *(brown)* soils ≈ Cambisols (Eutric)
- Acid brunatne *(brown)* soils ≈ Cambisols (Dystric)
- Plove soils (lessives) ≈ Luvisols

Bielicoziemne *(white earth)* soils ≈ Podzols / Arenosols
The following soil <u>types</u> are distinguished within the order:

- Rusty soils ≈ Entic Podzols / Brunic Arenosols
- Bielicowe soils ≈ Umbric Podzols
- Bielice ≈ Podzols

<u>Semihydrogenic soils</u> – soil <u>section</u>. ≈ Gleyic Podzols / Gleysols
The section is divided into three <u>orders</u>:
Glejobielicoziemne *(gley white earth)* soils ≈ Gleyic Podzols
The following soil <u>types</u> are distinguished within the order:

- Glejobielicowe ≈ Umbri-Gleyic Podzols
- Glejobielice ≈ Gleyic Podzols

Czarne ziemie (black soils) ≈ Mollic Gleysols / Umbric Gleysols
Only one soil <u>type</u> is distinguished within the order:

- Czarne ziemie ≈ Mollic Gleysols / Umbric Gleysols

Boggy soils ≈ Gleysols / Stagnosols
The following soil <u>types</u> are distinguished within the order:

- Surface-gleyed soils ≈ Stagnosols
- Groundwater gleyed soils ≈ Gleysols

<u>Hydrogenic soils</u> – soil <u>section</u>. ≈ Histosols / Gleysols
The section is divided into two <u>orders</u>:
Bog soils ≈ Histosols / Histic Gleysols
The following soil <u>types</u> are distinguished within the order:

- Clayey soils ≈ Histic Gleysols
- Peat soils ≈ Histosols

Bog-like soils ≈ Sapric Histosols / Mollic Gleysols
The following soil <u>types</u> are distinguished within the order:

- Mud soils ≈ Sapric Histosols
- Muddy soils ≈ Mollic Gleysols (Humic)

<u>Accumulated soils</u> – soil <u>section</u>. ≈ Fluvisols
The section is divided into two <u>orders</u>:
Alluvial soils ≈ Fluvisols
The following soil <u>types</u> are distinguished within the order:

- River clays ≈ Fluvisols
- Sea clays ≈ Salic Fluvisols / Tidalic Fluvisols

Deluvial soils ≈ (*Thaptic Colluvic*) Fluvisols
Only one soil <u>type</u> is distinguished within the order:

- Deluvial soils ≈ (*Thaptic Colluvic*) Fluvisols

<u>Salted soils</u> – soil <u>section</u>. ≈ Solonchaks / Solonetz / Hyposalic Regosols
The section contains only one <u>order</u> :
Salted-sodic soils ≈ Solonchaks / Solonetz
The following soil <u>types</u> are distinguished within the order:

- Solonczaki ≈ Solonchaks
- Solonczakowate soils ≈ Hyposalic Regosols
- Solonce ≈ Solonetz

<u>Anthropogenic soils</u> – soil <u>section</u>. ≈ Anthrosols / Technosols
The section is divided into two <u>orders</u>:
Kulturoziemne soils ≈ Anthrosols
The following soil <u>types</u> are distinguished within the order:

- *Hortisole* ≈ Hortic Anthrosols
- *Rigosole* ≈ Terric Anthrosols

Industroziemne and urbanoziemne soils ≈ Technosols
The following soil <u>types</u> are distinguished within the order:

- *Anthropogenic soils with undeveloped profile* ≈ (*Regic*) Technosols
- *Mudanthropogenic soils* ≈ Humic Technosols
- *Anthropogenic pararenzine* ≈ Technosols (Calcaric)
- *Anthropogenic salted soils* ≈ (*Salic*) Technosols

References

Anonymous (1956) 'Przyrodniczo-genetychna klasyfikocija gleb ze szczególnym uwzględnieniem gleb uprawnych', *Roczniki Nauk Rolniczych*, Warszawa, vol 74-D, 196pp
Charzyński, P., Hulisz, P. and Bednarek, R. (2005) 'Diagnostic subsurface horizons in Systematics of Polish Soils and their analogues in WRB classification', *Eurasian Soil Science*, vol 38, suppl 1, ppS55–S59
Charzyński, P. (2006) *Testing WRB on Polish soils*, Association of Polish Adult Educators, Toruń, 110pp
Polish Society of Soil Science (1989) 'Systematics of Polish Soils', *Roczniki Gleboznani*, vol 40, nos 3–4, 112pp

14
Soil Classification of Czech Republic

Objectives and scope

In the Czech Republic soil surveys cover almost all the national territory. To some extent the success of the soil inventory was due to the existence of a simple and adequate soil classification. The existing classifications were upgraded and harmonized with world classifications; in the case of the European Union it is the World Reference Base for Soil Resources (WRB) (IUSS Working Group WRB, 2006). Many European countries, even the smaller ones, developed their own soil classifications for large-scale soil survey. The Czech and Slovak republics, for example, after their peaceful separation, developed their own classifications. At the higher levels of taxonomy these soil classifications are harmonized with the WRB, and, thus, seem very similar at first glance. The Czech classification (Němeček et al, 2001) is a good example of such a tendency. It is similar to other new European classifications (like the Romanian, Slovak or Polish classifications), but has a number of particular features at lower levels, mainly used for large-scale soil mapping.

The objective of the new classification is to harmonize the existing taxonomy with the WRB, to integrate national databases to the European Union network, and to maintain the capacity of the classification for soil survey at different scales. Geographically, the classification covers the national territory (Table 14.1). Anthropogenically transformed soils are included in the reference class of *Anthrosols*: agricultural soils are included in *Kultizems*, and disturbed soils of industrial and urban areas including some transported materials are in the type *Anthrozems*. Bare rocks and underwater soils and sediments are not included in the classification.

Theoretical background

In general, the Czech classification is similar to several East European classifications which developed under the common influence of the Russian and European

Table 14.1 *The scope of soil classification of Czech Republic*

Superficial bodies	Representation in the system
Natural soils	National coverage
Urban soils	Partly included in the type of *Anthrozems*
Man-transported materials	Partly included in the type of *Anthrozems*
Bare rock	Not considered as soils
Subaquatic soils	Not considered as soils
Soils deeply transformed by agricultural activities	Included in a special group of cultivated soils (*Anthroposols*)

schools of soil science. The concepts of the main soil types (archetypes) are similar to those of German, Russian and other traditional schools. The types are grouped in such a way that most are comparable with the reference groups of the WRB. In the new version of the Czech soil taxonomy, the limits for soil taxa are defined quantitatively according to the demands for accuracy and comparability of soil diagnostics. The structure of the classification at the lower levels, being used for large-scale and detailed soil mapping, is strongly dependent on practical demands. A number of soil taxa at lower levels reflect the grouping of forest soils by their productivity, ecological evaluation and extent of land degradation.

Development

The development of soil classification in Czechoslovakia started in the early 1960s. The mapping of agricultural and forest lands was made separately by different institutions; thus, two separate classifications existed: Němeček et al (1967) and Houba (1970). In the 1980s, when many classifications were reformed according to the new concepts of diagnostic horizons and quantitative determination of soil properties, a uniform classification was prepared for the Czechoslovakian Republic (Hraško et al, 1987). Soon after that, the Czech Republic and Slovakia separated and each prepared its own new classification. Both used the previous classification system (Hraško et al, 1987) as a basis. In 2001 the new system was published (Němeček et al, 2001); an Internet version is also available.

Structure

The Czech soil classification has a complex hierarchical structure (Table 14.2). The upper, collective level of the taxonomy includes soil reference classes, which exist mainly for correlating the system with the WRB and other classifications. Soil types form the generic level of the taxonomy. Soil subtypes specify the morphology and properties of the types, some of them significantly. The level of varieties indicates less expressed qualitative soil characteristics. A specific feature of the Czech classification is the presence of a special taxonomic level of subvarieties, which is used only for forest soils and reflects their trophic status. It may be open to discussion whether a special taxonomic level should be used for a limited number of soils, but anyway it is an interesting experience; in most soil classifications forest soils are given much less attention than the agricultural ones. On the lower level ecological phases or degradation phases are used alternatively. At the lowest level there are soil forms. The classification is a complex hierarchical taxonomy with formalized borders.

Table 14.2 *The structure of soil classification of Czech Republic*

Level	Taxon name	Taxon characteristics	Borders between classes	Diagnostics	Terminology
0	Soils	Kingdom			
1	Reference class	Collective	Formal	Morphological	Artifical
2	Type	Generic	Formal	Chemico-morphological	Artificial / traditional
3	Subtype	Specific 1	Formal	Chemico-morphological	Artificial / traditional
4	Variety	Specific 2	Formal	Morphological	Traditional
5	Subvariety*	Varietal 1	Formal	Chemical	Traditional
6/7	Ecological phase	Varietal 2	Formal	Chemico-morphological	Traditional
6/7	Degradation phase	Varietal 2	Formal	Chemico-morphological	Traditional
8	Soil form	Specific / Varietal	Formalized	Geological and physico-morphological	Traditional

*This taxonomic level is used only for forest soils.

Diagnostics

In the new version of the Czech soil classification the taxa are defined by the presence or absence of certain diagnostic horizons and attributes. Water and temperature regimes are not regarded as diagnostic criteria as long as they do not cause certain changes in soil morphology and properties.

Terminology

The first collective level of the Czech classification repeats most terms of the WRB with a few exceptions. All the names of reference classes finish with –*sol*, and the names at the type level finish with –*zem*. Most type names are traditional soil names in Czech and European pedological schools but some of them are newly constructed words with the same roots as the WRB terms, such as Fluvizem or Andozem. There was a long discussion among the pedologists about whether it is possible to combine Greek, Latin and, for example, Russian or Czech roots in the same term. However, while the theoretical discussion goes on, most new classifications have terms with a variable mixture of Latin,

Greek, Russian, Japanese and other languages. At the level of subgroups names are partly traditional, and partly artificial; most of them repeat the modifiers used in the WRB. At lower levels the terminology is mainly traditional.

Correlation

The correlation of the Czech soil classification with the WRB was easy because the classification is specially harmonized with the international classification by Němeček and Kozák (2001). The correlation is presented only for the two upper levels. The subtypes are numerous and correspond with WRB modifiers; thus their correlation should not be difficult. The reference classes and their included types are listed in alphabetical order.

Andosoly – soil <u>reference class</u>. ≈ Andosols
The reference class includes only one <u>type</u>:

• Andozem ≈ Andosols

Anthroposoly – soil <u>reference class</u>. ≈ Anthrosols / Technosols
The reference class includes two <u>types</u>:

• Anthropozem ≈ Technosols
• Kultizems ≈ Anthrosols

Černosoly – soil <u>reference class</u>. ≈ Chernozems / Phaeozems
The reference class includes two <u>types</u>:

• Černice ≈ Phaeozems / Gleyic Chernozems
• Černozem ≈ Chernozems

Fluvisoly – soil <u>reference class</u>. ≈ Fluvisols
The reference class includes two <u>types</u>:

• Fluvizem ≈ Fluvisols
• Koluvizem ≈ (various reference groups with the modifier *Colluvic*)

Glejsoly – soil <u>reference class</u>. ≈ Gleysols
The reference class includes only one <u>type</u>:

• Glej ≈ Gleysols

Kambisoly – soil <u>reference class</u>. ≈ Cambisols / Umbrisols / Regosols
The reference class includes two <u>types</u>:

• Kambizem ≈ Cambisols / Umbrisols
• Pelozem ≈ Endogleyic Cambisols (Clayic) / Endogleyic Regosols (Clayic)

Leptosoly – soil <u>reference class</u>. ≈ Leptosols
The reference class includes four <u>types</u>:

- Litozem ≈ Lithic Leptosols
- Pararendzina ≈ Leptosols (Calcaric)
- Ranker ≈ Leptosols
- Rendzina ≈ Rendzic Leptosols

Luvisoly – soil <u>reference class</u>. ≈ Luvisols / Luvic Phaeozems
The reference class includes three <u>types</u>:

- Hnědozem ≈ Luvisols
- Luvizem ≈ Albic Luvisols
- Šedozem ≈ Luvic Greyic Phaeozems

Natrisoly – soil <u>reference class</u>. ≈ Solonetz
The reference class includes only one <u>type</u>:

- Slanec ≈ Solonetz

Organosoly – soil <u>reference class</u>. ≈ Histosols
The reference class includes only one <u>type</u>:

- Organozem ≈ Histosols

Podzosoly – soil <u>reference class</u>. ≈ Podzols
The reference class includes two <u>types</u>:

- Kryptopodzol ≈ Entic Podzols / Leptic Podzols
- Podzol ≈ Podzols

Regosoly – soil <u>reference class</u>. ≈ Regosols / Arenosols
The reference class includes only one <u>type</u>:

- Regozem ≈ Regosols / Arenosols

Salisoly – soil <u>reference class</u>. ≈ Solonchaks
The reference class includes only one <u>type</u>:

- Solončak ≈ Solonchak

Stagnosoly – soil <u>reference class</u>. ≈ Stagnosols / Stagnic Cambisols / Planosols
The reference class includes two <u>types</u>:

- Pseudoglej ≈ Stagnic Cambisols / Stagnosols
- Stagnoglej ≈ Stagnosols (Albic) / Planosols

Vertisoly – soil <u>reference class</u>. ≈ Vertisols / Vertic Phaeozems / Vertic Chernozems
The reference class includes only one <u>type</u>:

• Smonice ≈ Vertisols / Vertic Phaeozems / Vertic Chernozems

References

Houba, A. (1970) *Půdni typy a nižší taxonomické půdní jednotky typologického průzkumu půd (Soil types and lower taxonomic soil units of the typological soil survey)*, ÚHÚL Zvolen, pracoviště Brandys n. L., 20pp

Hraško, J., Němeček, R., Šály B. and Šurina M. (1987) *Morfogenetický klasifikačny systém pôd ČSFR*, VÚPÚ Bratislava, 107pp

IUSS Working Group WRB (2006) *World Reference Base for Soil Resources*, 2nd edition, World Soil Resources Reports no 103, UN Food and Agriculture Organization, Rome, 128pp

Němeček, J. and Kozak, J. (2001) 'The Czech taxonomic soil classification system and the harmonization of soil maps', in E. Micheli, F. O. Nachtergaele, R. J. A. Jones and L. Montanarella (eds) *Soil Classification 2001*, European Soil Bureau Research Report no 7, pp47–53

Němeček, J., Damaš, J., Hraško, J., Bedrna, Z., Zuska, V., Tomášek, M. and Kalenda, M. (1967) *Průzkum zemědělskych půd ČSSR (Soil survey of agricultural lands of Czechoslovakia)*, 1 dil, MZVŽ Praha, 246pp

Němeček, J., Macků, J, Vokoun, J., Vavříč, D. and Novák, P. (2001) *Taxonomický klasifikační system půd České Republiky*, ČZU, Prague, 180pp

15
Soil Classification of Slovakia

Objectives and scope

For a long time the Czech and Slovak nations were parts of the same state, and they had the same soil classification. Later, for administrative reasons, both countries needed official documents for soil inventory. The taxonomies of the Czech and Slovakia republics are somewhat different, and are discussed in separate chapters.

The objective of the Slovak classification (Sobocká, 2000) was to provide a basis for soil inventory and mapping within the country (Table 15.1). The classification is also used for educational purposes. Anthropogenically transformed soils are included in the group of *Anthropic soil*. The soils used in agriculture are *Kultizems*, and disturbed soils of industrial and urban areas, including some transported materials, are *Anthrozems*. The classification of human-transformed soils is very detailed in the Slovak classification at the subtype level. Bare rocks and underwater soils and sediments are not regarded as soils.

Table 15.1 *The scope of soil classification of Slovakia*

Superficial bodies	Representation in the system
Natural soils	National coverage
Urban soils	Partly included in *Anthrozems*
Transported materials	Partly included in *Anthrozems*
Bare rock	Not considered as soils
Subaquatic soils	Not considered as soils
Soils deeply transformed by agricultural activities	Included in *Kultizems*

Theoretical background

The Slovak classification has the same roots as the Czech classification and its basic ideas are similar. By definition, it is a morphogenetic system based on the soil properties diagnostic for soil genesis. Like many other soil classifications developed at the end of the 20th century, it uses a system of diagnostic horizons, but unlike the World Reference Base for Soil Resources (WRB) (IUSS Working Group WRB, 2006), diagnostic horizons cannot overlap. In the case of particular or intermediate features of a horizon, variants of diagnostic horizons are applied. For example, *cambic* horizons have the following variants: *luvic cambic, marble-like cambic, podzolic cambic, andic cambic* and *ranker-like cambic*.

Development

The classification of soils of Slovakia originated in the taxonomy published for the Czechoslovakian Republic in the late 1980s (Hraško et al, 1987). After separation the newly independent states developed their own systems of soil classification (Sobocká, 2000; Němeček and Kozak, 2001). On the higher levels of the taxonomy, the Slovak classification is closer to the initial version (Hraško et al, 1987); however, it was significantly updated and detailed at the subtype and lower levels of taxonomy, especially for anthropogenically transformed soils.

Structure

The Slovak soil classification has seven hierarchical levels (Table 15.2). The upper, collective level of the taxonomy includes soil <u>groups</u>, which are defined by the 'main' diagnostic horizon that reflects the major pedogenetic process. Soil <u>types</u> form the generic level of the taxonomy, and are characterized by variants of diagnostic horizons, or a combination of them. Soil <u>subtypes</u> specify the morphology and properties of the types, or transitions between them. The <u>variety</u> level indicates quantitative, mainly laboratory-determined soil characteristics. Soil <u>forms</u> define multiple features related to humus forms, erosion or accumulation and so forth. Lower levels indicate soil <u>texture</u> and geological <u>substrate</u>. The classification is a hierarchical taxonomy with formalized borders.

Diagnostics

In the Slovak soil classification the taxa are defined by the presence of diagnostic horizons and attributes. The water and temperature regimes are not used as diagnostic criteria. The Slovak soil classification is focused on field morphological diagnostics. Only on the level of varieties are chemical analyses required, and texture determination is desirable for the complete soil name.

Terminology

The names of soil types in the Slovak classification are almost the same as were used in the older Czechoslovak classification (Hraško et al, 1987). They have roots derived from both traditional European soil names and artificial

Table 15.2 *The structure of soil classification of Slovakia*

Level	Taxon name	Taxon characteristics	Borders between classes	Diagnostics	Terminology
0	Soils	Kingdom			
1	Group	Collective	Formal	Morphological	Artificial / traditional
2	Type	Generic	Formal	Morphological	Artificial / traditional
3	Subtype	Specific 1	Formal	Morphological	Artificial / traditional
4	Variety	Varietal 1	Formal	Chemical	Traditional
5	Form	Specific 2	Formal	Morphological	Traditional
6	Texture	Varietal 2		Physical	
7	Substrate	Specific 3	Formalized	Geological	Traditional

internationally approved terms (*ando-*, *luvi-*, *fluvi-*, *rego-*); most of the names finish with *–zem*. The groups have the same roots, but are more descriptive, and change according to the rules of Slovak language. In the correlation below they are given in English. At the level of subgroups soil names are mixed, and at lower levels the terminology is mainly traditional.

Correlation

The correlation of the Slovak soil classification with the WRB is only for the two upper levels of the taxonomy. For correlation of subtypes we recommend using the original text of the Slovak classification (Sobocká, 2000); unfortunately, that correlation was done with the draft version of the WRB (Spaargaren, 1994).

The reference groups and types are listed in alphabetical order.

Andozem soils – <u>soil group</u>. ≈ Andosols
Only one type is in this group:

- *Andozem* ≈ Andosols

Anthropic soils – <u>soil group</u>. ≈ Anthrosols / Technosols
Two types are included in this group:

- *Anthrozem* ≈ Technosols
- *Kultizem* ≈ Anthrosols

Brown soils – <u>soil group</u>. ≈ Cambisols
One type is in this group:

- *Cambizem* ≈ Cambisols / Brunic Arenosols

Hydromorphic soils – <u>soil group</u>. ≈ Stagnosols / Planosols / Gleysols / Histosols
Three types are in this group:

- *Gley* ≈ Gleysols
- *Organozem* ≈ Histosols
- *Pseudogley* ≈ Stagnosols / Planosols

Ilimeric (Illuviated) soils – <u>soil group</u>. ≈ Luvisols
Two types are in this group:

- *Hnedozem* ≈ Luvisols
- *Luvizem* ≈ Albic Luvisols / Albeluvisols

Initial soils – <u>soil group</u>. ≈ Leptosols / Regosols / Fluvisols / Arenosols
Four types are in this group:

- *Fluvizem* ≈ Fluvisols
- *Litozem* ≈ Lithic Leptosols
- *Ranker* ≈ Leptosols
- *Regozem* ≈ Regosols / Arenosols

Mollic soils – <u>soil group</u>. ≈ Chernozems / Phaeozems / Vertisols / Mollic Gleysols / Mollic Fluvisols
Three types are in this group:

- *Černozem* ≈ Chernozems / Phaeozems
- *Čiernica* ≈ Gleyic Phaeozems / Gleyic Chernozems / Mollic Gleysols / Mollic Fluvisols
- *Smonitza* ≈ Vertisols

Podzolic soils – <u>soil group</u>. ≈ Podzols
Only one type is in this group:

- *Podzol* ≈ Podzols

Rendzina soils – <u>soil group</u>. ≈ Rendzic Leptosols / Rendzic Phaeozems
Two types are in this group:

- *Pararendzina* ≈ Cambisols (Calcaric) / Phaeozems (Calcaric)
- *Rendzina* ≈ Rendzic Leptosols / Rendzic Phaeozems

Salted soils – <u>soil group</u>. ≈ Solonchaks / Solonetz
Two types are in this group:

- *Slanec* ≈ Solonetz
- *Slanisko* ≈ Solonchaks

References

Hraško, J., Němeček, R., Šály B. and Šurina M. (1987) *Morfogenetický klasifikačny systém pôd ČSFR*, VÚPÚ Bratislava, 107pp

IUSS Working Group WRB (2006) *World Reference Base for Soil Resources*, 2nd edition, World Soil Resources Reports no 103, UN Food and Agriculture Organization, Rome, 128pp

Němeček, J. and Kozak, J. (2001) 'The Czech taxonomic soil classification system and the harmonization of soil maps', in E. Micheli, F. O. Nachtergaele, R. J. A. Jones and L. Montanarella (eds) *Soil Classification 2001*, European Soil Bureau Research Report no 7, pp47–53

Sobocká, J. (ed) (2000) *Morfogenetický klasifikačný systém pôd Slovenska. Bazálna referenčná taxonómia*, Výskumný ústav pôdoznalectva a ochrany pôdy, Bratislava, 74pp

Spaargaren O. (ed) (1994) *World Reference Base for Soil Resources (Draft)*, World Soil Resources Reports no 84, UN Food and Agriculture Organization, Rome, 88pp

16
Soil Classification of Hungary

written together with Erika Michéli[1]

Objectives and scope

Hungary is one of the pedologically most studied countries in Europe. Though its territory is small, Hungary possesses a great variety of soils. A demand for soil classification was determined both by the needs of scientific soil research and practical soil mapping. The current official soil classification of Hungary is one of the last systems in Europe designed in the old style based mainly on the assumptions on soil genesis. It does not use diagnostic horizons and has practically no artificial terms. The lower taxa are developed in detail to be effective for soil survey at any scale. As there is no major soil survey activity going on, the current official system is applied mainly for land evaluation and soil conservation planning. Because of the need for harmonized soil information in the EU, education about the World Reference Base for Soil Resources (WRB) became necessary. The younger generation of soil scientists is familiar with both the national and WRB systems. The Hungarian soil classification was developed mostly for mapping soils of agricultural lands of the country, and does not include urban soils or technogenically disturbed substrates (Table 16.1). Bare rock and underwater sediments are not recognized as soils.

Theoretical background

The Hungarian soil classification has two main roots. First, it borrowed a pedogenetic basis from the Russian soil classification and the highest taxa are determined mainly according to pedogenetic concepts as primary criteria.

---

[1]Szent Istvan University, Gödölo, Hungary.

Table 16.1 *The scope of soil classification of Hungary*

Superficial bodies	Representation in the system
Natural soils	National coverage
Urban soils	Not included in the classification
Man-transported materials	Not included in the classification
Bare rock	Not considered as soils
Subaquatic soils	Not considered as soils
Soils deeply transformed by agricultural activities	Classified as if they are natural soils

Second, the Hungarian classification was influenced by the German school, especially Kubiëna's taxonomic system (1953). The 'dynamitic soil classification' developed by de Sigmond (1938) included diagnostic elements and influenced several modern classification systems and approaches (Jenny, 1941) but only slightly influenced the development of the Hungarian system.

Development

The development of the Hungarian soil classification started in the late 1950s (Stefanovits, 1963) but the early versions were too general to use for large-scale soil survey. Classification and mapping guidelines for large-scale mapping (Szabolcs, 1966; Jassó et al, 1988) were still genetically based but included more laboratory data requirements for the lower level taxonomic units. Influenced by modern diagnostic systems, mostly the WRB, the Hungarian Soil Classification is under major revision and modernization. The revised system is based on stricter definitions and criteria for describing soils and uses a key to define classes. The definitions and limits are compatible with the WRB (Michéli, 2008). Since the revised system is still under testing, in this publication only the current official system is described.

Structure

There are four levels in the soil classification of Hungary (Table 16.2). The highest level is soil <u>main type</u> which groups soils according to the 'most important soil-forming process'. The main types include soils with different properties depending on the pedogenetic process thought to be important; thus, the level of the main type is regarded as collective. Within each main type there are soil <u>types </u>that serve as a basic level of taxonomy with genetic definitions and traditional archetypes similar to those proposed by the Russian and German schools. <u>Subtypes</u> are distinguished by additional properties reflecting the soil processes such as the presence or absence of carbonates, the content of salts and so on. Soil <u>varieties</u> are distinguished by quantitative specifications of the depth of soil horizons, texture and content of organic carbon, salts, carbonates and some other measurable properties. The classification is a hierarchical taxonomy with mainly fuzzy borders.

Table 16.2 *The structure of soil classification of Hungary*

Level	Taxon name	Taxon characteristics	Borders between classes	Diagnostics	Terminology
0	Soils	Kingdom			
1	Main type	Collective	Fuzzy	Landscape-morphological	Traditional
2	Type	Generic	Fuzzy	Chemico-morphological	Traditional
3	Subtype	Specific	Fuzzy	Chemico-morphological	Traditional
4	Variety	Varietal	Mostly formal	Chemico-morphological	Traditional

Diagnostics

The characteristics of the levels of main types and types in the soil classification of Hungary are mainly qualitative. The main types in places also require landscape criteria: the type of ecosystem, sediments, geomorphology and so on. The types are defined by internal soil properties, mainly morphological ones. Subtypes require additional data on morphology and chemistry of soils. At the variety level, certain measured laboratory data and quantitative field observations are required. No climatic or soil regime criteria are included directly in the classification. The diagnostics are mostly qualitative landscape-chemico-morphological.

Terminology

The terminology of Hungarian soil classification is traditional and descriptive. Many soil names are long and seem more like definitions. It has an advantage in that nobody can mistake the name for another soil.

Correlation

The correlation of the Hungarian soil classification with the WRB (IUSS Working Group WRB, 2006) is not very precise because the classes in the soil taxonomy of Hungary are mostly fuzzy (Michéli et al, 2006). The archetypes of soils are recognizable, and easily fit certain WRB concepts, though their limits do not correspond completely. The correlation of soil names of the soil classification with the terms of the WRB is provided down to the type level. The names of soil main types and types are listed in the same order as in their text of classification. The Hungarian names for soils are given in parentheses.

Skeletal soils (váztalajok) – soil <u>main type</u>. ≈ Leptosols / Arenosols / Regosols / Calcisols
The following five soil <u>types</u> are distinguished within the main type:

- *Stony skeletal soils (köves-sziklás váztalajok)* ≈ Lithic Leptosols

- *Pebble skeletal soils (kavicsos váztalajok)* ≈ Hyperskeletic Leptosols
- *Shifting sand (futóhomok)* ≈ Protic Arenosols
- *Humic sandy soils (humuszos homok talajok)* ≈ Arenosols (*Humic*)
- *Barren Earth (barnaföldek)* ≈ Cambisols, Calcisols

Lithomorphic soils (kőzethatású talajok) – soil <u>main type</u>. ≈ Leptosols / Regosols / Phaeozems / Cambisols
The following four soil <u>types</u> are distinguished within the main type:

- *Humus-carbonate soils (humuszkarbonát talajok)* ≈ Rendzic Phaeozems / Calcisols / Calcaric Regosols
- *Rendzinas (rendzinák)* ≈ Rendzic Leptosols / Rendzic Phaeosems
- *Erubase soils (feketenyirok talajok)* ≈ Leptosols / Regosols / Umbrisols
- *Rankers (ranker talajok)* ≈ Leptosols / Regosols / Umbrisols

Brown forest soils (barna erdőtalajok) – soil <u>main type</u>. ≈ Cambisols / Luvisols / Umbrisols
The following seven soil <u>types</u> are distinguished within the main type:

- *Acidic, non-podzolic brown forest soils (savanyú nem podzolos barna erdőtalajok)* ≈ Cambisols (Dystric)
- *Podzolic brown forest soils (podzolos barna erdőtalajok)* ≈ Albic Luvisols
- *Brown forest soils with clay illuviation (agyagbemosódásos barna erdőtalajok)* ≈ Luvisols
- *Stagnating brown forest soils (pangóvizes barna erdőtalajok)* ≈ Stagnic Luvisols / Stagnic Cambisols
- *Ramann's brown earth (Ramann barnaföldek)* ≈ Cambisols / Umbrisols
- *Banded brown forest soils (kovárványos barna erdőtalajok)* ≈ Luvisols (*Lamellic*)
- *Chernozem brown forest soils (csernozjom barna erdőtalajok)* ≈ Luvic Chernozems

Chernozems (csernozjom talajok) – soil <u>main type</u>. ≈ Chernozems / Phaeozems / Kastenozems / Vertisols)
The following four soil <u>types</u> are distinguished within the main type:

- *Leached chernozems (kilúgzott csernozjom talajok)* ≈ Luvic Chernozems / Luvic Phaeozems
- *Pseudomyceliar chernozems (mészlepedékes csernozjomok)* ≈ Calcic Chernozems, Calcic Kastanozems
- *Meadow chernozems (réti csernozjom talajok)* ≈ Gleyic Chernozems / Vertisols
- *Alluvial chernozems (öntés csernozjom talajok)* ≈ Chernozems / Mollic Fluvisols

Salt-affected soils (szikes talajok) – soil <u>main type</u>. ≈ Solonchaks / Solonetz / Vertisols / Chernozems

The following four soil <u>types</u> are distinguished within the main type:

- *Solonchaks (szoloncsák talajok)* ≈ Solonchaks
- *Solonchaks-solonetz (szoloncsák-szolonyec talajok)* ≈ Sodic Solonchaks / Salic Solonetz
- *Meadow solonetz (rétiszolonyec talajok)* ≈ Gleyic Solonetz / Vertisols (Sodic)
- *Steppe meadow solonetz (sztyeppesedő rétiszolonyec)* ≈ Mollic Solonetz / Chernozems (Sodic)

Meadow soils (réti talajok) – soil <u>main type</u>. ≈ Humic Gleysols / Mollic Gleysols / Gleyic Phaeosems / Gleyic Chernozems
The following six soil <u>types</u> are distinguished within the main type:

- *Solonchak meadow soils (szoloncsákos réti talajok)* ≈ Endosalic Gleysols / Cambisols
- *Solonetz meadow soils (szolonyeces réti talajok)* ≈ Vertisols (Hyposodic) / Chernozems (Hyposodic) / Phaeozems (Hyposodic)
- *Meadow soils (típusos réti talajok)* ≈ Gleyic Phaeozems / Gleyic Chernozems / Gleyic Vertisols / Mollic Gleysols
- *Alluvial meadow soils (öntés réti talajok)* ≈ Gleyic Fluvisols (Humic)
- *Peat-meadow soils (lápos réti talajok)* ≈ Histic Gleysols / *Histic* Gleyic Phaeozems / *Histic* Gleyic Chernozems
- *Chernozem meadow soils (csernozjom réti talajok)* ≈ Gleyic Chernozems

Peat soils (láptalajok) – soil <u>main type</u>. ≈ Histosols
The following three soil <u>types</u> are distinguished within the main type:

- *Sphagnum peats (mohaláptalajok)* ≈ Fibric Histosols
- *Meadow bog soils (rétláptalajok)* ≈ Histosols
- *Ameliorated peats (lecsapolt réti talajok)* ≈ Histosols (Drainic)

Soils of swampy forests (moscári erdők talajai) – soil <u>main type</u>. ≈ Gleysols (Dystric) / Fluvisols (Dystric)
No further subdivision of the main type.

Soils developed in river, lake and slope sediments (öntés és lejtőhordalék talajok) – soil <u>main type</u>. ≈ Fluvisols / Regosols
The following three soil <u>types</u> are distinguished within the main type:

- *Raw (recent) alluvial soils (nyers öntés talajok)* ≈ Fluvisols
- *Humic alluvial soils (humuszos öntés talajok)* ≈ Fluvisols (Humic)
- *Soils of slope sediments (lejtőhordalék talajok)* ≈ Regosols

References

IUSS Working Group WRB (2006) *World Reference Base for Soil Resources*, 2nd edition, World Soil Resources Reports no 103, UN Food and Agriculture Organization, Rome, 128pp

Jassó, F. (ed) (1989) *Guidelines for the National Large-scale Soil Mapping (1:10,000)*, Agroinform, Budapest (in Hungarian)

Jenny, H. (1941) *Factors of Soil Formation. A System of Quantitative Pedology*, McGraw Hill Book Company, New York, NY, 281pp

Kubiëna, W. L. (1953) *Bestimmungsbuch und Systematic der Boden Europas*, Verlag Enke, Stuttgart, 392pp

Michéli, E., Fuchs, M., Hegymegi, P. and Stefanovits, P. (2006) 'Classification of the major soils of Hungary and their correlation with the World Reference Base for Soil Resources (WRB)', *Agrokémia és Talajtan*, vol 55, no 1, pp19–28

Michéli, E. (2008) *Improvement and International Correlation of the Hungarian Soil Classification System*, Research Report no 46513, Hungarian Scientific Research Foundation, 44pp

Sigmond, A. A. J. de (1938) *The Principles of Soil Science*, translated by A. B. Yolland, Thomas, Murby & Co, London, 362pp

Stefanovits, P. (1963) *The Soils of Hungary*, 2nd edition, Akadémiai Kiadó, Budapest, 224pp (in Hungarian)

Szabolcs, I. (ed) (1966) 'Methodology for genetic soil mapping in a farm scale', *Genetikus Talajtérképek*, ser 1, no 9, 112pp (in Hungarian)

17
Soil Classification of Romania

Objectives and scope

The main objective of this classification was to unify the basis for soil survey and soil interpretation in the country. The new classification also hopes to integrate soil information of the country with the European Union format. A special effort was undertaken to harmonize the Romanian classification with the World Reference Base for Soil Resources (WRB) which is used as a basic system for soil information storage in Europe. The classification is limited to Romania (Table 17.1). The soils of urban and industrial areas with minor development of profile constitute a special taxonomic group *Entianthroposols*, and there are also corresponding taxa at the subtype level for more developed soils. Bare rock is not included in the classification. Subaquatic soils, both mineral and organic ones, are included in a special soil type *Limnosols*. Soils deeply transformed by agricultural practices are included in the class *Anthrisols*, particularly in the type *Anthroposols*.

Table 17.1 *The scope of soil classification of Romania*

Superficial bodies	Representation in the system
Natural soils	National coverage
Urban soils	Partly included in the type *Entianthroposols*, partly distinguished at the subtype level
Man-transported materials	Partly included in the type *Entianthroposols*, partly distinguished at the subtype level
Bare rock	Not considered as soils
Subaquatic soils	Constitute a special soil type *Limnosols*
Soils deeply transformed by agricultural activities	Included in the *Anthrisols* class

Theoretical background

Three main sources were used as the basis of Romanian soil taxonomy: the genetic and geographic concepts of the Russian school, the system of diagnostic horizons and properties of the US Soil Taxonomy, and the principle of dividing the taxa borrowed from the Food and Agriculture Organization (FAO) soil map legend (Miclaus, 1980). Romanian soil classification is a successful compilation of these approaches. The taxonomic structure is similar to that of the old Russian classification, especially at the lower levels; however, the borders between classes are strict at all levels of the taxonomy. The grouping of types is similar to those used in the first edition of the WRB (FAO-ISRIC-ISSS, 1998). The first edition of the classification (Conea et al, 1980) had soil archetypes at the generic ('soil genetic type') level similar to those used by most East European pedological schools, but the new variant is much closer to the WRB way of clustering the soils. Another important source, not mentioned by the authors, is the evolutionary scheme for soil classification rooted in the ideas of Kubiëna (1953). This approach determined to a great extent soil groupings on a collective level.

Development

The history of soil classification in Romania started at the beginning of the 20th century with the classical works of G. M. Murgoci, the founder of Romanian soil science and one of the fathers of modern pedology (Munteanu and Florea, 2001). For a long period there was no official soil classification in Romania, and provisional schemes were used for soil research. At the end of the 1960s the work to create an official classification led towards the publication of a complete taxonomic classification of Romanian soils (Conea et al, 1980). Twenty years later, a new version of Romanian soil classification was published (Florea and Munteanu, 2000). It was designed to improve the taxonomy, and to harmonize it better with the WRB (FAO-ISRIC-ISSS, 1998). In comparison with the previous version, the new Romanian classification has more diagnostic horizons, classes and types. Some soil types, however, were moved down to the subtype level (Munteanu and Florea, 2001).

Structure

The taxonomy of soils of Romania is divided into two parts: the upper part includes three taxonomic categories, and the lower part includes four categories (Table 17.2). Soil class represents a group of soils with a common evolution and stage of pedogenesis as given by the presence of a specific diagnostic horizon or property. Soil type is an archetype, defined by the presence of a unique combination of diagnostic horizons and properties, or having 'a specific mode of expression' of the diagnostic elements (Munteanu and Florea, 2001). Soil subtype has some important additional diagnostic horizons and properties; sometimes it is a transition between the types. Soil variety has qualitative differences from the 'central image' of minor importance, or a quantitative variation of some soil characteristics. Soil species are named after the texture for mineral soils and peat decomposition for organic ones. Soil families represent lithological grouping of parent material, and soil variants indicate

Table 17.2 *The structure of soil classification of Romania*

Level	Taxon name	Taxon characteristics	Borders between classes	Diagnostics	Terminology
0	Soils	Kingdom			
1	Class	Collective	Formal	Chemico-morphological	Artificial
2	Type	Generic	Formal	Chemico-morphological	Artificial
3	Subtype	Specific 1	Formal	Chemico-morphological	Artificial
4	Variety	Specific / Varietal	Formal	Chemico-morphological	Artificial
5	Species	Varietal 1	Formal	Physical (texture)	Traditional scientific
6	Family	Specific 2	Mostly formal	Geological (parent material)	Traditional scientific
7	Variant	Varietal 2	Mostly formal	Morphological (use and degradation)	Traditional scientific

minor anthropogenic effect on soil properties. The Romanian soil classification is a hierarchical taxonomy with formal borders between taxa.

Diagnostics

Soil diagnostics are based on the presence of diagnostic horizons, materials and properties. In general, the diagnostics are similar to those used in the US Soil Taxonomy (Soil Survey Staff, 1999) or in the WRB (IUSS Working Group WRB, 2006), some of them with the same definitions. Consequently, the diagnostics in Romanian classification also require both field morphology and extensive chemical laboratory analyses. The source of diagnostics is a soil profile, with no information on water and temperature regimes required. The diagnostics are characterized as quantitative chemico-morphological.

Terminology

The terminology of the Romanian soil classification at higher levels is completely artificial. Most terms were borrowed from the legend of the WRB (FAO-ISRIC-ISSS, 1998) and the US Soil Taxonomy (Soil Survey Staff, 1999); one of the class names – *Hidrisols* – was adapted from the Australian classification (Isbell, 2002). Some of the soil names were slightly modified if the concepts were close to those of the WRB but the diagnostic criteria differed: for example, *Luvosols*, *Alosols* or *Eutricambosols* (compare with the FAO/WRB terms *Luvisols*, *Alisols* and *Eutric Cambisols*) A number of new terms were introduced. Some of them may cause confusion: for example, the type name *Turbosols* might be thought to relate to cryoturbated soils (*Turbic Cryosols* in WRB), however it was derived from the Latin word *turba* ('peat'). Some of the terms were proposed in the Romanian classification before the identical terms of the WRB – for example, the *Umbrisols* group (*Umbrisoluri* in Romanian) (Miclaus, 1980). It is curious that the term was not borrowed by the WRB Working Group but was invented independently. The same happened with the Romanian soil type name *Negrisols*; a similar term *Negrosols* was recently proposed for buried dark-coloured soils (Krasilnikov and García-Calderón, 2006) with no connection with the Romanian name.

In the previous version of the classification (Conea et al, 1980) the names were presented in Romanian writing (*Cambisoluri, Spodosoluri*), and the recent edition gives soil names in the form adapted for an English-speaking reader, as in the WRB (*Cambisols, Spodisols*). At the level of subtypes and varieties, modifiers are used, constructed from Greek and Latin roots; some of them were borrowed from the WRB, and the others were constructed in an identical way. At the lower levels of taxonomy, traditional scientific names are used for texture, parent material, and soil degradation characteristics.

Correlation

Since the Romanian soil classification was intentionally harmonized with the WRB, it is easy to correlate to the earlier version of the WRB. Also, a good correlation was presented recently by the authors of the classification themselves (Munteanu and Florea, 2001). However, the correlation had to be fitted to the recent edition of the WRB (IUSS Working Group WRB, 2006), and in places slightly corrected. Several soils in the Romanian classification were difficult to correlate, for example, strongly eroded soils.

The list of correlation of terms of the Romanian soil classification with the terms of the WRB is provided down to the level of genetic types. The names of classes are listed in alphabetical order and the names of types are in alphabetical order within classes.

Andisols – soil class. ≈ Andosols
Only one soil type is distinguished within the class:

- *Andosols* ≈ Andosols

Anthrisols – soil class. ≈ Anthrosols / Regosols
The following soil types are distinguished within the class:

- *Anthroposols* ≈ Anthrosols
- *Erodosols* ≈ (No direct equivalents in the WRB: eroded soils. Partly may correspond to Regosols)

Cambisols – soil class. ≈ Cambisols
The following soil types are distinguished within the class:

- *Dystricambosols* ≈ Cambisols (Dystric)
- *Eutricambosols* ≈ Cambisols (Eutric)

Cernisols – <u>soil class</u>. ≈ Chernozems / Phaeozems / Kastanozems / Rendzic Leptosols
The following soil <u>types</u> are distinguished within the class:

- *Cernoziom* ≈ Chernozems
- *Faeoziom* ≈ Phaeozems
- *Kastanoziom* ≈ Kastanozems
- *Rendzină* ≈ Rendzic Leptosols

Hidrisols – <u>soil class</u>. ≈ Gleysols / Stagnosols
The following soil <u>types</u> are distinguished within the class:

- *Gleiosols* ≈ Gleysols
- *Limnosols* ≈ Subaquatic Gleysols / Subacuatic Histosols
- *Stagnosols* ≈ Stagnosols / Stagnic Luvisols

Histisols – <u>soil class</u>. ≈ Histosols
The following soil <u>types</u> are distinguished within the class:

- *Foliosols* ≈ Folic Histosols
- *Turbosols* ≈ Histosols

Luvisols – <u>soil class</u>. ≈ Luvisols / Planosols / Alisols
The following soil <u>types</u> are distinguished within the class:

- *Alosols* ≈ Alisols
- *Luvosols* ≈ Luvisols (Albic) / Luvisols (Abruptic)
- *Planosols* ≈ Planosols
- *Preluvisols* ≈ Luvisols / Calcic Luvisols / *Cutanic* Cambisols

Pelisols – <u>soil class</u>. ≈ Vertisols / Vertic Regosols / Vertic Gleysols
The following soil <u>types</u> are distinguished within the class:

- *Pelosols* ≈ Vertic Regosols / Vertic Gleysols / Regosols (Clayic) / Gleysols (Clayic)
- *Vertosols* ≈ Vertisols

Protisols – <u>soil class</u>. ≈ Regosols / Arenosols / Fluvisols / Leptosols / Technosols
The following soil <u>types</u> are distinguished within the class:

- *Aluviosols* ≈ Fluvisols
- *Entianthroposols* ≈ Technosols
- *Lithosols* ≈ Leptosols
- *Psamosols* ≈ Arenosols
- *Regosols* ≈ Regosols

Salsodisols – <u>soil class</u>. ≈ Solonchaks / Solonetz
The following soil <u>types</u> are distinguished within the class:

* *Solonceak* ≈ Solonchak
* *Soloneţ* ≈ Solonetz

Spodisols – <u>soil class</u>. ≈ Podzols
The following soil <u>types</u> are distinguished within the class:

* *Cryptopodzols* ≈ Brunic Arenosols / *Spodic* Cambisols / *Spodic* Arenosols
* *Podzols* ≈ Podzols
* *Prepodzols* ≈ Entic Podzols / Leptic Podzols

Umbrisols – <u>soil class</u>. ≈ Umbrisols
The following soil <u>types</u> are distinguished within the class:

* *Humisiolsols* ≈ Umbrisols (Humic)
* *Nigrosols* ≈ Umbrisols

References

Conea, A., Florea, N., Puiu, Şt. (coord) (1980) *Sistemul Român de Clasificare a Solurilor (Romanian System of Soil Classification)*, ICPA, Bucure ti, 173pp

FAO-ISRIC-ISSS (1998) *World Reference Base for Soil Resources*, World Soil Resources Report no 84, UN Food and Agriculture Organization, Rome, 88pp

Florea, N. and Munteanu, I. (2000) *Sistemul Român de Taxonomie a Solurilor (Romanian System of Soil Taxonomy)*, University 'Al. I. Cuza', Iasi, 107pp

Isbell, R. F. (2002) *Australian Soil Classification*, revised edition, CSIRO Land & Water, Canberra, 144pp

IUSS Working Group WRB (2006) *World Reference Base for Soil Resources*, 2nd edition, World Soil Resources Reports no 103, UN Food and Agriculture Organization, Rome, 128pp

Krasilnikov, P. and García Calderón, N. E. (2006) 'A WRB-based buried paleosol classification', *Quaternary International*, vols 156–157, pp176–188

Kubiëna, W. L. (1953) *Bestimmungsbuch und Systematik der Boden Europas*, Verlag Enke, Stuttgart, 392pp

Miclaus, V. (1980) 'Considerati cu privire la sistemul Roman de clasificare a solurior', *Bulletin Instituto Agronomic Cluj-Napoca*, Ser. Agricultura, vol 34, pp9–15

Munteanu, I. and Florea, N. (2001) 'Present-day status of Soil Classification in Romania', in E. Michéli, F. O. Nachtergaele, R. J. A. Jones and L. Montanarella (eds), *Soil Classification 2001*, European Soil Bureau Research Report no 7, pp55–62

Soil Survey Staff (1999) *Soil Taxonomy: A Basic System of Soil Classification for Making and Interpreting Soil Surveys*, USDA, Handbook 436, 2nd edition, United States Government Printing Office, Washington DC, 696pp

18
Soil Classification of Bulgaria

Objectives and scope

Soil classification in Bulgaria was designed to serve as a basis for soil inventory and mapping. The new Bulgarian classification (Ninov, 2002) also aims at correlation and harmonization with soil databases and maps of the European Union. The classification has national coverage (Table 18.1). Although Bulgaria is a relatively small country it has diverse environmental conditions containing 20 of the 32 reference groups of the World Reference Base for Soil Resources (WRB). Compared with an older classification (Kojnov, 1987), the new one has broader scope and includes urban soils and soils deeply transformed by agriculture. Bare rock and underwater sediments are not recognized as soils in this classification. Because the Bulgarian language uses the Cyrillic alphabet, the soil names are provided in English.

Table 18.1 *The scope of soil classification of Bulgaria*

Superficial bodies	Representation in the system
Natural soils	National coverage
Urban soils	Included in the subtype of *Urbanogenic Anthropogenic soils*
Man-transported materials	Partially included in the subtype of *Recultivated Anthropogenic soils*
Bare rock	Not considered as soils
Subaquatic soils	Not considered as soils
Soils deeply transformed by agricultural activities	Included in the subtypes of *Agrogenic* and *Long-irrigated Anthropogenic soils*

Theoretical background

The new Bulgarian soil classification is similar in bases and terminology to an older version, but includes a novel system of diagnostics, similar to that used by the WRB (IUSS Working Group WRB, 2006). Although the archetypes recognized in this classification are similar to other European classifications, the scientific pedological basis is slightly different. It is based on a geographical view rather than on an evolutionary or hydrological approach to soil grouping. Soil types are believed to be related to certain climatic zones or mountain belts and are usually divided into 'zonal' and 'azonal' soils. To some extent this may be a residual influence of the Soviet school, but most likely results from diverse bioclimatic conditions and good correlations between actual environments and soil cover.

Development

A general soil classification was developed in Bulgaria in the early 1960s, but only in the 1970s was it developed in sufficient detail to be used for soil survey (Kojnov, 1987). The older version had been used a long time with minor modifications. The development of soil classification was pushed forward in 1980–1981, when Sofia hosted the meeting where the work on the International Reference Base for Soil Classification (later – WRB) was launched (IUSS Working Group WRB, 2006, p1). However, until the late 1990s the work on a new version of the Bulgarian classification had not started and the new soil classification (Ninov, 2002) was only recently published.

Structure

The Bulgarian classification (Ninov, 2002) has three levels (Table 18.2). The first level is generic consisting of soil types which represent the main soil-forming processes and an overall sequence of soil horizons. Subtypes is a specific level which has some particular properties, like secondary carbonates or gleying, present in the profile. The third level, varieties, indicates quantitative characteristics of soils, such as the depth of the A horizon, carbonates or organic C content. The classification is a simple taxonomy with formalized limits of the classes.

Diagnostics

The diagnostics in the new Bulgarian soil classification are based on the system of defined horizons and properties. The limits of soil taxa are mostly

Table 18.2 *The structure of soil classification of Bulgaria*

Level	Taxon name	Taxon characteristics	Borders between classes	Diagnostics	Terminology
0	Soils	Kingdom			
1	Type	Generic	Mostly formal	Chemico-morphological	Traditional
2	Subtype	Specific	Mostly formal	Morphologico-chemical	Traditional
3	Variety	Varietal	Formal	Morphologico-chemical	Traditional

quantitative. More attention is given to field soil morphology than to laboratory chemical analyses; however, the chemical properties are of major importance at the subtype and varieties levels. Hydrological and temperature regimes are not required for the diagnostics. In general, the diagnostics in the Bulgarian classification are chemico-morphological.

Terminology

The previous soil classification in Bulgaria used traditional scientific names, mainly derived from folk soil terminology. A special study of the terminology used by peasants showed that as many as 1000 soil names were used in the country (Stranski, 1954, 1956, 1957). Many folk terms are used in the new classification along with artificial terms borrowed from the Food and Agriculture Organization (FAO) soil map and its continuation in the WRB (IUSS Working Group WRB, 2006). The terminology of the Bulgarian classification actually may be characterized as mixed.

Correlation

The correlation of the terms of the Bulgarian classification with WRB terms is for the top two levels. The correlation was facilitated by Ninov's (2002) version of correlation. The correlation is not very strict due to different quantitative limits for soil groups. For example, the limit of soil depth for shallow soils in the Bulgarian classification is 40cm, whereas in the WRB the Leptosols have a depth limit of 25cm.

The soil names of the Bulgarian classification are given in English translation; the types are listed in alphabetical order, and subtypes are alphabetical within the types.

Accumulated soils – soil type. ≈ Fluvisols
The following subtypes are distinguished within the type:

- *carbonaceous* ≈ Calcic Fluvisols
- *dark* ≈ Fluvisols (Humic)
- *poor* ≈ Fluvisols (Dystric)
- *rich* ≈ Fluvisols (Eutric)
- *salted* ≈ Salic Fluvisols
- *wet* ≈ Gleyic Fluvisols

Andosols – soil type. ≈ Andosols
There are no subtypes within the type.

Anthropogenic soils – soil type. ≈ Anthrosols / Technosols
The following subtypes are distinguished within the type:

- *agrogenic* ≈ Anthrosols
- *long-irrigated* ≈ Irragric Anthrosols

- *recultivated* ≈ Terric Anthrosols / Technosols
- *urbanogenic* ≈ Urbic Technosols

Brown mountain forest soils – soil type. ≈ Cambisols
The following subtypes are distinguished within the type:

- *light-coloured* ≈ Cambisols
- *dark-coloured* ≈ Cambisols (Humic)

Chernozems (black earths) – soil type. ≈ Chernozems
The following subtypes are distinguished within the type:

- *carbonaceous* ≈ Calcic Chernozems
- *lessivated (degraded)* ≈ Luvic Chernozems
- *meadow (gleyed)* ≈ Gleyic Chernozems
- *ordinary* ≈ Chernozems

Chervenozems (red earths) – soil type. ≈ Nitisols
The following subtypes are distinguished within the type:

- *cinnamonic-red* ≈ Nitisols (Chromic)
- *lithic* ≈ *Leptic* Nitisols (Rhodic)
- *ordinary* ≈ Nitisols (Rhodic)

Cinnamonic soils – soil type. ≈ Cambisols / Kastanozems / Phaeozems
The following subtypes are distinguished within the type:

- *carbonaceous* ≈ *Calcic* Cambisols (Chromic) / Kastanozems (Chromic)
- *typical* ≈ Cambisols (Chromic) / Phaeozems (Chromic)

Dark mountain forest soils – soil type. ≈ Mollic Umbrisols
The following subtypes are distinguished within the type:

- *ordinary* ≈ Mollic Umbrisols
- *peaty* ≈ Mollic Histic Umbrisols

Deluvial soils – soil type. ≈ Fluvisols
The following subtypes are distinguished within the type:

- *deluvial* ≈ Fluvisols (Skeletic)
- *deluvial meadow* ≈ Gleyic Fluvisols (Skeletic)

Lessivated soils – soil type. ≈ Luvisols
The following subtypes are distinguished within the type:

- *carbonaceous* ≈ Calcic Luvisols

- *cinnamon-like* ≈ Luvisols (Chromic)
- *gleyed* ≈ Gleyic Luvisols
- *light-coloured* ≈ Albic Luvisols
- *ordinary* ≈ Cutanic Luvisols
- *red-coloured* ≈ Luvisols (Rhodic)
- *smolnitza-like* ≈ Vertic Luvisols

Mountain meadow soils – soil type. ≈ Umbrisols
The following subtypes are distinguished within the type:
- *chernozem-like* ≈ Umbrisols
- *ordinary* ≈ Folic Umbrisols
- *peaty* ≈ Histic Umbrisols

Phaeozems (dark chernozem-like) – soil type. ≈ Phaeozems / Chernozems
The following subtypes are distinguished within the type:

- *carbonaceous* ≈ Endocalcic Endogleyic Chernozems
- *lessivated (degraded)* ≈ Luvic Greyic Phaeozems
- *meadow (gleyed)* ≈ Gleyic Phaeozems
- *ordinary* ≈ Phaeozems

Planosols – soil type. ≈ Planosols (Albic)
The following subtypes are distinguished within the type:

- *saturated (neutral)* ≈ Planosols (Albic, Eutric)
- *unsaturated (acid)* ≈ Planosols (Albic, Dystric)

Regosols – soil type. ≈ Regosols
The following subtypes are distinguished within the type:

- *carbonaceous* ≈ *Calcic* Regosols / Regosols (Calcaric)
- *saturated* ≈ Regosols (Eutric)
- *unsaturated* ≈ Regosols (Dystric)

Sandy soils – soil type. ≈ Arenosols
There are no subtypes within the type.

Shallow soils – soil type. ≈ Leptosols / Umbrisols / Phaeozems
The following subtypes are distinguished within the type:

- *lithosols* ≈ Leptosols
- *rankers* ≈ Umbric Leptosols / Epileptic Umbrisols
- *rendzinas* ≈ Rendzic Leptosols / Rendzic Phaeozems

Smolnitzy – soil type. ≈ Vertisols
The following subtypes are distinguished within the type:

- *carbonaceous* ≈ Calcic Vertisols
- *gleyed* ≈ Gleyic Vertisols
- *gypsireous* ≈ Gypsic Vertisols
- *rich* ≈ Vertisols (Eutric)

Solonchatsi – soil type. ≈ Solonchaks
The following subtypes are distinguished within the type:

- *carbonaceous* ≈ Calcic Solonchaks
- *gleyed* ≈ Gleyic Solonchaks
- *ordinary* ≈ Solonchaks
- *sodic (solonchatsi-solontsi)* ≈ Sodic Solonchaks / Salic Solonetz

Solontsi – soil type. ≈ Solonetz
The following second level units are distinguished:

- *gleyed* ≈ Gleyic Solonetz
- *ordinary* ≈ Solonetz

Peat soils – soil type. ≈ Histosols
The following subtypes are distinguished within the type:

- *peat-moor (fibrous)* ≈ Fibric Histosols
- *peat-moor (earthy)* ≈ Hemic Histosols / Sapric Histosols

Wetland soils – soil type. ≈ Gleysols
The following subtypes are distinguished within the type:

- *carbonaceous* ≈ Calcic Gleysols
- *meadow-moor* ≈ Mollic Gleysols / Umbric Gleysols
- *peaty* ≈ Histic Gleysols
- *unsaturated* ≈ Gleysols (Dystric)

Zheltozems (yellow earths) – soil type. ≈ Alisols
The following subtypes are distinguished within the type:

- *gleyed* ≈ Stagnic Alisols
- *ordinary* ≈ Alisols

References

IUSS Working Group WRB (2006) *World Reference Base for Soil Resources*, 2nd edition, World Soil Resources Reports no 103, UN Food and Agriculture Organization, Rome, 128pp

Kojnov, V. (1987) 'Correlation of Bulgarian soils with the main international classifications', *Pochvoznanie, Agrohimiya i Rastitelna Zastshita*, no 5, pp5–13 (in Bulgarian)

Ninov, N. (2002) 'Taxonomic list of Bulgarian soils according to the FAO world soil system', *Geography 21*, no 5, pp4–20 (in Bulgarian, English summary)

Stranski, I. (1954) 'Bulgarian folk soil names based on their colour and humidity', *Transactions of Soil Science Institute*, Bulgarian Academy of Sciences Press, Sophia, vol 2, pp281–360 (in Bulgarian)

Stranski, I. (1956) 'Bulgarian folk soil names based on their physical properties', *Transactions of Soil Science Institute*, Bulgarian Academy of Sciences Press, Sophia, vol 3, pp329–410 (in Bulgarian)

Stranski, I. (1957) 'Bulgarian folk soil names based on their miscellaneous features', *Transactions of Soil Science Institute*, Bulgarian Academy of Sciences Press, Sophia, vol 4, pp307–409 (in Bulgarian)

19
Soil Classification and Diagnostics of the Former Soviet Union, 1977

Objectives and scope

The Soviet Union disappeared from the world political maps almost 20 years ago. However, we considered that the soil classification of the Soviet Union should be included in this volume, for several reasons. First, the new Russian soil classification (Shishov et al, 2004), discussed in the next chapter, is still not officially approved in the Russian Federation and the book *Classification and Diagnostics of Soils of USSR* (Egorov et al, 1977) remains the main document, regulating soil classification for soil survey. A review conducted by Moscow State University also showed that the majority of pedologists still use the older classification in their scientific work. Second, this classification served as a source system for the new independent states, formed after the fall of the Soviet Union. Basically, the classifications of the Baltic States and Caucasian republics use the same archetypes and terms as the older Soviet classification. In some states, like in Belorussia or in Kazakhstan, the older classification is still used officially. Third, the Soviet school generated many maps and publications that cannot be understood without understanding the system of soil classification that was used during that time.

The geographical scope of the classification is limited to the territory of the former Soviet Union (Table 19.1). However, the territory is huge and diverse, including soils from arctic to subtropical regions; only the US has a larger range of environmental conditions, including tropical soils in Hawaii and Puerto Rico. The Soviet soil classification was designed to support the mapping of the soils used in agriculture and led to a certain disproportion in the geographical scope of the system: the soils of enormous northern territories, especially in Siberia, were not represented in the classification because of their minor agricultural importance. The soils of urban and industrial territories were disregarded in the classification. Bare rock and even shallow soils are also absent in the classification because they are not used for agricultural purposes. On most maps the technogenically disturbed areas and bare rocks are shown

Table 19.1 *The scope of soil classification of USSR*

Superficial bodies	Representation in the system
Natural soils	National coverage
Urban soils	Not recognized as soils
Man-transported materials	Not recognized as soils
Bare rock	Not recognized as soils
Subaquatic soils	Not recognized as soils
Soils deeply transformed by agricultural activities	For some groups special types of 'cultivated' soils are recognized

as badlands or 'non-soils'. Agricultural transformation is recognized only for northern areas where deep ploughing led to the complete mixing of natural topsoil horizons, and for irrigated fields.

Theoretical background

The classification of soils of USSR is based on the ideas of soil genesis. Soil taxa are defined according to the presence and stage of development of the major and additional pedogenetic processes. However, the Soviet classification focuses mainly on the subjective interpretation of soil genesis by the researcher. To some extent it is the result of an underdeveloped system of diagnostics, but to a greater extent it results from the conceptual opinions of the authors. Accordingly, a particular soil profile may have disturbances or deviations from a modal soil; however, the researcher, while mapping, should show a typical, 'normal' situation, rather than a local random deviation. In doubtful cases the surveyor had to interpret soil genesis with the help of analysis of the regional factors of soil formation. In some cases this approach has led to disappointing results, including an inappropriate application of the zonal theory that the soils within a certain bioclimatic zone were included in the same 'zonal' type even though the profile was different from that. There have been many myths about the Soviet classification in the West. Information was very limited in international scientific literature with the book edited by Finkl (1982) as almost the only exception. Most information on this classification scheme was obtained from several early works of its authors, published in *Soviet Soil Science* journal (e.g. Ivanova and Rozov, 1967). These publications were presented together with interpretations of the geographical distribution of soils and seemed to mislead readers who considered that climatic conditions were included as diagnostics at the highest level of taxonomy. The classification itself is based on soil profile characteristics, and soil moisture and temperature regimes were not considered as major characteristics for classification. However, the classification has a geographical concept as a basis: soil groups correspond to big bioclimatic areas, soil zones, rather than to certain stages of soil profile development, like most European classifications (e.g. Kubiëna, 1953). Even the first official edition of the Soviet classification (Ministry of Agriculture of USSR, 1967) was published in five separate volumes, each one corresponding to a certain bioclimatic zone. It was considered that a soil surveyor would work within a specific zone, and should consult mainly a corresponding volume. To a great extent this approach reflects the real distribution of major soils of the country, where vast areas are occupied by more or less uniform soils of the same age, and the most important gradient for soil distribution is climate.

Development

The development of soil classification started in Russia at the end of the 19th century with the early works of Dokuchaev (e.g. Dokuchaev, 1950). However, the first soil classifications were very general, and divided soils only at the highest level. Unlike in the US, where the classification from the very beginning was suited for soil survey at a detailed scale, and then developed 'from bottom to top', from soil series to higher taxonomic levels, in Russia the situation was somewhat the reverse. Soil mapping initially was made at a general scale, and time passed before large-scale maps came into existence. Large-scale mapping required much more detailed soil characteristics, and lower levels were introduced. For a long time these lower levels were provisional, and every soil map included its own legend, usually different from others. When a systematic soil survey started in the Soviet Union, only internal circulars of the Ministry of Agriculture existed for unifying the cartographic units. A complete detailed system of classification was prepared for the Ministry of Agriculture by the leading specialists of the Dokuchaev Soil Science Institute, E. I. Ivanova and N. N. Rozov (Ministry of Agriculture of USSR, 1967). Later on, the Dokuchaev Institute reworked the classification, and edited the final version of the Soviet classification (Egorov et al, 1977). The latter was adjusted to the newer concepts of pedology, included a number of new types, and corrected some errors of the 1967 edition. It was smaller compared to the earlier version by eliminating soils of the arctic and tundra regions which had been considered to be of lesser importance for agriculture.

Structure

The Soviet classification is based on the idea of a <u>soil type</u>, which is defined as a group of soils formed under common conditions and regimes, having the same major soil-forming process, and with a similar structure to the profile (Table 19.2). In the taxa of a type there are <u>subtypes</u> that are defined by a superimposed soil-forming process, or are transitional between different soil types. Also in the latest version of that classification (Egorov et al, 1977) there is a group of 'facial' subtypes, defined by the regional climatic conditions. Within a subtype there are soil <u>genera</u> which are characterized by some important local conditions of soil formation (e.g. the type of parent rocks, or the chemical composition of groundwater). Within a soil genus there are soil <u>classes,</u> distinguished by the rate of the development of the main soil-forming processes (the diagnostic for this level is formal: it is based on the depth of horizons, organic matter content, etc.). Within a class there are <u>subclasses</u> distinguished by texture, and within a subclass are <u>sorts</u> distinguished by the origin of parent material. On the highest levels the borders between the taxa are fuzzy, whereas on the levels of classes and subclasses the borders are formal. The structure of the latest version of the classification (Egorov et al, 1977) is somewhat confusing because of the presence of 'groups' of soils at different levels that are not included in the taxonomic structure. For example, the text mentions a 'group of types' of alluvial soils, and then the presence of three 'groups of types' within the group of alluvial soils. For purposes of illustration we will consider the Alluvial and Irrigated soils as separate branches. Within the podzolic type there are two groups of soils: one for soils with a dominant process of clay illuviation, and the other for soils with dominant iron, aluminium and organic matter migration.

The structure of the classification may be called a hierarchical taxonomy with mainly fuzzy borders.

Table 19.2 *The structure of soil classification of USSR*

Level	Taxon name	Taxon characteristics	Borders between classes	Diagnostics	Terminology
0	Soils	Kingdom			
1	Type	Generic	Fuzzy	Morphological	Traditional
2	Subtype	Specific 1	Fuzzy	Climatic-chemico-morphological	Traditional
3	Genus	Specific 2	Fuzzy	Chemico-morphological	Traditional scientific
4	Class	Varietal 1	Formal	Chemico-morphological	Traditional scientific
5	Subclass	Varietal 2	Formal	Textural	Traditional scientific
6	Sort	Specific 3	Formal	Geological	Traditional

Diagnostics

The diagnostics of the highest taxa in the Russian classification are based on qualitative characteristics of the profile. The data on chemical properties used for the characteristics of the 'central images' are not very strict. The source of diagnostics is a soil profile with a more or less determined sequence of horizons. At the subtype level soils may have additional horizons or features, or may just vary in the expression of their features. A special case is the 'facial' criteria for some subtype determinations. They more or less correspond to the classes of soil temperature regimes in the US Soil Taxonomy (Soil Survey Staff, 1999), but are more detailed. 'Facial subtypes' were introduced when it was found that under the same bioclimatic conditions the appearance of soils might vary depending on the regional climatic regimes. For the diagnostics of 'facial subtypes' the following data were used: the sum of air temperatures above 10°C, the sum of soil temperatures above 10°C at a depth of 20cm, and the duration of the period of soil temperatures below zero at a depth of 20cm. In practice, the separation of soils into facial subtypes was seldom used either in science or practical soil survey. Chemical criteria are used mainly as additional characteristics for types and subtypes. At lower levels of the taxonomy both morphological criteria (depth of horizons) and chemical and physical attributes are evaluated quantitatively.

At the higher levels of the taxonomy almost all the limits of classes are fuzzy which allows more flexibility in definitions. In classifications with strict formal borders between the taxa, sometimes very close objects are separated. For example, in the World Reference Base for Soil Resources (WRB) (IUSS Working Group WRB, 2006), base saturation of 49 or 51 per cent in an *argic* horizon leads to the separation of soils at the reference group level, even though the difference is within the limit of error. However, the lack of strict criteria leads to a subjective approach to classification, useless discussions and possible errors.

The diagnostics in this classification may be called climatic-chemico-morphological.

Terminology

The Soviet classification maintains the traditional terminology introduced by Dokuchaev and even earlier workers (see Krasilnikov, 1999). Most often this terminology is the result of borrowing some folk terms and some stylized names. It is important to remember that borrowing a name from a folk lexicon usually means a significant transformation and narrowing of its meaning. For example, the folk name 'podzol' (having ash colour underneath – in a word-by-word translation) meant any soil having a bright horizon under a cultivated one, but when transformed into a scientific term meant a light-textured soil with a redistribution of organic matter and iron and aluminium hydroxides in the profile having a number of specific interconnected characteristics. The tradition of stylizing scientific terms also originated from the early works by Dokuchaev, who knew folk soil terms well (Dokuchaev, 1953). This tradition was carried over in the Soviet classification: one might compare folk terms 'chernozem' (black earth) and 'podzol' (having ash underneath) and terms, invented by soil scientists, such as 'serozem' (grey earth), 'krasnozem' (red earth), 'zheltozem' (yellow earth) and 'podbel' (having white colour underneath). This shows the importance of morphochromatic characteristics in the diagnostics of soils in Russian classification. Perceptive terms, such as 'forest', 'desert', 'mountainous' and 'irrigated', which are associated with the importance of factors of soil formation as diagnostics of many taxa, are widely used in this classification.

Correlation

A list of soils included in *Classification and Diagnostics of Soils of USSR* (Egorov et al, 1977) down to the level of subtypes is given, except for facial subtypes because they are very numerous (sod-carbonate soils have 21 facial subtypes), and their profiles are similar. The names of Russian soils of tundra and Arctic regions according to the Ministry of Agriculture of USSR (1967) are presented at the end of this chapter.

The soil names are listed alphabetically for convenience. Within the types the subtypes are listed in the same order as in the text of the classification, thereby indicating some opinion of authors of the system.

Folk (or stylised) terms are transcribed, and other characteristics that can be directly translated into English are translated. In some cases it leads to curiosities: for example, 'kastanovyje pochvy' in Russian would be translated into English as 'chestnut soils' and correlated with a WRB term 'Kastanozems' formed from the same root as the Russian term. Some Russian soil names are often translated into English erroneously. For example, the name *dernovo-podzolistye pochvy* is often translated into English as 'sod-podzolic soils', because *dern* in Russian means 'sod' or 'turf'. However, the Russian name refers not to the presence of turf but to the development of the *dernovy process*, which is the process of humus accumulation. Thus, a more correct translation is 'humus podzolic soils', not 'sod-podzolic soils', because the presence of a sod layer does not always mean the development of a humus-enriched horizon.

A group of types of alluvial soils is a special, somewhat particular group of soils, forming in river plains and deltas, which are characterized by a periodic flooding and sedimentation of fresh alluvial layers on the surface of soils. Though in many classifications alluvial soils are considered to be azonal, in the Soviet classification they were included in the general scheme of soil zonality. The alluvial soils were divided into three subgroups according to their drainage, and then subdivided into types according to the zone of their main occurrence. The 'acid' types were considered to develop in the taiga zone, the 'saturated' types in the forest-steppe and steppe zones, and 'carbonate' types in the dry steppe, semidesert and desert zones. According to the drainage they are separated into the groups of *alluvial humus*, *alluvial meadow* and *alluvial moor soils*. Each of this group corresponds to a certain geomorphologic element of a flood plain. Since the grouping of alluvial soils is complex, we used a hierarchical numbering with capital A especially for this group. ≈ Fluvisols / Histosols

A1. A group of types of alluvial meadow soils – soils forming in the conditions of moistening by floods and groundwater, lying at a depth of 1–2 metres; the capillary water rises to the soil profile. The accumulation of matter occurs due to both rich alluvial sediments and the effect of groundwater. The group is separated into three types. ≈ Gleyic Fluvisols

A1.1. Alluvial meadow acid soils – soil type. ≈ Gleyic Fluvisols (Dystric)
The type is separated into three subtypes:

- *alluvial meadow layered primitive acid soils* ≈ Gleyic Fluvisols (Dystric, Arenic)
- *alluvial meadow acid layered soils* ≈ Gleyic Fluvisols (Dystric)

- *alluvial meadow acid soils* ≈ Umbric Gleyic Fluvisols

A1.2. Alluvial meadow carbonate soils – <u>soil type</u>. ≈ Calcic Gleyic Fluvisols
The type is separated into three <u>subtypes</u>:

- *alluvial meadow carbonate layered soils* ≈ Calcic Gleyic Fluvisols
- *alluvial meadow carbonate 'tugai' soils* ≈ Calcic Mollic Gleyic Fluvisols
- *alluvial meadow carbonate soils* ≈ Calcic Mollic Gleyic Fluvisols

A1.3. Alluvial meadow saturated soils – <u>soil type</u>. ≈ Gleyic Fluvisols (Eutric)
The type is separated into four <u>subtypes</u>:

- *alluvial meadow saturated layered primitive soils* ≈ Gleyic Fluvisols (Eutric, Arenic)
- *alluvial meadow saturated layered soils* ≈ Gleyic Fluvisols (Eutric)
- *alluvial meadow saturated soils* ≈ Mollic Gleyic Fluvisols
- *alluvial meadow saturated dark-coloured soils* ≈ Mollic Gleyic Fluvisols (Humic)

A2. A <u>group of types</u> of alluvial moor soils – alluvial soils 'forming in the conditions of long flooding and sustainable excess atmospheric and groundwater moistening; are characterized by accumulation of poorly decomposed vegetative residues, as well as organic materials originating from flood and groundwater' (Egorov et al, 1977). Three types are distinguished in this group. ≈ Gleyic Histic Fluvisols / Histosols
A2.1. Alluvial meadow moor soils – <u>soil type</u>. ≈ Gleyic Fluvisols
The type is separated into two <u>subtypes</u>:

- *alluvial meadow moor soils* ≈ Gleyic Fluvisols (Humic)
- *alluvial peaty meadow moor soils* ≈ Gleyic Histic Fluvisols

A2.2. Alluvial moor clay-mud gleyic soils – <u>soil type</u>. ≈ Sapric Histosols
The type is separated into two <u>subtypes</u>:

- *alluvial moor clay gleyic soils* ≈ Gleyic Histic Fluvisols (*Sapric*)
- *alluvial moor mud gleyic soils* ≈ Gleyic Histic Fluvisols (*Hemic*)

A2.3. Alluvial moor mud-peat soils – <u>soil type</u>. ≈ Histosols / Gleyic Histic Fluvisols
The type is separated into two <u>subtypes</u>:

- *alluvial moor mud-peat soils* (peat layer depth > 50cm) ≈ Sapric Histosols
- *alluvial moor mud-peat gleyic soils* (peat layer depth < 50cm) ≈ Histic Gleyic Fluvisols

A3. A <u>group of types of</u> alluvial humus soils – soils forming in the conditions of short moistening with annual floods. The groundwater table lays much deeper than the soil profile; alluvial sediments have mostly light texture. The group is separated into three types. ≈ Mollic Fluvisols / Umbric Fluvisols

A3.1. Alluvial humus acid soils – <u>soil type</u>. ≈ Fluvisols (Dystric)
The type is separated into four <u>subtypes</u>:

- *alluvial humus acid layered primitive soils* ≈ Fluvisols (Dystric, Arenic)
- *alluvial humus acid layered soils* ≈ Fluvisols (Dystric)
- *alluvial humus acid soils* ≈ Umbric Fluvisols
- *alluvial humus acid podzolized soils* ≈ Umbric Fluvisols (*Albic*)

A3.2. Alluvial humus desertified carbonate soils – <u>soil type</u>. ≈ Calcic Fluvisols (Aridic)
The type is separated into three <u>subtypes</u>:

- *alluvial humus desertified carbonate layered primitive soils* ≈ Calcic Fluvisols (Aridic, Arenic)
- *alluvial humus desertified carbonate layered soils* ≈ Calcaric Fluvisols (Aridic)
- *alluvial humus desertified carbonate soils* ≈ Calcaric Fluvisols (Aridic)

A3.3. Alluvial humus saturated soils – <u>soil type</u>. ≈ Fluvisols (Eutric)
The type is separated into four <u>subtypes</u>:

- *alluvial humus saturated layered primitive soils* ≈ Fluvisols (Eutric, Arenic)
- *alluvial humus saturated soils* ≈ Fluvisols (Eutric)
- *alluvial humus saturated soils* ≈ Mollic Fluvisols
- *alluvial humus saturated 'steppified' soils* ≈ Mollic Calcic Fluvisols

Brown forest gley soils – <u>soil type</u>. ≈ Gleyic Cambisols / Stagnosols
The type is separated into the following <u>subtypes</u>:

- *brown forest surface-gleyed podzolized soils* ≈ Stagnic Cambisols
- *brown forest surface-gley podzolized soils* ≈ Stagnosols
- *brown forest gleyed soils* ≈ Endogleyic Cambisols
- *brown forest gley soils* ≈ Gleyic Cambisols

Brown forest soils (burozems) – <u>soil type</u>. ≈ Cambisols
The type is separated into six <u>subtypes</u> (not speaking of 'facial' subtypes):

- *brown forest acid raw-humus soils* ≈ Folic Cambisols (Dystric)
- *brown forest acid raw-humus podzolized soils* ≈ Folic Cambisols (Dystric, *Albic*)
- *brown forest acid soils* ≈ Cambisols (Dystric)
- *brown forest acid podzolized soils* ≈ Cambisols (Dystric, *Albic*)

- *brown forest weakly unsaturated soils* ≈ Cambisols (Eutric)
- *brown forest weakly unsaturated podzolized soils* ≈ Cambisols (Eutric, *Albic*)

Brown semidesert soils – <u>soil type</u>. ≈ Luvic Calcisols / Hypoluvic *Calcic* Arenosols
Only facial <u>subtypes</u> are distinguished.
Chernozems (black earths) – <u>soil type</u>. ≈ Chernozems / Phaeozems
The type is separated into five <u>subtypes</u>:

- *podzolized chernozems* ≈ Greyi-Luvic Phaeozems
- *leached chernozems* ≈ Luvic Chernozems
- *typical chernozems* ≈ Chernic Chernozems
- *ordinary chernozems* ≈ Chernozems
- *southern chernozems* ≈ Calcic Glossic Chernozems

Chestnut (kashtanovyje) soils – <u>soil type</u>. ≈ Kastanozems / Calcisols
The type is separated into three <u>subtypes</u>:

- *dark chestnut soils* ≈ Kastanozems
- *chestnut soils* ≈ Kastanozems
- *light chestnut soils* ≈ Calcisols

Cinnamonic soils – <u>soil type</u>. ≈ Luvic Kastanozems
The type is separated into three <u>subtypes</u>:

- *cinnamonic leached soils* ≈ Luvic Kastanozems
- *cinnamonic typical soils* ≈ Luvic Endocalcic Kastanozems
- *cinnamonic carbonate soils* ≈ Luvic Calcic Kastanozems

Gley zheltozems (gley yellow earths) – <u>soil type</u>. ≈ Gleyic Lixisols
The type is separated into three <u>subtypes</u>:

- *surface-gleyed zheltozems* ≈ Stagnic Lixisols
- *gleyed zheltozems* ≈ Endogleyic Lixisols
- *gley zheltozems* ≈ Gleyic Lixisols

Grey-brown desert soils – <u>soil type</u>. ≈ Calcic Gypsisols
Only facial <u>subtypes</u> are distinguished within this type.
Grey-cinnamonic soils – <u>soil type</u>. ≈ Luvic Calcisols / Luvic Calcic Kastanozems
The type is separated into three <u>subtypes</u>:

- *dark grey-cinnamonic soils* ≈ Luvic Calcic Kastanozems
- *ordinary grey-cinnamonic soils* ≈ Luvic Calcisols
- *light grey-cinnamonic soils* ≈ Luvic Calcisols

Grey forest gley soils – <u>soil type</u>. ≈ Luvic Gleyic Phaeozems (Albic) / Luvic Stagnic Phaeozems (Albic)
The type is separated into three <u>subtypes</u>:

- *grey forest surface-gleyed soils* ≈ Luvic Stagnic Phaeozems (Albic)
- *grey forest groundwater-gleyed soils* ≈ Luvic Endogleyic Phaeozems (Albic)
- *grey forest groundwater-gleyic soils* ≈ Luvic Gleyic Phaeozems (Albic)

Grey forest soils – <u>soil type</u>. ≈ Albic Luvisols / Luvic Greyic Phaeozems
The type is separated into three <u>subtypes</u>:

- *light-grey forest soils* ≈ Albic Luvisols
- *grey forest soils* ≈ Albic Luvisols
- *dark-grey forest soils* ≈ Luvic Greyic Phaeozems

Humus-carbonate soils – <u>soil type</u>. ≈ Rendzic Leptosols / Rendzic Phaeozems
The type is separated into three <u>subtypes</u>:

- *humus-carbonate typical soils* ≈ Rendzic Leptosols / Rendzic Phaeozems
- *humus-carbonate leached soils* ≈ Rendzic Leptosols / Rendzic Phaeozems
- *humus-carbonate podzolized soils* ≈ Rendzic Leptosols / Rendzic Phaeozems

Humus-gley soils – <u>soil type</u>. ≈ Mollic Gleysols / Mollic Stagnosols
The type is separated into four <u>subtypes</u>:

- *humus surface-gley soils* ≈ Mollic Histic Stagnosols
- *humus surface-gleyed soils* ≈ Mollic Stagnosols
- *humus groundwater-gley soils* ≈ Mollic Histic Gleysols
- *humus groundwater-gleyed soils* ≈ Mollic Gleysols

Irrigated soils – a <u>group of types</u> of soils, which are characterized with a special irrigation type of water regime, bringing with it carbonates, salts and small particles of mineral and organic materials. Both soils having a thick horizon of irrigation-originated solid material (Irragric Anthrosols), and soils without such a horizon are included in this group.
Irrigated brown soils of semideserts – <u>soil type</u>. ≈ *Irragric* Luvic Calcisols
No <u>subtypes</u> are distinguished within this type.
Irrigated grey-brown soils of deserts – <u>soil type</u>. ≈ *Irragric* Calcic Gypsisols
No <u>subtypes</u> are distinguished within this type.
Irrigated meadow brown soils of semideserts – <u>soil type</u>. ≈ *Irragric* Luvic Gleyic Calcisols
No <u>subtypes</u> are distinguished within this type.
Irrigated meadow desert soils – <u>soil type</u>. ≈ *Irragric* Gleyic Calcisols / Gleyic Irragric Anthrosols

The type is separated into two <u>subtypes</u>:

- *irrigated meadow desert soils* ≈ *Irragric* Gleyic Calcisols
- *long-irrigated meadow desert soils* ≈ Gleyic Irragric Anthrosols

Irrigated meadow serozem soils – <u>soil type</u>. ≈ *Irragric* Gleyic Calcisols
The type is separated into two <u>subtypes</u>:

- *irrigated meadow serozem soils* ≈ *Irragric* Endogleyic Calcisols
- *irrigated serozem meadow soils* ≈ *Irragric* Gleyic Calcisols

Irrigated meadow soils of deserts and semideserts – <u>soil type</u>. ≈ *Irragric* Gleyic Calcisols / *Irragric* Calcic Gleysols / Irragric Gleyic Anthrosols
The type is separated into three <u>subtypes</u>:

- *irrigated meadow soils of deserts and semideserts* ≈ *Irragric* Gleyic Calcisols
- *irrigated wet meadow soils of deserts and semideserts* ≈ *Irragric* Calcic Gleysols
- *long-irrigated meadow soils of deserts and semideserts* ≈ Irragric Gleyic Anthrosols

Irrigated moor soils of deserts and semideserts – <u>soil type</u>. ≈ *Irragric* Gleyic Anthrosols
No <u>subtypes</u> are distinguished within this type.

Irrigated serozems – <u>soil type</u>. ≈ *Irragric* Luvic Calcisols
The type is separated into four <u>subtypes</u>:

- *irrigated light serozems* ≈ *Irragric* Luvic Calcisols (Hyperochric)
- *irrigated serozems* ≈ *Irragric* Luvic Calcisols
- *irrigated dark serozems* ≈ *Irragric* Luvic Calcisols
- *long-irrigated serozems* ≈ Irragric Anthrosols

Irrigated takyr-like soils of deserts – <u>soil type</u>. ≈ *Irragric* Takyric Regosols
The type is separated into two <u>subtypes</u>:

- *irrigated takyr-like soils* ≈ *Irragric* Takyric Regosols
- *long-irrigated takyr-like soils of deserts* ≈ Irragric Anthrosols

Krasnozems (red earths) – soil type. ≈ Acrisols (Rhodic) / Nitisols (Rhodic)
The type is separated into two <u>subtypes</u>:

- *typical krasnozems* ≈ Acrisols (Rhodic) / Nitisols (Rhodic)
- *podzolized krasnozems* ≈ Stagnic Acrisols (Rhodic) / Acrisols (Albic, Rhodic)

Meadow brown semidesert soils – soil type. ≈ Luvic Endogleyic Calcisols
The type is separated into two subtypes:

- *meadow-like brown semidesert soils* ≈ Luvic *Stagnic* Calcisols
- *meadow brown semidesert soils* ≈ Luvi-Gleyic Calcisols

Meadow chernozem-like soils – soil type. ≈ Stagnic Phaeozems
The type contains only one subtype:

- *meadow chernozem-like soils with surface moistening* ≈ Stagnic Phaeozems

Meadow chernozemic soils – soil type. ≈ Gleyic Chernozems
The type is separated into two subtypes:

- *meadow-like chernozemic soils* ≈ Endogleyic Chernozems
- *meadow chernozemic soils* ≈ Gleyic Chernozems

Meadow chestnut soils – soil type. ≈ Gleyic Kastanozems
The type is separated into two subtypes:

- *meadow-like chestnut soils* ≈ Epigleyic Kastanozems
- *meadow chestnut soils* ≈ Gleyic Kastanozems

Meadow cinnamonic soils – soil type. ≈ Luvic Gleyic Kastanozems
The type is separated into three subtypes:

- *surface-gleyed meadow-like cinnamonic soils* ≈ Luvic Stagnic Kastanozems
- *meadow-like cinnamonic soils* ≈ Luvic Endogleyic Kastanozems
- *meadow cinnamonic soils* ≈ Luvic Gleyic Kastanozems

Meadow dark chernozem-like soils – soil type. ≈ Gleyic Phaeozems / Mollic Gleysols
The type is separated into two subtypes:

- *meadow dark chernozem-like soils* ≈ Gleyic Phaeozems
- *wet meadow dark chernozem-like soils* ≈ Mollic Gleysols (Humic)

Meadow desert soils – soil type. ≈ Gleyic Calcisols / Endogleyic Arenosols
The type is separated into five subtypes:

- *meadow-like desert soils* ≈ Endogleyic Calcisols (Takyric)
- *meadow desert soils* ≈ *Gleyic* Calcisols (Takyric)
- *meadow desert soils with additional surface moistening* ≈ *Stagnic* Calcisols

- *meadow desert grey-brown soils* ≈ *Gleyic* Endosalic Calcisols
- *meadow desert sandy soils* ≈ *Gleyic Calcic* Arenosols

Meadow grey cinnamonic soils – soil type. ≈ Luvic Gleyic Calcisols
The type is separated into three subtypes:

- *surface-gleyed meadow-like grey cinnamonic soils* ≈ Luvic *Stagnic* Calcisols
- *meadow-like grey cinnamonic soils* ≈ Luvic Endogleyic Calcisols
- *meadow grey cinnamonic soils* ≈ Luvic *Gleyic* Calcisols

Meadow grey forest soils – soil type. ≈ Gleysols (Humic, Eutric)
The type is separated into two subtypes:

- *meadow forest grey soils* ≈ Gleysols (Humic, Eutric)
- *wet meadow forest grey soils* ≈ Gleysols (Humic, Eutric)

Meadow moor soils – soil type. ≈ Histic Gleysols
The type is separated into two subtypes:

- *mud meadow moor soils* ≈ Histic Gleysols
- *clay meadow moor soils* ≈ Mollic Histic Gleysols

Meadow podbels – soil type. ≈ Planosols (Albic, Ferric)
The type is separated into two subtypes:

- *meadow podzolized podbels* ≈ Planosols (Albic, Ferric)
- *meadow podzolized gleyed podbels* ≈ Gleyic Planosols (Albic, Ferric)

Meadow serozemic soils (meadow grey earths) – soil type. ≈ Gleyic Calcisols
The type is separated into two subtypes:

- *meadow-like serozemic soils* ≈ Endogleyic Calcisols
- *meadow serozemic soils* ≈ *Gleyic* Calcisols

Meadow soils – soil type. ≈ Gleyic Phaeozems / Mollic Gleysols
The type is separated into two subtypes:

- *meadow soils* ≈ Gleyic Phaeozems
- *wet meadow soils* ≈ Mollic Gleysols

Meadow soils of deserts and semi-deserts – soil type. ≈ Calcic Gleysols
The type is separated into two subtypes:

- *meadow (typical) soils of deserts and semi-deserts* ≈ Calcic Gleysols

- *wet meadow (moor-meadow) soils of deserts and semi-deserts* ≈ Calcic Gleysols (Humic)

Moor podzolic soils – <u>soil type</u>. ≈ Planosols / Stagnosols / Gleyic Podzols
According to the character of moistening and the organic profile type this type is separated into six <u>subtypes</u>:

- *peaty-podzolic surface-gleyed soils* ≈ Histic Planosols / Histic Stagnosols
- *humus-podzolic surface-gleyed soils* ≈ Umbric Planosols / Umbric Stagnosols
- *mud-podzolic surface-gleyed soils* ≈ Histic Planosols / Umbric Stagnosols
- *peaty-podzolic groundwater-gleyed soils* ≈ Gleyic Histic Podzols
- *humus-podzolic groundwater-gleyed soils* ≈ Umbric Gleyic Podzols
- *mud-podzolic groundwater-gleyed soils* ≈ Gleyic Histic Podzols

Moor soils of deserts and semideserts – <u>soil type</u>. ≈ Eutric Histosols / Calcic Histic Gleysols
The type is separated into two <u>subtypes</u>:

- *peat-moor soils of deserts and semideserts* ≈ Eutric Histosols
- *clay-moor soils of deserts and semideserts* ≈ Calcic Histic Gleysols / Calcic Humic Gleysols

Mountain meadow chernozem-like soils – <u>soil type</u>. ≈ Phaeozems / Chernozems
The type is separated into three <u>subtypes</u>:

- *mountain meadow chernozem-like typical soils* ≈ Chernozems
- *mountain meadow chernozem-like leached soils* ≈ Phaeozems
- *mountain meadow chernozem-like carbonate soils* ≈ Calcic Chernozem

Mountain meadow soils – <u>soil type</u>. ≈ Umbrisols
The type is separated into two <u>subtypes</u>:

- *mountain meadow alpine soils* ≈ Umbrisols
- *mountain meadow subalpine soils* ≈ Umbrisols (Humic)

Mountain meadow-steppe soils – <u>soil type</u>. ≈ Phaeozems
The type is separated into two <u>subtypes</u>:

- mountain meadow-steppe alpine soils ≈ Phaeozems
- mountain meadow-steppe subalpine soils ≈ Phaeozems

Peat moor soils – a <u>group of soil types</u> characterized by the accumulation of a thick layer of organic residues in various stages of decomposition. ≈ Histosols
Soil types are distinguished according to the type of water supply and the

presence of drainage and cultivation.

Peat high (ombrotrophic) moor soils – soil type. ≈ Ombri-Fibric Histosols / Histic Gleysols (Dystric)

The type is separated into two subtypes:

- *peat-gleyic high moor soils* ≈ Histic Gleysols (Dystric)
- *peat high moor soils* ≈ Ombric Fibric Histosols

Peat low (minerotrophic) moor soils – soil type. ≈ Rheic Histosols / Histic Gleysols (Eutric)

The type is separated into four subtypes:

- *peat-gleyic impoverished low moor soils* ≈ Histic Gleysols
- *peat impoverished low moor soils* ≈ Rheic Histosols (Dystric)
- *peat-gleyic low moor soils* ≈ Histic Gleysols (Eutric)
- *peat low moor soils* ≈ Rheic Histosols (Eutric)

Peat high (oligotrophic) moor cultivated soils – soil type. ≈ Ombric Histosols (Drainic)

No subtypes are distinguished within this type.

Peat low (minerotrophic) moor cultivated soils – soil type. ≈ Rheic Histosols (Drainic)

Only facial subtypes are distinguished within this type.

Podzolic-brown forest gleyic soils – soil type. ≈ Stagnosols / Gleyic Luvisols (Albic)

The type is separated into four subtypes:

- *podzolic-brown forest surface-gleyed soils* ≈ Stagnic Luvisols (Albic)
- *podzolic-brown forest surface-gley soils* ≈ Stagnosols
- *podzolic-brown forest gleyed soils* ≈ Endogleyic Luvisols (Albic)
- *podzolic-brown forest gley soils* ≈ Gleyic Luvisols (Albic)

Podzolic-brown forest soils – soil type. ≈ Stagnic Luvisols

The type is separated into two subtypes:

- *podzolic-brown forest unsaturated soils* ≈ Stagnic Luvisols (Dystric)
- *podzolic-brown forest weakly unsaturated soils* ≈ Stagnic Luvisols (Eutric)

Podzolic soils – soil type. All the subtypes of this type are separated into two groups (the taxonomic meaning of this level is not defined in the classification):

A. Soils having an illuvial horizon enriched mainly with clay; developed predominately on clay or loam, and in rare cases on loamy sand parent materials ≈ Albeluvisols

B. Soils having an illuvial horizon enriched mainly with iron, aluminium and humus; developed predominately on sandy, loamy sand or gravelly parent material (podzols) ≈ Podzols
The type is separated into three <u>subtypes</u>:

- *gleyed podzolic soils* ≈ Stagnic Albeluvisols / Stagnic Podzols
- *podzolic soils* ≈ Albeluvisols / Podzols
- *humus-podzolic soils* ≈ Umbric Albeluvisols / Umbric Podzols

Cultivated podzolic soils – <u>soil type</u>. ≈ Podzols (Anthric) / Albeluvisols (Anthric)
The type is separated into three <u>subtypes</u>:

- *gleyed podzolic cultivated soils* ≈ Epigleyic Podzols (Anthric) / Epigleyic Albeluvisols (Anthric)
- *podzolic cultivated soils* ≈ Podzols (Anthric) / Albeluvisols (Anthric)
- *humus-podzolic cultivated soils* ≈ Podzols (Anthric) / Albeluvisols (Anthric)

Podzolic-zheltozem gleyic soils (yellow-podzolic gleyic soils) – <u>soil type</u>. ≈ Gleyic Lixisols (Albic)
The type is separated into three <u>subtypes</u>:

- *podzolic-zheltozem surface-gleyed soils* ≈ Stagnic Lixisols (Albic)
- *podzolic-zheltozem gleyed soils* ≈ Endogleyic Lixisols (Albic)
- *podzolic-zheltozem gleyic soils* ≈ Gleyic Lixisols (Albic)

Podzolic-zheltozem soils (yellow-podzolic soils) – <u>soil type</u>. ≈ Acrisols (Albic) / Lixisols (Albic)
The type is separated into two <u>subtypes</u>:

- *podzolic-zheltozem unsaturated soils* ≈ Albic Acrisols (Albic)
- *podzolic-zheltozem saturated soils* ≈ Albic Lixisols (Albic)

Sandy desert soils – <u>soil type</u>. ≈ Protic *Calcic* Arenosols
No subtypes are distinguished within this type.
Serozems (grey earths) – <u>soil type</u>. ≈ Calcisols
The type is separated into three <u>subtypes</u>:

- *light serozems* ≈ Calcisols (Hyperochric)
- *typical serozems* ≈ Calcisols
- *dark serozems* ≈ Calcisols

Solod – <u>soil type</u>. ≈ Solodic Planosols
The type is separated into three <u>subtypes</u>:

- *meadow-steppe (humus gley) solod* ≈ Calcic Mollic Endogleyic Solodic Planosols
- *meadow (humus gley) solod* ≈ Mollic Endogleyic Solodic Planosols
- *meadow-moor solod* ≈ Mollic Endogleyic Histic Solodic Planosols

Solonchaks – a <u>group of soil types</u> characterized by a high concentration of soluble salts in surface horizons. ≈ Solonchaks
Types are distinguished according to water regime.
Automorphic solonchaks – <u>soil type</u>. ≈ Solonchaks
The type is separated into two <u>subtypes</u>:

- *automorphic typical solonchaks* ≈ Solonchaks
- *automorphic takyr-like solonchaks* ≈ Solonchaks (Takyric)

Hydromorphic solonchaks – <u>soil type</u>. ≈ Gleyic Solonchaks
The type is separated into six <u>subtypes</u>:

- *typical solonchaks* ≈ Gleyic Solonchaks
- *meadow solonchaks* ≈ Mollic Gleyic Solonchaks
- *moor solonchaks* ≈ Gleyic Histic Solonchaks
- *'sor' solonchaks* ≈ Petrosalic Gleyic Solonchaks
- *mud-volcanic solonchaks* ≈ Gleyic Solonchaks
- *uneven solonchaks* ≈ Gleyic Solonchaks (Aridic)

Solonetz – a <u>group of soil types</u>, characterized by a high exchangeable sodium content in the upper horizons; as a result the reaction in these horizons is alkaline, and the physical properties are unfavourable (these soils are compact when dry, and sticky when moist, etc.) Types are distinguished according to water regimes. ≈ Solonetz
Automorphic solonetz – <u>soil type</u>. ≈ Solonetz
The type is separated into three <u>subtypes</u>:

- *chernozemic solonetz* ≈ Mollic Solonetz
- *chestnut solonetz* ≈ Mollic Solonetz
- *semidesert solonetz* ≈ Salic Solonetz (Aridic)

Hydromorphic solonetz – <u>soil type</u>. ≈ Gleyic Solonetz
The type is separated into four <u>subtypes</u>:

- *chernozemic-meadow solonetz* ≈ Mollic Gleyic Solonetz
- *chestnut-meadow solonetz* ≈ Mollic Gleyic Solonetz
- *meadow-moor solonetz* ≈ Gleyic Solonetz
- *meadow-frost solonetz* ≈ *Gelic* Gleyic Solonetz

Semihydromorphic solonetz – <u>soil type</u>. ≈ Endogleyic Solonetz
The type is separated into four <u>subtypes</u>:

- *meadow-chernozemic solonetz* ≈ Mollic Endogleyic Solonetz
- *meadow-chestnut solonetz* ≈ Mollic Endogleyic Solonetz
- *meadow-semidesert solonetz* ≈ Salic Endogleyic Solonetz
- *semihydromorphic frost solonetz* ≈ *Gelic* Endogleyic Solonetz

Takyr-like desert soils – <u>soil type</u>. ≈ Regosols (Takyric)
Only facial <u>subtypes</u> are distinguished within this type.
Takyrs – <u>soil type</u>. ≈ Calcisols (Takyric)
No <u>subtypes</u> are distinguished within this type.
Zheltozems (yellow earths) – <u>soil type</u>. ≈ Acrisols / Lixisols
The type is separated into four <u>subtypes</u>:

- *unsaturated zheltozems* ≈ Acrisols
- *unsaturated podzolized zheltozems* ≈ Stagnic Acrisols
- *weakly unsaturated zheltozems* ≈ Lixisols
- *weakly unsaturated podzolized zheltozems* ≈ Stagnic Lixisols

Soil types, included in the first edition of the classification (Ministry of Agriculture of USSR, 1967), but not included in the second one (Egorov et al, 1977).
Arctic semi-moor soils – <u>soil type</u>. ≈ *Gleyic* Leptic Histic Cryosols (Aridic)
No <u>subtypes</u> are distinguished within this type.
Arctic soils – <u>soil type</u>. ≈ Leptic Cryosols (Aridic, Skeletic)
No <u>subtypes</u> are distinguished within this type.
Arctic solonchak soils – <u>soil type</u>. ≈ Salic Leptic Cryosols (Aridic, Skeletic)
No <u>subtypes</u> are distinguished within this type.
Frost meadow-forest soils – <u>soil type</u>. ≈ Umbrisols / Phaeozems
The type is separated into two <u>subtypes</u>:

- *frost gleyed meadow-forest soils* ≈ Endogleyic Umbrisols (Gelic)
- *frost steppe-like meadow-forest soils* ≈ (*Endocalcic*) Umbrisols / Phaeozems

Humus (mud) lithogenic soils – <u>soil type</u>. ≈ Umbric Leptosols / Mollic Leptosols
The type is separated into three <u>subtypes</u>:

- *humus lithogenic saturated soils* ≈ Mollic Leptosols
- *humus lithogenic acid soils* ≈ Umbric Leptosols
- *humus lithogenic podzolized soils* ≈ Umbric Leptosols (*Albic*)

Moor tundra soils – <u>soil type</u>. ≈ Cryic Histosols
No <u>subtypes</u> are distinguished within this type.
Tundra semi-moor soils – <u>soil type</u>. ≈ *Gleyic* Histic Cryosols
No <u>subtypes</u> are distinguished within this type.
Tundra soils – <u>soil type</u>. ≈ Cryosols

The type is separated into three <u>subtypes</u>:

- *arctic tundra soils* ≈ Cryosols (Aridic)
- *typical tundra soils* ≈ Umbric Cryosols
- *podzolized tundra soils* ≈ Umbric Cryosols (*Albic*)

References

Dokuchaev, V. V. (1950) 'Mapping of Russian soils', in V. V. Dokuchaev, *Complete Set of Works*, USSR Academy of Sciences Publisher, Moscow, vol 2, pp69–241 (in Russian)

Dokuchaev, V. V. (1953) 'On the use of the study of local names of Russian soils', in V. V. Dokuchaev, *Complete Set of Works*, USSR Academy of Sciences Publisher, Moscow, vol 7, pp332–340 (in Russian)

Egorov, V. V., Fridland, V. M., Ivanova, E. N., Rozov, N. N., Nosin, V. A. and Fraev, T. A. (1977) *Classification and Diagnostics of Soils of USSR*, Kolos Press, Moscow, 221pp (in Russian)

Finkl, C. W. Jr (ed) (1982) 'Soil classification', *Benchmark Papers in Soil Science*, vol 1, Hutchinson Ross Publishing Company, Stroudsburg, PA, 391pp

IUSS Working Group WRB (2006) *World Reference Base for Soil Resources*, 2nd edition, World Soil Resources Reports no 103, UN Food and Agriculture Organization, Rome, 128pp

Ivanova, E. N. and Rozov, N. N. (1967) 'Soil classification of USSR', *Soviet Soil Science*, no 2, pp3–11

Krasilnikov, P. V. (1999) 'Early studies on folk soil terminology', *Eurasian Soil Science*, vol 32, no 10, pp1147–1150

Kubiëna, W. L. (1953) *Bestimmungsbuch und Systematic der Boden Europas*, Verlag Enke, Stuttgart, 392pp

Ministry of Agriculture of USSR (1967) *Instructions on Soils Diagnostics and Classification*, Kolos Press, Moscow, in five volumes (in Russian)

Shishov, L. L., Tonkonogov, V. D., Lebedeva, I. I. and Gerasimova, M. I. (2004) *Classification and Diagnostics of Soils of Russia*, Oykumena, Smolensk, 342pp (in Russian)

Soil Survey Staff (1999) *Soil Taxonomy: A Basic System of Soil Classification for Making and Interpreting Soil Surveys*, USDA, Handbook 436, 2nd edition, United States Government Printing Office, Washington DC, 696pp

20
Russian Soil Classification, 2004

Objectives and scope

A new classification of the soils of the Russian Federation (Shishov et al, 2004) was made to replace the older taxonomic system used in the Soviet Union (Egorov et al, 1977). Initially planned as a tool for soil survey in the country, ironically, it was published when Russian Giprozems (State Project Institutes for Land Resources) practically stopped soil survey. These institutes were transformed into commercial enterprises mainly focused on the arranging of land ownership documentation, and the soil survey divisions were closed. Currently the new classification of Russian soils has a strange status: it is published, but not approved as an official one for Russia because there are no institutions that can approve it. Due to this undefined status of the classification, it is not included as obligatory material in university courses on soil science. It is included in some university courses by a decision of the departments, or particular professors. According to an on-line survey organized by the Faculty of Soil Science of Moscow State University, less than 20 per cent of university scientists and students use the new classification in their work.

The classification is used in Russia at least by some members of the scientific community. It has many new ideas which have no analogies in other taxonomies, and is interesting from a theoretical point of view. The classification represents a major effort of a group of leading researchers, and reflects their views on soil genesis, classification and systematics. The geographical scope of the new Russian classification is narrower than that of the previous version (Egorov et al, 1977): it is limited to the borders of the Russian Federation (Table 20.1). The authors of the classification decided to classify human-transported materials and substrates of urban and industrial areas apart from the main soil classification. Bare rock and underwater soils are also disregarded in the classification. Soils transformed by agricultural activities are widely represented in the Russian classification: in fact, this classification pays major attention especially to the degree and type of agricultural transformation of soils. The most transformed soils are classified at the level of soil section (collective level), the less transformed are at the level of soil types (generic level), and

Table 20.1 *The scope of the Russian soil classification*

Superficial bodies	Representation in the system
Natural soils	National coverage
Urban soils	Classified as technogenic substrates apart from the main classification
Man-transported materials	Classified as technogenic substrates apart from the main classification
Bare rock	Not included in the classification
Subaquatic soils	Not included in the classification
Soils deeply transformed by agricultural activities	Included at various levels of taxonomy depending on the degree of transformation

those with minor agricultural disturbance are classified at the level of subtypes (specific level).

Theoretical background

Russian classification declares four principles as its theoretical basis: genetic approach, historical continuity, reproducibility and openness. These principles are common for many soil classification schemes. In practice these principles are not strictly followed but serve as guidelines.

Completely new is the attempt to use a system of diagnostic horizons and properties. However, in the Russian classification these diagnostic genetic horizons have somewhat different meanings than in most Western schools. The latter ones are stricter, but often embrace broader concepts. Thus the same horizon in a real profile may fit the concepts of several diagnostic horizons, properties or materials: they are considered to be overlapping or coinciding in part. For example, in the World Reference Base for Soil Resources (WRB) (IUSS Working Group WRB, 2006) a single horizon can fit the definitions of *gypsic*, *calcic* and *argic* horizons, and also a number of other properties – all of which may be taken into account for classifying soils. In the Russian classification the designated horizons are mutually exclusive which provides more order to the taxonomy; for example, this classification has a key for diagnostic horizons (Anonymous, 2008) that is impossible in other classifications. Any new combination of properties usually results in the introduction of a new horizon and, according to the logic of classification, results in a new soil type.

The other important feature of the Russian classification is its special attention to the degree of agricultural transformation of soils. No other classification has such a large number of horizons and properties related to agricultural transformations. The basic concepts of the new Russian classification, including the system of naming agrogenic soil horizons and agricultural soils, are described in more detail in some journal articles (e.g. Lebedeva et al, 1996; Shishov et al, 2005).

Development

The 1977 classification and diagnostics of soils of USSR (Egorov et al, 1977) did not satisfy both scientists and practical workers. Just five years after its publication, the Dokuchaev Soil Science Institute initiated work on a new version of soil classification (Fridland, 1982). One of the approximations of the classification was used for the legend of the Soil Map of Russia at a scale of 1:2,500,000. A complete version was published in 1997 (Shishov et al, 1997) and caused extensive discussion in the scientific community. For an English-speaking reader this version is also available, in an adapted and improved form (Arnold, 2001). The discussion of the classification resulted in its further revision, and soon after, a new version was published (Shishov et al, 2004). In fact, the recent changes in the Russian classification are sometimes difficult to follow: the last guidebook (Anonymous, 2008) also contains significant modifications in the structure and terminology of the taxonomy. In this book the correlations are for the text published in 2004 (Shishov et al, 2004).

Structure

In general the Russian soil classification follows the structure of the classification of soils of USSR (Egorov et al, 1977). The basis unit of this classification (generic level) is soil type (Table 20.2), but its meaning has changed somewhat. Soil types have become much narrower; for example, in previous classifications textural differentiation in chernozems was used only for distinguishing soils on the level of subtypes, whereas in the new classification this feature is used to separate chernozems from clay-illuviated chernozems. As a result the number of types has increased. Accordingly, the subtypes became narrower and mostly separate soils having slight gleying or minor agricultural transformations. The soils differing by the composition of rocks and groundwater are distinguished at the genus level. The concepts of classes, subclasses and sorts are the same as in the 1977 version.

A new feature is the presence of two taxa on levels above type, namely trunks and sections. There are three trunks: postlithogenic, synlithogenic and organogenic soils. The first two are distinguished on the basis of the ratio between lithogenesis and soil formation, and the soils of the third trunk develop in peat. In the latter case the peculiarity of parent material determines all the soil properties.

The structure of the classification is a hierarchical taxonomy with fuzzy borders between taxa.

Diagnostics

The diagnostics of soil taxa in the new Russian classification are derived from the composition of a soil profile; the authors define this classification as 'substantial-genetic'. A soil profile is defined as a sequence of diagnostic genetic horizons. The characteristics of horizons are determined using both

Table 20.2 *The structure of the Russian soil classification*

Level	Taxon name	Taxon characteristics	Borders between classes	Diagnostics	Terminology
0	Soils	Kingdom			
1	Trunk	Collective	Fuzzy	Morphological	Semiartificial
2	Section	Collective	Fuzzy	Morphological	Semiartificial
3	Type	Generic	Fuzzy	Chemico-morphological	Traditional and scientific
4	Subtype	Specific	Fuzzy	Chemico-morphological	Traditional and scientific
5	Genus	Specific	Fuzzy	Chemical	Scientific
6	Class	Varietal	Formal	Chemico-morphological	Traditional
7	Subclass	Varietal	Formal	Physical (texture)	Traditional
8	Sort	Specific	Formal	Morphological	Traditional (geological)

field morphological criteria and laboratory analysis. However, the diagnostics of the genetic horizons are not very strict, and very few quantitative criteria are used. Thus, though at a first glance the diagnostics of Russian classification seem to be similar to other classifications, in fact its fuzziness occurs with the diagnostic horizons. The horizons are defined by the description of 'central images' rather than by clearly defined limits. The authors try to avoid formal quantitative criteria and appeal more to the genetic interpretation of properties than to their formal use. In practice the diagnostics are still subjective for the most part.

Climatic conditions of soil formation and the mineralogical composition of soils are classified independently of soil profiles. In general the diagnostics in this classification can be called qualitative chemico-morphological.

Terminology

A significant part of the terminology in the new Russian classification is preserved from the Soviet classification; this terminology is mainly traditional (e.g. *chernozems, podzols, burozems*). The other part consists of terms that are stylized names (e.g. *agrozems, podburs, stratozems and agroabrazems*). Most Russian specialists disliked this part, because it was quite unusual terminology, combining Greek and Latin roots with Russian ones. The third part is based mainly on scientific terms indicating some processes and features of soils; this terminology is widely used on the levels of sections and subtypes (e.g. sections of *metamorphic, Al-Fe-humus, texture-differentiated* soils, and subtypes such as *surface-turbated, clay-illuvial* soils, etc.). Thus, the terminology of the new Russian classification can be called mixed. Most soil terms at the level of soil types are long descriptive names.

Correlation

Correlating soil classifications based on different basic principles is not an easy task; however, the central concepts of soil taxa are surprisingly close in the Russian classification and in the WRB. There is no correspondence for deeply ploughed or turbated soils.

In the translation of the terms we followed the same rules as in the previous chapter. The word *dernovy* was translated as 'humus', not 'sod'. The classification uses two synonyms for a relatively light-coloured surface humus-enriched horizon AY: *dernovy* and *serogumusovy* ('grey-humus'), and both of them are used as formative elements for soil names. We used the term 'grey-humus' in all cases to avoid confusion.

Only correlations of trunks, sections and types are provided. The classification has a complex hierarchical structure: thus, the types are presented in the order proposed by the authors of the classification.

Postlithogenic soils – a trunk of soils – bringing together soils where soil-formation processes occur on a previously formed parent material, and modern accumulation of matter on the surface is negligible. (No identical units in WRB.)

Texture differentiated soils – soil section. ≈ Albeluvisols / Luvic Phaeozems / Planosols / Stagnosols / Luvisols

- podzolic soils – soil type. ≈ Albeluvisols
- gleyic podzolic soils – soil type. ≈ Gleyic Albeluvisols
- peat gleyic podzolic soils – soil type. ≈ Gleyic Histic Albeluvisols
- grey-humus podzolic soils – soil type. ≈ Umbric Albeluvisols
- grey-humus gleyic podzolic soils – soil type. ≈ Gleyic Umbric Albeluvisols
- grey soils – soil type. ≈ Luvic Phaeozems (Albic)
- gleyic grey soils – soil type. ≈ Luvic Gleyic Phaeozems (Albic)
- dark-grey soils – soil type. ≈ Luvic Greyic Phaeozems
- dark-grey gleyic soils – soil type. ≈ Luvic Gleyic Greyic Phaeozems
- dark-humus podbels – soil type. ≈ Umbric Stagnosols
- grey-humus brown podzolic soils – soil type. ≈ *Greyic* Luvisols
- grey-humus solods – soil type. ≈ Solodic Planosols
- dark-humus solods – soil type. ≈ Mollic Solodic Planosols
- gleyic grey-humus solods – soil type. ≈ Endogleyic Solodic Planosols
- mud-dark-humus hydrometamorphic solods – soil type. ≈ Mollic Endogleyic Solodic Planosols
- agric peat gleyic podzolic soils – soil type. ≈ Gleyic-Histic Albeluvisols (Anthric)
- agric grey-humus podzolic soils – soil type. ≈ Umbric Albeluvisols (Anthric)
- agric grey-humus gleyic podzolic soils – soil type. ≈ Umbric Gleyic Albeluvisols (Anthric)
- agric grey soils – soil type. ≈ Luvic Phaeozems (Anthric, Albic)

- agric dark-grey soils – <u>soil type</u>. ≈ Luvic Greyic Phaeozems (Anthric)
- agric gleyic dark-grey soils – <u>soil type</u>. ≈ Luvic Gleyic Greyic Phaeozems (Anthric)
- agric dark-humus podbels – <u>soil type</u>. ≈ Umbric Stagnosols (*Anthric*)
- agric dark-humus gleyic podbels – <u>soil type</u>. ≈ Umbric Endogleyic Stagnosols (*Anthric*)
- agric solods – <u>soil type</u>. ≈ Solodic Planosols (*Anthric*)
- agric dark solods – <u>soil type</u>. ≈ Mollic Solodic Planosols (*Anthric*)
- agric gleyic grey-humus solods – <u>soil type</u>. ≈ Endogleyic Solodic Planosols (*Anthric*)
- agric dark hydrometamorphic solods – <u>soil type</u>. ≈ Mollic Endogleyic Solodic Planosols (*Anthric*)

<u>Al-Fe-humus soils</u> – <u>soil section</u>. ≈ Podzols

- podburs – <u>soil type</u>. ≈ Entic Podzols
- gleyic podburs – <u>soil type</u>. ≈ Gleyic Entic Podzols
- dry peat podburs – <u>soil type</u>. ≈ Histic Entic Podzols
- gleyic peat podburs – <u>soil type</u>. ≈ Gleyic Histic Entic Podzols
- grey-humus podburs – <u>soil type</u>. ≈ Umbric Entic Podzols
- gleyic grey-humus podburs – <u>soil type</u>. ≈ Umbric Gleyic Entic Podzols
- podzols – <u>soil type</u>. ≈ Podzols
- gleyic podzols – <u>soil type</u>. ≈ Gleyic Podzols
- dry peat podzols – <u>soil type</u>. ≈ Folic Podzols
- grey-humus podzols – <u>soil type</u>. ≈ Umbric Podzols
- gleyic peat podzols – <u>soil type</u>. ≈ Gleyic Histic Podzols
- gleyic grey-humus podzols – <u>soil type</u>. ≈ Umbric Gleyic Podzols
- agric grey-humus podzols – <u>soil type</u>. ≈ Umbric Podzols (Anthric)
- agric grey-humus gleyic podzols – <u>soil type</u>. ≈ Umbric Gleyic Podzols (Anthric)
- agric peat gleyic podzols – <u>soil type</u>. ≈ Gleyic Histic Podzols (Anthric)

<u>Iron metamorphic soils</u> – <u>soil section</u>. ≈ Brunic Arenosols

- rzhavozems (rusty earths) – <u>soil type</u>. ≈ Brunic Arenosols
- raw-humus rzhavozems – <u>soil type</u>. ≈ Brunic Arenosols
- organic rzhavozems – <u>soil type</u>. ≈ Folic Brunic Arenosols

<u>Structural metamorphic soils</u> – <u>soil section</u>. ≈ Umbrisols / Cambisols / Phaeozem

- burozems (brown earths) – <u>soil type</u>. ≈ Cambisols
- raw-humus burozems – <u>soil type</u>. ≈ *Greyic* Cambisols
- dark burozems – <u>soil type</u>. ≈ *Cambic* Phaeozems / Cambic Mollic Umbrisols
- grey metamorphic soils – <u>soil type</u>. ≈ Cambisols

- dark-grey metamorphic soils – <u>soil type</u>. ≈ *Cambic* Greyic Phaeozems
- eluvial metamorphic soils – <u>soil type</u>. ≈ *Cambic* Stagnosols / Stagnic Cambisols / Cambisols (*Albic*)
- grey-humus eluvial metamorphic soils – <u>soil type</u>. ≈ *Cambic* Stagnosols / Stagnic Cambisols / Cambisols (*Albic*)
- cinnamonic soils – <u>soil type</u>. ≈ *Cambic* Kastanozems (Chromic)
- agric eluvial metamorphic soils – <u>soil type</u>. ≈ *Cambic* Stagnosols / Stagnic Cambisols / Cambisols (*Albic*)
- agric grey metamorphic soils – <u>soil type</u>. ≈ Cambisols
- agric dark-grey metamorphic soils – <u>soil type</u>. ≈ *Cambic* Greyic Phaeozems
- agric cinnamonic soils – <u>soil type</u>. ≈ *Cambic* Kastanozems (Chromic)

<u>Cryometamorphic soils</u> – <u>soil section</u>. ≈ Cryosols

- cryometamorphic soils – <u>soil type</u>. ≈ Cambic Cryosols
- mud cryometamorphic soils – <u>soil type</u>. ≈ Cambic Histic Cryosols
- raw-humus cryometamorphic soils – <u>soil type</u>. ≈ *Greyic* Cryosols
- svetlozems (light earth) – <u>soil type</u>. ≈ Cambic Cryosols (*Albic*)
- iron-illuviated svetlozems – <u>soil type</u>. ≈ Spodic Cryosols (*Albic*)
- texture-differentiated svetlozems – <u>soil type</u>. ≈ *Luvic* Cryosols (*Albic*)

<u>Pale metamorphic soils</u> – <u>soil section</u>. ≈ Cryosols

- pale soils – <u>soil type</u>. ≈ Calcic Cryosols
- dark-humus pale soils – <u>soil type</u>. ≈ Calcic Mollic Cryosols
- cryoarid soils – <u>soil type</u>. ≈ Calcic Mollic Cryosols
- agric pale soils – <u>soil type</u>. ≈ Calcic Cryosols (*Anthric*)

<u>Cryoturbated soils</u> – <u>soil section</u>. ≈ Cryosols

- cryozems – <u>soil type</u>. ≈ Turbic Cryosols
- raw-humus cryozems – <u>soil type</u>. ≈ (*Greyic*) Turbic Cryosols
- peat cryozems – <u>soil type</u>. ≈ Histic Turbic Cryosols

<u>Gleyic soils</u> – <u>soil section</u>. ≈ Gleysols

- gleezems – <u>soil type</u>. ≈ Gleysols
- cryometamorphic gleezems – <u>soil type</u>. ≈ Gleysols (Gelic)
- peat gleezems – <u>soil type</u>. ≈ Histic Gleysols
- dark-humus gleyic soils – <u>soil type</u>. ≈ Mollic Gleysols
- mud gleyic soils – <u>soil type</u>. ≈ Mollic Gleysols (Humic)
- agric cryometamorphic gleezems – <u>soil type</u>. ≈ Gleysols (Gelic, *Anthric*)
- agric peat gleezems – <u>soil type</u>. ≈ Histic Gleysols (*Anthric*)
- agric dark-humus gleyic soils – <u>soil type</u>. ≈ Mollic Gleysols (*Anthric*)
- agric mud gleyic soils – <u>soil type</u>. ≈ Mollic Gleysols (Humic, *Anthric*)

Humus-accumulating soils – soil section. ≈ Chernozems / Phaeozems / Kastanozems

- clay-illuvial chernozems – soil type. ≈ Luvic Chernozems / Greyic Luvic Phaeozems
- chernozems – soil type. ≈ Chernozems
- textural carbonaceous chernozems – soil type. ≈ Vertic Chernozems (Glossic)
- black slitizated soils – soil type. ≈ Vertisols (Pellic)
- chernozem-like soils – soil type. ≈ Phaeozems
- agric clay-illuvial chernozems – soil type. ≈ Luvic Chernozems / Greyic Luvic Phaeozems
- agric chernozems – soil type. ≈ Chernozems
- agric textural carbonaceous chernozems – soil type. ≈ Vertic Chernozems (Glossic)
- agric slitizated soils – soil type. ≈ Vertisols (Pellic)
- agric chernozem-like soils – soil type. ≈ Anthric Phaeozems

Light-humus carbonate-accumulating soils – soil section. ≈ Calcisols

- kashtanovye (chestnut) soils – soil type. ≈ *Cambic* Calcisols (*Glossic*)
- brown (arid) soils – soil type. ≈ *Cambic* Calcisols
- light-humus carbonate-accumulating soils – soil type. ≈ Calcisols

Alkaline clay-differentiated soils – soil section. ≈ Solonetz

- dark solonetz – soil type. ≈ Mollic Solonetz
- light solonetz – soil type. ≈ Solonetz
- dark hydrometamorphic solonetz – soil type. ≈ Mollic Gleyic Solonetz
- light hydrometamorphic solonetz – soil type. ≈ Gleyic Solonetz
- agric dark solonetz – soil type. ≈ Mollic Solonetz
- agric light solonetz – soil type. ≈ Solonetz
- agric dark hydrometamorphic solonetz – soil type. ≈ Mollic Gleyic Solonetz (Humic)
- agric light hydrometamorphic solonetz – soil type. ≈ Gleyic Solonetz (Humic)

Halomorphic soils – soil section. ≈ Solonchaks

- solonchaks – soil type. ≈ Solonchaks
- gleyic solonchaks – soil type. ≈ Gleyic Solonchaks
- sulphidic solonchaks – soil type. ≈ Solonchaks (*Protothionic*)
- dark solonchaks – soil type. ≈ Mollic Solonchaks
- peat solonchaks – soil type. ≈ Gleyic Histic Solonchaks
- secondary solonchaks – soil type. The type includes soils, where strong secondary salinization is overlaid on a developed profile of any other soil.

Subtypes are distinguished according to the primary profile, e.g. *chernozemic secondary solonchaks* or *chestnut secondary solonchaks*. (Solonchaks with corresponding qualifiers or *salic* qualifier in the reference groups that are listed in the key before Solonchaks in WRB.)

Hydrometamorphic soils – soil section. ≈ Gleysols

- humus hydrometamorphic soils – soil type. ≈ Mollic Gleysols
- mud hydrometamorphic soils – soil type. ≈ Histic Gleysols (Eutric)
- agric hydrometamorphic soils – soil type. ≈ Mollic Gleysols
- agric mud hydrometamorphic soils – soil type. ≈ Histic Gleysols (Eutric)

Organic matter accumulating soils – soil section. ≈ Phaeozems / Umbrisols / Regosols

- grey-humus soils – soil type. ≈ Umbrisols / Regosols (Dystric)
- dark-humus soils – soil type. ≈ Phaeozems
- mud soils – soil type. ≈ Histic Umbrisols
- mud dark-humus soils – soil type. ≈ Histic Phaeozems / Mollic Histic Umbrisols
- light-humus soils – soil type. ≈ Regosols (Eutric)
- agric humus soils – soil type. ≈ Umbrisols (Anthric)
- agric dark-humus soils – soil type. ≈ Phaeozems

Eluvial soils – soil section. ≈ Planosols

- eluvozems – soil type. ≈ Planosols
- gleyic eluvozems – soil type. ≈ Endogleyic Planosols
- grey-humus eluvozems – soil type. ≈ Planosols / Umbric Planosols
- grey-humus gleyic eluvozems – soil type. ≈ Endogleyic Planosols / Umbric Endogleyic Planosols
- peat gleyic eluvozems – soil type. ≈ Endogleyic Histic Planosols
- podzol-eluvozems – soil type. ≈ Planosols (Albic, Ruptic)
- peat gleyic podzol-eluvozems – soil type. ≈ Histic Planosols (Albic, Ruptic)
- agric grey-humus eluvozems – soil type. ≈ Planosols / Umbric Planosols
- agric grey-humus gleyic eluvozems – soil type. ≈ Endogleyic Planosols / Umbric Endogleyic Planosols
- agric peat gleyic eluvozems – soil type. ≈ Endogleyic Histic Planosols

Lithozems – soil section. ≈ Leptosols / Leptic Histosols

- peat lithozems – soil type. ≈ Leptic Histosols
- dry peat lithozems – soil type. ≈ Leptic Folic Histosols
- raw-humus lithozems – soil type. ≈ *Greyic* Leptosols
- mud lithozems – soil type. ≈ Umbric Leptosols

- mud carbo-lithozems – <u>soil type</u>. ≈ Umbric Leptosols (Calcaric)
- grey-humus lithozems – <u>soil type</u>. ≈ Leptosols (Dystric) / Umbric Leptosols
- mud dark-humus lithozems – <u>soil type</u>. ≈ Mollic Leptosols
- dark-humus lithozems – <u>soil type</u>. ≈ Mollic Leptosols
- dark-humus carbo-lithozems (rendzinas) – <u>soil type</u>. ≈ Rendzic Leptosols
- light-humus lithozems – <u>soil type</u>. ≈ Leptosols (Eutric)
- agric dark-humus lithozems – <u>soil type</u>. ≈ Mollic Leptosols

<u>Weakly developed soils</u> – <u>soil section</u>. ≈ Regosols, Arenosols, Leptosols

- pelozems – <u>soil type</u>. ≈ Regosols (Clayic)
- humus pelozems – <u>soil type</u>. ≈ Regosols (Clayic)
- psammozems – <u>soil type</u>. ≈ Arenosols
- humus psammozems – <u>soil type</u>. ≈ Arenosols
- petrozems – <u>soil type</u>. ≈ Leptosols
- humus petrozems – <u>soil type</u>. ≈ Leptosols
- carbo-petrozems – <u>soil type</u>. ≈ Leptosols (Calcaric)
- gypso-petrozems – <u>soil type</u>. ≈ Leptosols (Gypsiric)
- humus gypso-petrozems – <u>soil type</u>. ≈ Leptosols (Gypsiric)

<u>Abrazems</u> – <u>soil section</u>. ≈ (partially Regosols, but in fact any soil group may enter, depending on the exposed horizon) – strongly eroded soils; no direct equivalent in WRB.

- texture differentiated abrazems – <u>soil type</u>. ≈ *Nudiargic* Luvisols
- Al-Fe-humus abrazems – <u>soil type</u>. ≈ Entic Podzols
- iron metamorphic abrazems – <u>soil type</u>. ≈ Brunic Arenosols
- structural metamorphic abrazems – <u>soil type</u>. ≈ Cambisols
- cryometamorphic abrazems – <u>soil type</u>. ≈ Cambic Cryosols
- pale metamorphic abrazems – <u>soil type</u>. ≈ Cambic Calcic Cryosols
- clay-illuvial abrazems – <u>soil type</u>. ≈ *Nudiargic* Luvisols
- carbonate-accumulative abrazems – <u>soil type</u>. ≈ Calcisols / Calcic Cambisols / Calcic Regosols
- textural carbonaceous abrazems – <u>soil type</u>. ≈ Vertic Calcisols / Calcic Vertic Cambisols / Calcic Vertic Regosols
- solonetz abrazems – <u>soil type</u>. ≈ Sodic Regosols

<u>Agrozems</u> – <u>soil section</u>. ≈ Anthrosols

- light agrozems – <u>soil type</u>. ≈ Regic Anthrosols
- dark agrozems – <u>soil type</u>. ≈ Regic Anthrosols
- dark gleyic agrozems – <u>soil type</u>. ≈ Gleyic Anthrosols
- dark hydrometamorphic agrozems – <u>soil type</u>. ≈ Gleyic Anthrosols
- peat agrozems – <u>soil type</u>. ≈ *Histic* Anthrosols
- peat-mineral agrozems – <u>soil type</u>. ≈ *Histic* Anthrosols
- texture differentiated agrozems – <u>soil type</u>. ≈ Luvic Anthrosols

- texture differentiated gleyic agrozems – <u>soil type</u>. ≈ Luvic Gleyic Anthrosols
- Al-Fe-humus agrozems – <u>soil type</u>. ≈ Spodic Anthrosols
- Al-Fe-humus gleyic agrozems – <u>soil type</u>. ≈ Spodic Gleyic Anthrosols
- structural metamorphic light agrozems – <u>soil type</u>. ≈ *Cambic* Anthrosols
- structural metamorphic dark agrozems – <u>soil type</u>. ≈ *Cambic* Anthrosols (Humic)
- clay-illuvial agrozems – <u>soil type</u>. ≈ Luvic Anthrosols
- dark carbonate-accumulative agrozems – <u>soil type</u>. ≈ *Calcic* Anthrosols
- textural carbonaceous agrozems – <u>soil type</u>. ≈ Vertic Anthrosols
- dark solonetz agrozems – <u>soil type</u>. ≈ *Sodic* Anthrosols
- light solonetz agrozems – <u>soil type</u>. ≈ *Sodic* Anthrosols
- dark solonetz hydrometamorphic agrozems – <u>soil type</u>. ≈ *Sodic* Gleyic Anthrosols
- light solonetz hydrometamorphic agrozems – <u>soil type</u>. ≈ *Sodic* Gleyic Anthrosols

<u>Agroabrazems</u> – <u>soil section</u>. Cultivated soils, previously affected with intensive erosional processes; no direct equivalents in WRB. ≈ Anthrosols

- agroabrazems – <u>soil type</u>. ≈ Anthrosols
- gleyic agroabrazems – <u>soil type</u>. ≈ Gleyic Anthrosols
- hydrometamorphic agroabrazems – <u>soil type</u>. ≈ Gleyic Anthrosols
- texture differentiated agroabrazems – <u>soil type</u>. ≈ *Nudiargic* Luvic Anthrosols
- texture differentiated gleyic agroabrazems – <u>soil type</u>. ≈ Anthrosols
- Al-Fe-humus agroabrazems – <u>soil type</u>. ≈ Spodic Anthrosols
- Al-Fe-humus gleyic agroabrazems – <u>soil type</u>. ≈ Spodic Gleyic Anthrosols
- structural metamorphic agroabrazems – <u>soil type</u>. ≈ *Cambic* Anthrosols
- structural metamorphic gleyic agroabrazems – <u>soil type</u>. ≈ *Cambic* Gleyic Anthrosols
- clay-illuvial agroabrazems – <u>soil type</u>. ≈ *Nudiargic* Luvic Anthrosols
- carbonate-accumulative agroabrazems – <u>soil type</u>. ≈ *Calcic* Anthrosols
- textural carbonaceous agroabrazems – <u>soil type</u>. ≈ *Calcic Vertic* Anthrosols
- solonetz agroabrazems – <u>soil type</u>. ≈ *Sodic* Anthrosols

<u>Turbated soils</u> – <u>soil section</u>. ≈ Sodic Plaggic Anthrosols

- dark post-solonetz turbozems – <u>soil type</u>. ≈ Mollic Sodic Plaggic Anthrosols
- light post-solonetz turbozems – <u>soil type</u>. ≈ Sodic Plaggic Anthrosols
- dark post-solonetz hydrometamorphic turbozems – <u>soil type</u>. ≈ Mollic Sodic Gleyic Plaggic Anthrosols
- light post-solonetz hydrometamorphic turbozems – <u>soil type</u>. ≈ Sodic Gleyic Plaggic Anthrosols

<u>Sinlithogenic soils</u> – a <u>trunk of soils</u>, gathering together the soils, where soil formation occurs in the condition of periodical or continuous accumulation of new material on the surface. ≈ Fluvisols / Andosols / Cambisols / Regosols / Arenosols

<u>Alluvial soils</u> – <u>soil section</u>. ≈ Fluvisols

- grey-humus alluvial soils – <u>soil type</u>. ≈ Fluvisols (Dystric) / Umbric Fluvisols
- dark-humus alluvial soils – <u>soil type</u>. ≈ Mollic Fluvisols
- alluvial peat gleyic soils – <u>soil type</u>. ≈ Gleyic Histic Fluvisols
- mud gleyic alluvial soils – <u>soil type</u>. ≈ Mollic Fluvisols / Umbric Fluvisols
- light-humus alluvial soils – <u>soil type</u>. ≈ Fluvisols (Eutric)
- dark-humus hydrometamorphic alluvial soils – <u>soil type</u>. ≈ Mollic Gleyic Fluvisols
- alluvial slitizated soils – <u>soil type</u>. ≈ Vertic Fluvisols
- agric humus alluvial soils – <u>soil type</u>. ≈ Umbric Fluvisols
- agric dark-humus alluvial soils – <u>soil type</u>. ≈ Mollic Fluvisols
- agric peat gleyic alluvial soils – <u>soil type</u>. ≈ Gleyic Histic Fluvisols
- agric humus gleyic alluvial soils – <u>soil type</u>. ≈ Umbric Gleyic Fluvisols
- agric humus hydrometamorphic alluvial soils – <u>soil type</u>. ≈ Mollic Gleyic Fluvisols (Anthric)
- agric slitizated alluvial soils – <u>soil type</u>. ≈ Vertic Fluvisols

<u>Volcanic soils</u> – <u>soil section</u>. ≈ Andosols

- volcanic ochreous soils – <u>soil type</u>. ≈ Andosols
- mud ochreous volcanic soils – <u>soil type</u>. ≈ Umbric Andosols
- ochreous podzolic volcanic soils – <u>soil type</u>. ≈ *Spodic* Andosols
- agric ochreous soils – <u>soil type</u>. ≈ Andosols (Anthric)

<u>Stratozems</u> – <u>soil section</u>. Includes all the soils with periodical aeolian or fluvial accumulation of sediments (including long-term irrigation) with no or weak development of soil profile on the surface. (No direct equivalents in WRB; mainly correspond to Irragric or Terric Anthrosols, and in some cases any other soil with a modifier *colluvic*.)

- grey-humus stratozems – <u>soil type</u>. ≈ Irragric Anthrosols (Dystric)
- dark-humus stratozems – <u>soil type</u>. ≈ Irragric Anthrosols (Humic)
- light-humus stratozems – <u>soil type</u>. ≈ Irragric Anthrosols (Eutric)
- grey-humus stratozems on a buried soil – <u>soil type</u>. ≈ Irragric Anthrosols (Dystric)
- dark-humus stratozems on a buried soil – <u>soil type</u>. ≈ Irragric Anthrosols (Humic)
- light-humus stratozems on a buried soil – <u>soil type</u>. ≈ Irragric Anthrosols (Eutric)
- agric humus stratozems – <u>soil type</u>. ≈ Irragric Anthrosols (Humic)

- agric dark-humus stratozems – <u>soil type</u>. ≈ Irragric Anthrosols (Humic) / Terric Anthrosols
- agric humus stratozems on a buried soil – <u>soil type</u>. ≈ Irragric Anthrosols (Humic)
- agric dark-humus stratozems on a buried soil – <u>soil type</u>. ≈ Irragric Anthrosols (Humic) / Terric Anthrosols

<u>Weakly developed soils</u> – <u>soil section</u>. ≈ Fluvisols, Regosols, Arenosols

- stratified alluvial soils – <u>soil type</u>. ≈ *Protic* Fluvisols
- stratified ash volcanic soils – <u>soil type</u>. ≈ Tephric Regosols / Tephric Arenosols

<u>Organogenic soils</u> – a <u>trunk of soils</u>, including the soils formed in organic material, mainly in peat. ≈ Histosols
<u>Peat soils</u> – <u>soil section</u>. ≈ Histosols

- peat oligotrophic soils – <u>soil type</u>. ≈ Dystri-Fibric Histosols
- peat eutrophic soils – <u>soil type</u>. ≈ Eutric Histosols
- dry peat soils – <u>soil type</u>. ≈ Folic Histosols

<u>Torfozems</u> – <u>soil section</u>. ≈ Histosols (Drainic, *Anthric*)

- torfozems – <u>soil type</u>. ≈ Histosols (Dystric, Drainic, *Anthric*)
- agro-mineral torfozems – <u>soil type</u>. ≈ Histosols (Eutric, Drainic, *Anthric*)

References

Anonymous (2008) *Field Guide for Russian Soils*, Dokuchaev Soil Science Institute, Moscow, 182pp

Arnold, R. W. (ed) (2001) *Russian Soil Classification System*, Oykumena, Smolensk, 214pp

Egorov, V. V., Fridland, V. M., Ivanova, E. N., Rozov, N. N., Nosin, V. A. and Fraev, T. A. (1977) *Classification and Diagnostics of Soils of USSR*, Kolos Press, Moscow, 221pp (in Russian)

Fridland, V. M. (1982) *The Main Principles and Elements of Basic Soil Classification and a Working Plan for its Construction*, Dokuchaev Soil Science Institute, Moscow, 149pp (in Russian)

IUSS Working Group WRB (2006) *World Reference Base for Soil Resources*, 2nd edition, World Soil Resources Reports no 103, UN Food and Agriculture Organization, Rome, 128pp

Lebedeva, I. I., Tonkonogov, V. D. and Shishov, L. L. (1996) 'Agrogenically transformed soils: Evolution and taxonomy', *Eurasian Soil Science*, vol 29, no 3, pp214–219

Shishov, L. L., Tonkonogov, V. D. and Lebedeva, I. I. (1997) *Classification of Soils of Russia*, Dokuchaev Soil Science Institute, Moscow, 236pp

Shishov, L. L., Tonkonogov, V. D., Lebedeva, I. I. and Gerasimova, M. I. (2004) *Classification and Diagnostics of Soils of Russia*, Oykumena, Smolensk, 342pp (in Russian)

Shishov, L. L., Tonkonogov, V. D., Gerasimova, M. I. and Lebedeva, I. I. (2005) 'New classification system of Russian soils', *Eurasian Soil Science*, vol 38, suppl 1, pp6–12

21
Soil Classifications of the New Independent States

The reasons for multiplication of classifications

Many new independent states of the post-Soviet territory published their own classifications. The source system for all these classifications was, to a great extent, *Classification and Diagnostics of Soils of USSR* (Egorov et al, 1977); however, both the structure and the list of soil groups differed between countries. It depended on several factors. First, the soil cover of every country is different, and the lists of soils are narrower than that of the Soviet classification. Second, each region in the former Soviet Union had a particular school of pedology, and some regional schools had different points of view on soil genesis and classification from those of the main Moscow institutes and universities. Finally, after the fall of the Soviet Union the pedologists of the new independent states tried to accept some novel ideas of the Western classifications, such as the United States Department of Agriculture (USDA) Soil Taxonomy (Soil Survey Staff, 1999) and the World Reference Base for Soil Resources (WRB) (IUSS Working Group WRB, 2006).

Not all the new independent states developed their own classifications; thus, 'by default' the older Soviet classification (Egorov et al, 1977) is still used in places. The countries that published their own classifications are Azerbaijan, Belorussia, Estonia, Latvia, Lithuania and Ukraine. The countries which did not develop their own classification systems, or at least did not publish any official documents on their national classifications, are Armenia, Georgia, Moldova, Kazakhstan, Kirgizia, Tadzhikistan, Turkmenia and Uzbekistan.

Brief reviews of the new classifications are provided but because most archetypes are borrowed from the old Soviet classification discussed in Chapter 19, a complete correlation of terms is not made with the WRB (IUSS Working Group WRB, 2006). In doubtful cases additional comments are made. For further details one may consult the original texts of classification as most of them include a correlation of terms with those of the WRB.

Azerbaijan

The Azerbaijani soil classification has a long history, starting from the 1920s (Babaev et al, 2006). A number of regional classification schemes had been proposed in the Soviet period, but most of them did not develop to the stage of a real classification. A systematic list of the soils of Azerbaijan was made in the 1960s, and it was used in preparing the Soviet soil classification (Egorov et al, 1977) and as a legend of the 1:200,000 soil map of Azerbaijan. In the period of independence, a complete version of soil classification was published (Salaev, 1999). It was based mainly on the same map legend but included more detailed division of taxa and a system of diagnostics and was successfully used for large-scale soil mapping. Recently, a new soil classification was proposed for Azerbaijan (Babaev et al, 2006). The new classification has a structure similar to that of the new Russian taxonomy (Shishov et al, 2004), because both taxonomies are rooted in the early 1980s, when the work on a new USSR soil classification started. The Azerbaijani classification has two collective levels. On the highest level there are three soil classes; *naturally evolved, anthropogenically transformed* and *technogenically transformed* soils. In fact, it is the only classification where agricultural and technogenic transformations are equivalent to pedogenesis at the highest level. The high priority of technogenic transformation and contamination is related to vast areas transformed by mining and petroleum industries in the country. Even some names of soil types reflect contamination by various products of the petroleum industry.

The class of naturally evolved soils includes the sections of *organic matter accumulative, textural-differentiated, humus-accumulative, carbonate-accumulative low in humus, metamorphic, alluvial* and *halomorphic* soils. The class of anthropogenically transformed soils includes irrigated soils with the following sections: *textural-differentiated, alluvial, humus-accumulative, carbonate-accumulative* and *irrigation-accumulative soils*. The class of technogenically transformed soils includes the sections of *technogenically disturbed* and *mining* soils. In every section there are from one to five soil types. The names of soil types are traditional, similar to those used in the Soviet classification (Egorov et al, 1977). The differentiating characteristics are based on the morphology of the soil with almost no chemical laboratory analyses needed. A system of diagnostic horizons and properties is applied but the definitions are mainly qualitative. Landscape, climatic criteria and water regimes are disregarded; the only exception is artificial irrigation which is diagnostic for anthropogenically transformed soils. The authors say that the classification is still under development (Babaev, 2006), and more research is needed, especially on technogenically transformed soils.

Belorussia

During the Soviet Union period, Belorussia had a strong regional school which had been developing since the 1920s. The study of peat soils was of special importance with focus on practical questions of peatland drainage and improvement. Soil classification was of minor importance. Recently the inventory of soil resources and scientific research has required a classification system, and a new Belorussian soil classification was published (Romanova, 2004). This classification was proposed by Romanova but has not yet been officially approved for soil mapping or for education. This classification, unlike most authorial schemes, has diagnostics and a complete taxonomic structure.

The Belorussian classification is unique in that it is based mostly on soil hydrology. At a domain level the soils are divided into alluvial and non-alluvial ones; this separation is not a hierarchical level as these groups are classified separately. The classification of each group includes eight taxonomic levels: soil classes, types, subtypes, genus, subgenus, species, subspecies and variants. The three *classes* are automorphous, subhydromorphous and hydromorphous soils that resemble the division proposed by Kubiëna (1953). Soil types are traditional, but also include a characteristic of their water regime: *underdeveloped soils*, *humus-carbonate soils*, *brown forest soils*, *pale soils*, *humus-carbonate waterlogged soils*, *pseudopodzolic soils*, *mud-gley soils*, *peat minerotrophic soils*, *peat transitional soils*, *peat oligotrophic soils*, *podzolic soils* and *humus waterlogged soils*. The subtypes are characterized by the water saturation period of the 0–20cm topsoil layer. The lower taxonomic levels reflect soil texture, mineralogical composition and additional characteristics of soil use and degradation. The text of the classification (Romanova, 2004) includes an approximate correlation of terms with the WRB.

Estonia

The Baltic States traditionally had developed independent schools of soil science, even in the Soviet period. Estonia was well known for the studies of soil biochemistry, the chemistry of microelements and also of soil genesis and geography. Soil classification was of minor importance for Estonian pedologists. For soil survey, the uniform Soviet classification was used (Egorov et al, 1977); however, in the scientific community a different classification had been applied. In the 1970s an extensive discussion was launched in Europe and in the Soviet Union about the relation between podzolization, pseudopodzolization, gleying and pseudogleying processes (e.g. Reintam, 1974), and a number of new taxa were proposed. The 'mainstream' Soviet school represented by the universities and institutes of Moscow and St Petersburg did not accept the new concepts, but in the Baltic States where the soils with surface gleying and clay illuviation features were widespread, the concepts of 'pseudogley' and 'pseudopodzolic soil' were used in their regional classification, thereby differing from the official one (Egorov et al, 1977). Until recently Estonia used the classification developed in the Soviet period, but since it was not officially approved, different institutions had slightly different versions of classifications (Kõlli, 1998). During the last decade, much was done to unify and formalize a classification in Estonia (Reintam, 2002; Reintam and Köster, 2006) though an official publication still is not available. The current classification has two or three taxonomic levels depending on the version. The first level includes soil types, and the lower levels divide the soils according their depth, humus content and so on. The WRB is widely used for soil research and mapping in Estonia, both as a correlation tool and as a map legend (Reintam and Köster, 2006).

Latvia

Latvia traditionally had a particular school of soil science; the classification used in Latvian universities was also different from that officially used for soil survey in the Soviet Union (see comments in Chapter 31). As an independent state, Latvia started work to develop its own soil classification which had to follow the traditional school but also correlate well with the European system. A group of researchers of the Latvian National University gradually improved the national system, and recently published a final complete version of it (Kārkliņš et al, 2009). The Latvian taxonomy has three hierarchical levels: the highest collective level is <u>class</u>, separated by the criterion of hydromorphism, the second – generic – level is soil <u>type</u>, followed by <u>subtypes</u> (specific level). The terminology is completely traditional. The differentiating characteristics are mostly qualitative, and much more flexible than that in the WRB (IUSS Working Group WRB, 2006); however, similar criteria are used that permit successful correlation of terms. The main difficulty for the Latvian scientists was to have a classification that worked well under the umbrella of the WRB, but also made possible the use of older soil maps without confusing the users of soil information (farmers and advisors). Generally the classification is simple, logical and easy to use, and may be considered a good example of an adequate classification for a relatively small state.

The class of automorphic soils includes the types of *Sod calcareous soils*, *Brown soils*, *Podzolic soils*, *Podzols*, *Weakly developed soils* and *Anthrosols*. The class of semihydromorphic soils includes *Gley*, *Podzolic-gley* and *Alluvial soils*. The class of hydromorphic soils includes *Fen peat*, *Transitional mire* and *Raised bog soils*. The correlation of all the terms down to the subtype level with the WRB is provided in the text of the classification (Kārkliņš et al, 2009).

Lithuania

Lithuania has long traditions in soil classification starting in the 16th century when a land reform was made in the Grand Duchy of Lithuania (Buivydaite, 2002). In the Soviet period, the Lithuanian soil science school was strong especially in research about forest soils. As an independent state, Lithuania started a discussion of a national soil classification system which was adopted in 1999 (Buivydaite, 2002). The system has two priorities: to reflect adequately the soil cover of the country and to correlate well with the umbrella WRB system to fit the format of the European soil databases. The WRB reference groups are integrated at the highest level of the national taxonomy, thus the national system is not replaced by the WRB but is harmonized at the highest taxonomic level. In Lithuania, there are 12 <u>major soil groups</u>, which correspond to the WRB groups and almost completely repeat the names of reference groups: *Histosols*, *Anthroposols* (the only group name differing from the original name *Anthrosols* in the WRB), *Leptosols*, *Fluvisols*, *Gleysols*, *Podzols*, *Planosols*, *Albeluvisols*, *Luvisols*, *Cambisols*, *Arenosols* and *Regosols*. On the second level of taxonomy there are 46 <u>subgroups</u>. On the lower levels the taxonomy becomes flexible and fuzzy: the basic concept of the classification is *soil typological unit*, which may be found at the third, fourth and even fifth levels of the hierarchy. The hierarchical levels relate to the number of modifiers used for characterizing soils. The modifiers are mostly the same as recommended by the WRB, but some of them have different forms and are combined in a way different from that recommended by the WRB (IUSS Working Group WRB, 2006), for example, *hypergleyic*, *hypostagnic* or *orthieutric*. Two parallel terminologies are used: each soil typological unit has a WRB-like 'international' name and a name in Lithuanian. It is interesting to note that soil names should be approved by the Lithuanian parliament. The differentiating soil characteristics are close to those used by the WRB, and are based on quantitative morphological and chemical properties of soils.

Ukraine

The Ukrainian school of pedology was always different from mainstream Soviet soil science. Soil research in the republic was based on early works of Sokolovsky (1932), which included a number of original concepts, and even the indexes for soil horizons were different. After the fall of the Soviet Union, there was a period of stagnation of soil research in Ukraine until the end of the 1990s. The Ukrainian pedologists published a new national soil classification in 2005 (Polupan et al, 2005). The classification is very detailed because Ukraine is a large country with many highly productive soils. The classification is a hierarchical taxonomy with six levels: <u>types</u>, <u>subtypes</u>, <u>genus</u>, <u>species</u>, <u>variants</u> and <u>lithological series</u>. Soil type is a generic taxon, subtype is a specific level, and lower taxa are varietal. Lithological series do not correspond to soil series in the US Soil Taxonomy (Soil Survey Staff, 1999) but reflect the origin of parent material, thus resembling a similar level in the Soviet classification (Egorov et al, 1977) or of substrate or series in the German classification (Ad-hoc-AG Boden, 2005). There are 54 types in the Ukrainian classification, and for convenience they are grouped into provisional units which are not regarded as a special taxonomic level although they operate as one. They are *automorphic with humus-accumulative profile*, *podzolized with humus-accumulative profile*, *podzolized with textural-differentiated profile*, *halogenic with textural-differentiated profile*, *soils with surface and combined surface and ground gleying*, *soils with surface gleying and humus-accumulative profile*, *soils with ground gleying and humus-accumulative profile*, *alluvial*, *organic natural and natural-anthropogenic soils* and *replanted soils*. The characteristic soil properties at the highest levels of the Ukrainian taxonomy are semiquantitative based on a system of diagnostic genetic horizons; at the lower levels they are quantitative. Field morphology and basic chemical and physical analyses are required; special attention is given to humus content and distribution in the profile. Water regime is used at the subtype level because waterlogging does not produce morphologically evident gleying in soils rich in dark humus. The terminology of the classification is completely traditional.

References

Ad-hoc-AG Boden (2005) *Bodenkundliche Kartieranleitung*, herausgegeben von der Bundesanstalt für Geowissenschaften und Rohstoffe und den Geologischen Landesämtern in der Bundesrepublik Deutschland, Hannover, 5 Auflage, 438pp (in German)

Babaev, M. P., Dzhafarova, Ch. M. and Gasanov, V. G. (2006) 'Modern Azerbaijani soil classification system', *Eurasian Soil Science*, vol 39, no 11, pp1176–1182

Buivydaite, V. V. (2002) 'Classification of soils of Lithuania based on FAO-UNESCO soil classification system and WRB', *Transactions of the 17th World Congress of Soil Science*, Bangkok, Thailand, 14–21 August 2002, CD, pp2189-1–2189-13

Egorov, V. V., Fridland, V. M., Ivanova, E. N., Rozov, N. N., Nosin, V. A. and Fraev, T. A. (1977) *Classification and Diagnostics of Soils of USSR*, Kolos Press, Moscow, 221pp (in Russian)

IUSS Working Group WRB (2006) *World Reference Base for Soil Resources*, 2nd edition, World Soil Resources Reports no 103, UN Food and Agriculture Organization, Rome, 128pp

Kārkliņš, A., Gemste, I., Mežals, H., Nikodemus, O. and Skujāns, R. (2009) *Latvijas augšņu noteicējs (Taxonomy of Latvian soils)*, LLU, Jelgava, 240pp (in Latvian, English summary)

Kõlli, R. (1998) 'Problems of Estonian soil classification', *Problems of Estonian Soil Classification*, Transactions of Estonian Agricultural University, vol 198, pp9–23 (in Estonian, English summary)

Kubiëna, W. L. (1953) *Bestimmungsbuch und Systematik der Böden Europas*, Verlag Enke, Stuttgart, 392pp (in German)

Polupan, M. I., Solovey, V. B. and Velichko, V. A. (2005) *Classification of Ukrainian Soils*, Agrarna nauka, Kiev, 300pp (in Ukrainian)

Reintam, L. (1974) 'Burozem formation and pseudopodzolization in the soils of the Estonian SSR', in S. V. Zonn, (ed.) *Burozem Formation and Pseudopodzolization in the Soils of the Russian Plain*, Nauka Publishers, Moscow, pp118–161 (in Russian)

Reintam, L. (2002) 'Correlation of the diagnostic properties of soil genetic units for harmonization of soil map units', in E. Michéli, F. O. Nachtergaele, R. J. A. Jones and L. Montanarella (eds), *Soil Classification 2001*, European Soil Bureau Research Report no 7, pp205–210

Reintam, E. and Köster, T. (2006) 'The role of chemical indicators to correlate some Estonian soils with WRB and Soil Taxonomy criteria', *Geoderma*, vol 136, no 2, pp199–209

Romanova, T. A. (2004) *Diagnostics of Soils in Belarus and their Classification in FAO-WRB System*, Institute for Soil Science and Agrochemistry, National Academy of Sciences of Belarus, Minsk, 428pp (in Russian)

Salaev, M. E. (1999) *Diagnostics and Classification of Soils of Azerbaijan*, Elm Publishers, Baku, 238pp (in Russian)

Shishov, L. L., Tonkonogov, V. D., Lebedeva, I. I. and Gerasimova, M. I. (2004) *Classification and Diagnostics of Soils of Russia*, Oykumena, Smolensk, 342pp (in Russian)

Soil Survey Staff (1999) *Soil Taxonomy: A Basic System of Soil Classification for Making and Interpreting Soil Surveys*, USDA, Handbook 436, 2nd edition, United States Government Printing Office, Washington DC, 696pp

Sokolovsky, A. N. (1932) 'The nomenclature of the genetic horizons of the soil', in *Proceeding Papers of the Second International Congress of Soil Science, Commission V, Classification, Geography and Cartography of Soils*, State Publishing House of Agricultural, Cooperative and Collective Farm Literature, Moscow, pp153–154

22
Soil Classification of Israel

Objectives and scope

Israel is a country where a desert environment has been converted into flourishing gardens. The task required good knowledge of soil resources. Soil science in Israel is well known all over the world both for developments in basic soil research and in applied studies. The soil classification of Israel is very pragmatic. It has been used both for scientific purposes and for soil survey at the national scale (Table 22.1) for almost half a century with minor corrections. The soils and sediments not used in agriculture, such as bare rock, underwater deposits and soils of urban and industrial areas, are not included in the classification. Soils deeply transformed by agriculture are not regarded as a separate soil group.

Table 22.1 *The scope of soil classification of Israel*

Superficial bodies	Representation in the system
Natural soils	National coverage
Urban soils	Not included in the classification
Man-transported materials	Not included in the classification
Bare rock	Not included in the classification
Subaquatic soils	Not included in the classification
Soils deeply transformed by agricultural activities	Classified as if they are natural soils

Theoretical background

Soil classification of Israel is based on soil genesis. However, from the very beginning it used measurable objective characteristics for differentiating soils, mainly those of B horizons. An important feature of the approach is that it follows folk tradition, and many archetypes of soils are traditional ones, derived from local historical soil names and concepts. Something resembling the soil series concept is used, though the Israeli concept is somewhat broader.

The lower level of the classification, soil types, is regarded as an open system, that is, depending on the recognition of additional types in soil surveys.

Development

More than 40 years ago two proposals for the soil classification of Israel were published (Ravikovitch, 1960; Dan et al, 1962) and since then two slightly different soil classification systems have existed in the country. The official classification was developed by a committee of the Ministry of Agriculture chaired by D. H. Yaalon in the 1970s. It was used in soil surveys, and the national soil map was produced on its basis (Dan et al, 1976). Its latest version was published 30 years ago (Dan and Koyumdjisky, 1979). The alternative classification was used unofficially until the 1990s, and even some soil maps of Israel were produced on its basis (Ravikovitch, 1992); the later publication was, in fact, a reprint of an old book initially published in Hebrew. The official version developed by the Ministry of Agriculture is used in Israel both for soil mapping and in scientific research (Singer, 2007).

Structure

The structure of the soil classification has four levels (Table 22.2). The upper generic level is <u>great groups</u> that include the main archetypes. The level of <u>subgroups</u> is specific and certain properties of soils indicate intergrades between two great groups or are related to an overlapping soil process. The <u>family</u> level provides additional, mostly qualitative, information about soil morphology and chemistry. <u>Soil types</u> are recognized by texture, carbonate content and topographical position; this level is open to the inclusion of new entities if such are found during soil surveys.

Table 22.2 *The structure of soil classification of Israel*

Level	Taxon name	Taxon characteristics	Borders between classes	Diagnostics	Terminology
0	Soils	Kingdom			
1	Great group	Generic 1	Semiquantitative	Chemico-morphological	Traditional
2	Subgroup	Specific	Semiquantitative	Chemico-morphological	Traditional
3	Family	Varietal	Semiquantitative	Chemico-morphological	Traditional
4	Soil type	Generic 2	Semiquantitative	Landscape-physico-chemical	Traditional

Diagnostics

The main source of soil diagnostics in the Israeli soil classification is field morphology. For some groups, laboratory data such as texture, carbonate content and salinity are essential. For some horizons, micromorphological study is recommended, but because the preparation of thin sections is expensive and time-consuming it is applied only for selected reference profiles. At the level of soil types certain external criteria, for example, the position in the landscape, are taken into account. In general the diagnostic characteristics can be called chemico-morphological.

Terminology

The names of the highest levels of Israel's soil taxonomy are traditional. Some terms are folk Israeli soil names, such as *Hamra, Husmas* or *Nazzaz*. Some soil names derived from the Russian school of pedology are used (*Krasnozem, Sierozem*), and some other terms are borrowed from various European classifications (*Grumosols, Terra Rossa, Rendzina,* etc.). Chromatic characteristics are widely used at the second level of the taxonomy.

Correlation

The correlation of the soil classification of Israel with the World Reference Base for Soil Resources (WRB) is not very easy because most archetypes are different in the two classifications. A lot of overlapping occurs, and for many great groups the correlation is multiple, making understanding of the concept difficult. This correlation is based on the text prepared by the Office of Arid Lands Studies of the University of Arizona, Tucson, AZ, USA. This text, presented on the Internet (website http://ag.arizona.edu/OALS/IALC/soils/israel/correlation.html), includes a correlation table with the US Soil Taxonomy and the 1975 legend of the FAO-UNESCO World Soil Map. Professor D. H. Yaalon checked this table and we did not make any radical changes; rather we adjusted the correlation to the latest WRB edition (IUSS Working Group WRB, 2006).

Alluvial soils (weathered) ≈ Fluvic Cambisols
There are four subgroups within the great group:

- *Brown Alluvial soils* ≈ Fluvic Cambisols (Calcaric)
- *Grumusolic Alluvial soils* ≈ Fluvic Vertic Cambisols
- *Hamric Alluvial soils* ≈ Fluvic Cambisols (Eutric)
- *Weathered Alluvium* ≈ Fluvic Cambisols

Brown (and dark brown) soils ≈ Luvisols / Cambisols / Fluvisols
There are six subgroups within the great group:

- *Dark brown soils* ≈ Calcic Vertic Luvisols
- *Grumusolic dark brown soils* ≈ Vertic Luvisols
- *Husmas* ≈ Calcic Luvisols
- *Light brown soils* ≈ Calcic Luvisols / Cambisols (Calcaric, Yermic) / Cambisols (Calcaric, Aridic)

- *Saline Siltic Alluvial brown soils* ≈ Fluvic Solonchaks (Siltic) / Salic Fluvisols (Siltic)
- *Siltic Alluvial brown soils* ≈ Fluvisols (Siltic) / Fluvic Cambisols (Siltic)

Brown forest soils ≈ Phaeozems / Kastanozems
There is one subgroup within the great group:

- *Brown forest soils* ≈ Phaeozems / Kastanozems

Brown Lithosol ≈ Lithic Leptosols (Calcaric)
There is one subgroup within the great group:

- *Brown Lithosol* ≈ Lithic Leptosols (Calcaric)

Coarse Alluvium ≈ Fluvisols
There are four subgroups within the great group:

- *Coarse Chalky and Marly Alluvium* ≈ Fluvisols (Calcaric, Arenic)
- *Coarse Desert Alluvium* ≈ Fluvisols (Calcaric, Arenic)
- *Coarse Regosolic Alluvium* ≈ Fluvisols (Arenic)
- *Coarse Silty and Clayey Stony Alluvium* ≈ Fluvisols (Skeletic)

Colluvial Alluvial soils (weathered) ≈ Fluvic Cambisols / Fluvisols
There are six subgroups within the great group:

- *Brown Basaltic Colluvial Alluvial soils* ≈ Fluvic Cambisols (Eutric) / Fluvisols (Eutric)
- *Brown Rendzinic Colluvial Alluvial soils* ≈ Fluvic Cambisols (Calcaric) / Mollic Fluvisols (Calcaric)
- *Light-coloured Chalky Colluvial Alluvial soils* ≈ Fluvic Cambisols (Calcaric) / Fluvisols (Calcaric)
- *Light-coloured Kurkaric Colluvial Alluvial soils* ≈ Fluvic Cambisols (Calcaric) / Mollic Fluvisols (Calcaric) / Fluvisols (Calcaric)
- *Red Colluvial Alluvial soils* ≈ Fluvic Cambisols (Rhodic) / Mollic Fluvisols (*Rhodic*) / Fluvisols (Eutric, *Rhodic*)
- *Yellow Colluvial Alluvial soils* ≈ *Fluvic* Phaeozems (*Xanthic*) / Mollic Fluvisols (*Xanthic*)

Dark Rendzina ≈ Rendzic Leptosols / Cambisols (Calcaric, Eutric) / Rendzic Phaeozems
There are two subgroups within the great group:

- *Brown Rendzina* ≈ Rendzic Leptosols / Cambisols (Calcaric, Eutric) / Rendzic Phaeozems
- *Dark Pararendzina* ≈ Rendzic Leptosols / Rendzic Phaeozems

Desert Lithosols ≈ Regosols (Calcaric) / Leptosols (Calcaric)
There are four subgroups within the great group:

- *Gypsiferous Desert Lithosols* ≈ Leptosols (Gypsiric) / Regosols (Gypsiric)
- *Marly and Chalky Desert Lithosols* ≈ Regosols (Calcaric) / Leptosols (Calcaric)
- *Rendzic Desert Lithosols* ≈ Leptosols (Calcaric)
- *Siliceous Desert Lithosols* ≈ Leptosols (Calcaric)

Gley ≈ Gleysols
There are five subgroups within the great group:

- *Calcareous Gley* ≈ Calcic Gleysols / Gleysols (Calcaric)
- *Gley* ≈ Mollic Gleysols
- *Grumusolic Gley* ≈ Gleyic Vertisols / *Vertic* Gleysols
- *Saline Calcareous Gley* ≈ Calcic Endosalic Gleysols / Gleyic Solonchaks
- *Sandy Gley* ≈ Mollic Gleysols (Arenic) / Eutric Gleysols (Arenic)

Grumusol (Vertisol) ≈ Vertisols
There are five subgroups within the great group:

- *Brown Grumusol* ≈ Vertisols (Chromic)
- *Calcareous Grumusol* ≈ Calcic Vertisols
- *Hydromorphic Grumusol* ≈ Gleyic Vertisols
- *Reddish-Brown Grumusol* ≈ Vertisols (Chromic)
- *Solonezic Grumosol* ≈ Vertisols (Hyposodic)

Hamra ≈ Luvisols
There are three subgroups within the great group:

- *Brown Hamra* ≈ Vertic Luvisols
- *Hamra (Orthohamra)* ≈ Luvisols (Chromic)
- *Nazzaric Hamra* ≈ Gleyic Luvisols

Krasnozem ≈ Nitisols (Rhodic) / Luvisols (Rhodic) / Cambisols (Rhodic)
There are two subgroups within the great group:

- *Krasnozem (Orthokrasnozem)* ≈ Nitisols (Rhodic) / Cambisols (Rhodic)
- *Podzolic Krasnozem* ≈ Albic Luvisols (Rhodic)

Light Rendzina ≈ Regosols (Calcaric) / Leptosols (Calcaric) / Cambisols (Calcaric) / Arenosols (Calcaric)
There are four subgroups within the great group:

- *Brown Light Rendzina* ≈ Lithic Leptosols (Calcaric) / Epileptic Cambisols (Calcaric)

- *Grumosolic Light Rendzina* ≈ Regosols (Calcaric, Clayic) / Vertic Leptosols / Vertic Cambisols
- *Light Rendzina (Ortholight Rendzina)* ≈ Regosols (Calcaric) / Lithic Leptosols (Calcaric) / Cambisols (Calcaric)
- *Light-coloured Pararendzina* ≈ Arenosols (Calcaric) / Leptosols (Calcaric) / Cambisols (Calcaric)

Loess and Fine Desert Alluvial soils ≈ Fluvisols
There are five subgroups within the great group:

- *Chalky and Marly Alluvial soils* ≈ Fluvisols (Calcaric)
- *Gravelly and Stony Loessial Alluvial soils* ≈ Fluvisols (Calcaric, Skeletic)
- *Gravelly Desert Alluvial soils* ≈ Fluvisols (Calcaric, Yermic, Skeletic)
- *Loess and Fine Desert Alluvium* ≈ Fluvisols (Calcaric, Aridic)
- *Sandy Loessial Alluvial soils* ≈ Fluvisols (Calcaric, Arenic)

Nazzaz ≈ Planosols (Eutric)
There is one subgroup within the great group:

- *Nazzaz* ≈ Planosols (Eutric)

Non-Desertic Lithosols on siliceous rocks ≈ Leptosols (Eutric)
There is one subgroup within the great group:

- *Basaltic Lithosols* ≈ Leptosols (Eutric)

Organic soils ≈ Histosols
There are two subgroups within the great group:

- *Organic Mineral soils* ≈ Sapric Histosols
- *Peat* ≈ Histosols

Protogrumosol ≈ Vertic Cambisols / Vertic Leptosols / Vertic Phaeozems
There is one subgroup within the great group:

- *Protogrumosol* ≈ Vertic Cambisols / Vertic Leptosols / Vertic Phaeozems

Reg ≈ Leptosols / Calcisols / Gypsisols
There are three subgroups within the great group:

- *Lithosolic Reg* ≈ Leptosols (Calcaric) / Leptosols (Gypsiric)
- *Reg (Orthoreg)* ≈ Endosalic Calcisols (Yermic) / Endosalic Gypsisols (Yermic)
- *Regosolic Reg* ≈ Calcisols (Yermic) / Gypsisols (Yermic)

Regosols ≈ Regosols / Arenosols

There are seven subgroups within the great group:

- *Brown Fine-Textured Regosols* ≈ Regosols (Calcaric)
- *Hamric Regosols* ≈ Regosols (Calcaric, Arenic) / Regosols (Eutric, Arenic)
- *Loessial Regosols* ≈ Regosols (Calcaric, Siltic)
- *Red Fine-Textured Regosols* ≈ Regosols (Calcaric, *Rhodic*) / Regosols (Eutric, *Rhodic*)
- *Sandy Regosols* ≈ Arenosols (Calcaric) / Arenosols (Eutric)
- *Stony Regosols* ≈ Regosols (Calcaric, Yermic)
- *Tuffic Regosols* ≈ Regosols (Tephric)

Sandy soils ≈ Arenosols / Fluvisols
There are two subgroups within the great group:

- *Aeolian sand* ≈ Arenosols (Calcaric) / Arenosols (Eutric)
- *Alluvial sand* ≈ Fluvisols (Calcaric, Arenic) / Fluvisols (Eutric, Arenic)

Shallow brown Mediterranean soils ≈ Luvic Phaeozems / Luvisols
There is one subgroup within the great group:

- *Shallow brown Mediterranean soils* ≈ Luvic Phaeozems / Luvisols

Sierozem ≈ Calcisols / Gypsisols
There are six subgroups within the great group:

- *Calcareous Sierozem* ≈ Hypercalcic Calcisols
- *Hydromorphic Calcareous Sierozem* ≈ Endogleyic Hypercalcic Calcisols
- *Marly Sierozem* ≈ Calcisols (Calcaric)
- *Saline-Gypsipherous Calcareous Sierozem* ≈ Calcic Endosalic Gypsisols
- *Sierozem (Orthosierozem)* ≈ Calcisols
- *Stony Sierozem* ≈ Calcisols (Yermic) / Calcisols (Skeletic)

Solonchak ≈ Solonchaks
There are five subgroups within the great group:

- *Aeolian and Alluvial Solonchak* ≈ Salic Fluvisols / *Fluvic* Solonchaks
- *Gley Solonchak* ≈ Gleyic Solonchaks
- *Marly Solonchak* ≈ Calcic Solonchaks
- *Organic Solonchak* ≈ Histic Solonchaks / Mollic Solonchaks
- *Sterile Solonchak* ≈ Hypersalic Solonchaks

Terra Rossa ≈ Luvisols / Cambisols / Phaeozems / Leptosols (Calcaric)
There are three subgroups within the great group:

- *Hamric Terra Rossa* ≈ Luvisols (Skeletic, Rhodic) / Leptosols (Calcaric, Rhodic)

- *Red Terra Rossa* ≈ Luvisols (Rhodic) / Cambisols (Rhodic) / Phaeozems (*Rhodic*)
- *Reddish Brown Terra Rossa* ≈ Luvisols (Chromic) / Cambisols (Chromic) / Phaeozems (Chromic)

Yellow soils ≈ Luvic Phaeozems (*Xanthic*)
There is one subgroup within the great group:

- *Yellow soils* ≈ Luvic Phaeozems (*Xanthic*)

Acknowledgement

The authors with to thank Professor D. H. Yaalon for his worthwhile comments.

References

Dan, J. and Koyumdjisky, H. (eds) (1979) *The Classification of Israel Soils*, Committee on Soil Classification in Israel, Special publication no 137, Institute of Soils and Water ARO, Bet Dagan, Israel, 95pp (in Hebrew with English abstract)

Dan, J., Koyumdjisky, H. and Yaalon, D. H. (1962) 'Principles of a proposed classification for the soils of Israel', *Transactions of the International Soil Conference*, Commissions IV and V, New Zealand, pp410–421

Dan Y., Yaalon, D. H., Koymdjisky, H. and Raz, Z. (1976) '*The soils of Israel*', Bulletin of the Volcanic Institute of Agricultural Research, Beit Dagan, Israel, vol 168, 28pp (in Hebrew)

IUSS Working Group WRB (2006) *World Reference Base for Soil Resources*, 2nd edition, World Soil Resources Reports no 103, UN Food and Agriculture Organization, Rome, 128pp

Ravikovitch, S. (1960) *Soils of Israel. Classification of the Soils of Israel*, Rehovot, Israel, 86pp

Ravikovitch S. (1992) *The Soils of Israel: Formation, Nature, and Properties,* Hakibbuts Hamehuchad Publishing House, Israel, 489pp (in Hebrew)

Singer, A. (2007) *The Soils of Israel*, Springer-Verlag, Berlin, 306pp

23
Soil Classification of People's Republic of China

Objectives and scope

The new Chinese soil classification (CRG-CST, 2001) was developed to replace an older genetic classification of soils of the People's Republic of China. It is supposed to be used for soil surveys at all scales and for scientific soil research and interpretation. The main objectives of the classification were to update it in accordance with recent pedogenetic findings and harmonize it with the main world soil classification systems. The geographical coverage of the classification is limited to China (Table 23.1). The soils of urban and industrial areas, as well as underwater soils and bare rocks, are not included in this classification.

Table 23.1 *The scope of the Chinese soil classification*

Superficial bodies	Representation in the system
Natural soils	National coverage
Urban soils	Partly included in the *Anthric Primosols* group
Transported materials	Partly included in the *Anthric Primosols* group
Bare rock	Not recognized as soils
Subaquatic soils	Not recognized as soils
Soils deeply transformed by agricultural activities	Form the order of *Anthrosols*

Theoretical background

The ideology of the new Chinese soil classifications, as well as the structure and some terminology, are similar to the US Soil Taxonomy (Soil Survey Staff, 1999). The taxa are recognized by their genesis, but the limits are strictly defined by the definitions of diagnostic horizons and properties. The classification uses water and temperature regimes for dividing soils at high taxonomic levels (suborder level). The Chinese taxonomy is not a copy of the American but reflects the peculiarity of the soil cover of the country. Much more attention is given to soils transformed by agriculture, because in China there are regions where soils have been cultivated for thousands of years (Gong Zitong, 1994). When the first draft version of the World Reference Base for Soil Resources (WRB) was in preparation (Spaargaren, 1994), the Chinese classification of anthropogenic soils was almost completely imported into the WRB.

Development

The first soil classification was documented in China more than 2500 YBP in an ancient book, *Yugong* (Gong Zitong, 1994). The early variant of scientific soil classification was proposed by Thorp (1939), and systematic work on the development of soil classification started in the 1950s (Gong Zitong, 1989). Chinese pedology was strongly influenced by the Soviet soil science school at that time and the classification resembled that of the soils of the USSR. In some instances the pedogenetic concepts developed for temperate and cold regions were mechanically used for soils of the tropical zones, causing confusion (Gerasimova, personal communication). Until recently the genetic soil classification of China served as a basis for soil survey and research in the country. In the 1980s leading pedologists of China understood that the classification system should be updated according to newer pedogenetic and classification concepts. After a series of internal documents a first proposal in English (Gong Zitong, 1994) was widely discussed internationally and within the country, and then a final version was published in 2001 (CRG-CST, 2001). Many older pedologists use the previous genetic classification, while the younger researchers use the new system.

Structure

The structure of Chinese soil taxonomy is similar to the upper levels of the US Soil Taxonomy (Soil Survey Staff, 1999), but the concept of these taxa is somewhat different. The highest level of the taxonomy, soil order, is determined by properties 'resulting from the leading soil-forming processes' (Table 23.2). An order is divided into suborders, defined by properties thought to reflect current soil-forming processes, or to be the main limiting properties for soil formation. Soil groups are defined as having 'differences in soil properties reflecting the intensity of soil formation, or are secondary soil-forming processes' (CRG-CST, 2001) determined by the presence and depth of diagnostic horizons and properties, and colours of soil horizons. In most cases the group level is generic. Soil subgroups are described as modifications of 'central images' of groups. They are recognized by the presence of additional diagnostic horizons and properties, and by lithological characteristics of parent material. The structure of the Chinese soil taxonomy is a hierarchical taxonomy with formal borders.

Table 23.2 *The structure of the Chinese soil classification*

Level	Taxon name	Taxon characteristics	Borders between classes	Diagnostics	Terminology
0	Soils	Kingdom			
1	Order	Collective	Formal	Chemico-morphological	Artificial
2	Suborder	Generic / Collective	Formal	Regimes and morphology	Artificial
3	Group	Generic / Specific	Formal	Chemico-morphological	Artificial, partially traditional
4	Subgroup	Specific	Formal	Chemico-morphological	Artificial

Diagnostics

The Chinese soil taxonomy uses both static soil characteristics and dynamic water and temperature regimes. The limits of soil taxonomic units are strict: a system of diagnostic horizons and properties is applied. At the order level, taxa are identified by the presence of certain diagnostic horizons and properties. An aridic moisture regime is considered to be a diagnostic soil property for the Aridisol order. At the suborder level some taxonomic units are characterized on the basis of internal static soil properties, and the others on the basis of soil moisture regimes. At the lower levels of taxonomy soil morphology, texture and chemical properties are the main soil properties selected as diagnostic indicators. The procedures for determining diagnostic horizons are much simpler than in the US Soil Taxonomy. More emphasis is given to morphological criteria: soil colour, horizon depth, quantity and form of pedogenic nodules and concentrations and so forth. Analytical criteria are simplified and often it is sufficient to determine texture, pH, soluble salts and organic carbon contents. For a *spodic* horizon and for *ferrallic, fersiallitic* and *siallic* properties more extensive chemical analyses are required. The diagnostics in Chinese soil taxonomy can be considered quantitative chemico-climatic-morphological.

Terminology

The terminology of Chinese soil taxonomy is almost completely artificial. It is a compilation of terms: some are borrowed directly from Soil Taxonomy, plus others with minor modifications (*Histosols, Spodosols, Aridosols, Vertosols*); some are from the reference group names of the WRB, including some slightly modified (*Andosols, Cambosols, Gleyosols*); and some are new artificial terms constructed in similar manner (*Halosol, Argosols, Isohumisols*, etc.). On the levels of groups and subgroups the terminology is also completely artificial, with occasional inclusion of folk terms like *Shajiang* soils. Earlier drafts of classification had mostly traditional Chinese names (Kanno, 1978) which were gradually replaced by artificial ones (Gong Zitong 1989, 1994; CRG-CST, 2001).

Correlation

For most soil taxa in the Chinese soil classification it is easy to find analogues in the WRB because many concepts are similar in these two classifications. Some groups, especially those related to intermediate weathering of mineral soil material (*Ferrosols*), were more difficult to correlate. Chinese terms down to the group level are correlated with the WRB terms. The soil orders are listed alphabetically, as are the suborders within orders, and the names of groups within suborders.

Andosols – soil order. ≈ Andosols / Cryosols
There are three suborders within the order:
Cryic Andosols – soil suborder. ≈ *Andic* Cryosols / Andosols (Gelic)
The following groups are distinguished within the suborder:

- *Geli-Cryic Andosols* ≈ Andosols (Gelic)
- *Hapli-Cryic Andosols* ≈ *Andic* Cryosols

Udic Andosols – soil suborder. ≈ Andosols
The following groups are distinguished within the suborder:

- *Hapli-Udic Andosols* ≈ Andosols
- *Humi-Udic Andosols* ≈ Melanic Andosols / Fulvic Andosols

Vitric Andosols – soil suborder. ≈ Vitric Andosols
The following groups are distinguished within the suborder:

- *Udi-Vitric Andosols* ≈ Vitric Andosols
- *Usti-Vitric Andosols* ≈ Vitric Andosols

Anthrosols – soil order. ≈ Anthrosols
The order is divided into two suborders:
Orthic Anthrosols – soil suborder. ≈ Anthrosols
The following groups are distinguished within the suborder:

- *Earth-cumuli-Orthic Anthrosols* ≈ Terric Anthrosols
- *Fumi-Orthic Anthrosols* ≈ Hortic Anthrosols
- *Mud-cumuli-Orthic Anthrosols* ≈ Hortic Anthrosols
- *Siltigi-Orthic Anthrosols* ≈ Anthrosols

Stagnic Anthrosols – soil suborder. ≈ Hydragric Anthrosols / Stagnic Anthrosols
The following groups are distinguished within the suborder:

- *Fe-leachi-Stagnic Anthrosols* ≈ Hydragric Anthrosols / Stagnic Anthrosols
- *Fe-accumuli-Stagnic Anthrosols* ≈ Hydragric Anthrosols (*Ferric*) / Stagnic Anthrosols (*Ferric*)

- *Gleyi-Stagnic Anthrosols* ≈ Gleyic Hydragric Anthrosols / Stagnic Gleyic Anthrosols
- *Hapli-Stagnic Anthrosols* ≈ Hydragric Anthrosols / Stagnic Anthrosols

Argosols – soil order. ≈ Luvisols
The order is divided into four suborders:
Boric Argosols – soil suborder. ≈ Luvisols / Luvic Phaeozems
The following groups are distinguished within the suborder:

- *Albi-Boric Argosols* ≈ Albic Luvisols
- *Hapli-Boric Argosols* ≈ Luvisols
- *Molli-Boric Argosols* ≈ Luvic Phaeozems

Perudic Argosols – soil suborder. ≈ Luvisols / Alisols
The following groups are distinguished within the suborder:

- *Ali-Perudic Argosols* ≈ Alisols
- *Carbonati-Perudic Argosols* ≈ Luvisols (*Calcaric*)
- *Hapli-Perudic Argosols* ≈ Luvisols

Udic Argosols – soil suborder. ≈ Luvisols / Lixisols / Acrisols / Alisols
The following groups are distinguished within the suborder:

- *Acidi-Udic Argosols* ≈ Acrisols / Alisols
- *Albi-Udic Argosols* ≈ Albic Luvisols
- *Ali-Udic Argosols* ≈ Alisols
- *Carbonati-Udic Argosols* ≈ Luvisols (*Calcaric*)
- *Claypani-Udic Argosols* ≈ Luvisols (Fragic)
- *Ferri-Udic Argosols* ≈ Acrisols / Lixisols
- *Hapli-Udic Argosols* ≈ Luvisols

Ustic Argosols – soil suborder. ≈ Luvisols / Lixisols
The following groups are distinguished within the suborder:

- *Calci-Ustic Argosols* ≈ Calcic Luvisols
- *Carbonati-Ustic Argosols* ≈ Luvisols (*Calcaric*)
- *Ferri-Ustic Argosols* ≈ Lixisols
- *Hapli-Ustic Argosols* ≈ Luvisols

Aridosols – soil order. ≈ Cryosols / Calcisols / Solonchaks / Gypsisols / Regosols
The order is divided into two suborders:
Cryic Aridosols – soil suborder. ≈ Cryosols (Aridic)
The following groups are distinguished within the suborder:

- *Argi-Cryic Aridosols* ≈ *Luvic* Cryosols (Aridic)

- *Calci-Cryic Aridosols* ≈ Calcic Cryosols (Aridic)
- *Gypsi-Cryic Aridosols* ≈ Gypsic Cryosols (Aridic)
- *Hapli-Cryic Aridosols* ≈ Cryosols (Aridic)

Orthic Aridosols – soil <u>suborder</u>. ≈ Calcisols
The following <u>groups</u> are distinguished within the suborder:

- *Argi-Orthic Aridosols* ≈ Luvisols (*Aridic*)
- *Calci-Orthic Aridosols* ≈ Calcisols
- *Gypsi-Orthic Aridosols* ≈ Gypsisols
- *Hapli-Orthic Aridosols* ≈ Regosols (Aridic)
- *Sali-Orthic Aridosols* ≈ Endosalic Cambisols (Aridic) / Endosalic Regosols (Aridic)

<u>Cambosols</u> – <u>soil order</u>. ≈ Cambisols / Phaeozems / Cryosols
The order is divided into five <u>suborders</u>:
Aquic Cambosols – soil <u>suborder</u>. ≈ Endogleyic Cambisols / Gleyic Phaeozems
The following <u>groups</u> are distinguished within the suborder:

- *Dark Aquic Cambosols* ≈ Humic Cambisols
- *Littery-Aquic Cambosols* ≈ Endogleyic Folic Cambisols
- *Ochri-Aquic Cambosols* ≈ Endogleyic Cambisols
- *Shajiang-Aquic Cambosols* ≈ Gleyic Phaeozems (Calcaric)

Gelic Cambosols – soil <u>suborder</u>. ≈ Cryosols
The following soil <u>groups</u> are distinguished:

- *Aqui-Gelic Cambosols* ≈ Cryosols (Oxiaquic) / Cryosols (Reductiaquic)
- *Hapli-Gelic Cambosols* ≈ Cryosols
- *Matti-Gelic Cambosols* ≈ Folic Cryosols
- *Molli-Gelic Cambosols* ≈ Mollic Cryosols
- *Permi-Gelic Cambosols* ≈ Cryosols
- *Umbri-Gelic Cambosols* ≈ Umbric Cryosols

Perudic Cambosols – soil <u>suborder</u>. ≈ Cambisols (Dystric)
The following soil <u>groups</u> are distinguished:

- *Acidi-Perudic Cambosols* ≈ Cambisols (Hyperdystric)
- *Ali-Perudic Cambosols* ≈ Cambisols (Alumic)
- *Bori-Perudic Cambosols* ≈ Cambisols (Dystric)
- *Carbonati-Perudic Cambosols* ≈ Cambisols (Calcaric, Dystric)
- *Hapli-Perudic Cambosols* ≈ Cambisols (Dystric)
- *Stagni-Perudic Cambosols* ≈ Stagnic Cambisols (Dystric)

Udic Cambosols – soil <u>suborder</u>. ≈ Cambisols / Umbrisols / Luvisols / Enthic Podzols

The following groups are distinguished within the suborder:

- *Acidi-Udic Cambosols* ≈ Cambisols (Dystric)
- *Ali-Udic Cambosols* ≈ Cambisols (Alumic)
- *Bori-Udic Cambosols* ≈ Cambisols
- *Carbonati-Udic Cambosols* ≈ Cambisols (Calcaric)
- *Ferri-Udic Cambosols* ≈ Cambisols (Chromic)
- *Hapli-Udic Cambosols* ≈ Cambisols
- *Purpli-Udic Cambosols* ≈ Cambisols (Rhodic)

Ustic Cambosols – soil suborder. ≈ Cambisols / Phaeozems
The following soil groups are distinguished:

- *Endorusti-Ustic Cambosols* ≈ Cambisols (Eutric)
- *Ferri-Ustic Cambosols* ≈ Cambisols (Chromic)
- *Hapli-Ustic Cambosols* ≈ Cambisols (Eutric)
- *Molli-Ustic Cambosols* ≈ Phaeozems
- *Siltigi-Ustic Cambosols* ≈ Cambisols (Eutric)

Gleyosols – soil order. ≈ Gleysols / Cryosols / Gleyic Fluvisols / Stagnosols
The order is divided into three suborders:
Orthic Gleyosols – soil suborder. ≈ Haplic Gleysols / Gleyic Fluvisols
Three soil groups are distinguished:

- *Hapli-Orthic Gleyosols* ≈ Haplic Gleysols / Gleyic Fluvisols
- *Histi-Orthic Gleyosols* ≈ Histic Gleysols / Gleyic Histic Fluvisols
- *Molli-Orthic Gleyosols* ≈ Mollic Gleysols / Mollic Gleyic Fluvisols

Permagelic Gleyosols – soil suborder. ≈ Cryosols (Reductaquic)
Two soil groups are distinguished:

- *Hapli-Permagelic Gleyosols* ≈ Cryosols (Reductaquic)
- *Histi-Permagelic Gleyosols* ≈ Histic Cryosols (Reductaquic)

Stagnic Gleyosols – soil suborder. ≈ Stagnosols / Planosols
Two groups are distinguished within the suborder:

- *Hapli-Stagnic Gleyosols* ≈ Stagnosols / Planosols
- *Histi-Stagnic Gleyosols* ≈ Histic Stagnosols / Histic Planosols

Ferrallosols – soil order. ≈ Ferralsols
The order has only one suborder with three groups:
Udic Ferrallisols – soil suborder. ≈ Ferralsols

- *Hapli-Udic Ferralosols* ≈ Ferralsols
- *Rhodi-Udic Ferralosols* ≈ Ferralsols (Rhodic)
- *Xanthi-Udic Ferralosols* ≈ Ferralsols (Xanthic)

<u>Ferrosols</u> – <u>soil order</u>. ≈ Cambisols / Luvisols
The order is divided into three suborders:
Perudic Ferrosols – soil <u>suborder</u>. ≈ Cambisols
The following <u>groups</u> are distinguished within the suborder:

- *Alliti-Perudic Ferrosols* ≈ *Gibbsic* Cambisols
- *Carbonati-Perudic Ferrosols* ≈ Cambisols (Calcaric, Chromic)
- *Hapli-Perudic Ferrosols* ≈ Cambisols (Chromic)

Udic Ferrosols – soil <u>suborder</u>. ≈ Cambisols
The following <u>groups</u> are distinguished within the suborder:

- *Argi-Udic Ferrosols* ≈ Lixisols
- *Alliti-Perudic Ferrosols* ≈ *Gibbsic* Cambisols
- *Carbonati-Udic Ferrosols* ≈ Cambisols (Calcaric, Chromic)
- *Hapli-Udic Ferrosols* ≈ Cambisols (Chromic)
- *Hiweatheri-Udic Ferrosols* ≈ Ferralic Cambisols (Chromic)

Ustic Ferrosols – soil <u>suborder</u>. ≈ Cambisols / Lixisols
The following <u>groups</u> are distinguished within the suborder:

- *Argi-Ustic Ferrosols* ≈ Lixisols
- *Hapli-Ustic Ferrosols* ≈ Cambisols

<u>Halosols</u> – <u>soil order</u>. ≈ Solonchaks / Solonetz / Salic Fluvisols
The order is divided into two suborders:
Alkalic Halosols – soil <u>suborder</u>. ≈ Solonetz
The following <u>groups</u> are distinguished within the suborder:

- *Aqui-Alkalic Halosols* ≈ Gleyic Solonetz
- *Hapli-Alkalic Halosols* ≈ Solonetz
- *Takyri-Alkalic Halosols* ≈ Solonetz (Takyric)

Orthic Halosols – soil <u>suborder</u>. ≈ Solonchaks / Salic Fluvisols
The following <u>groups</u> are distinguished within the suborder:

- *Aqui-Orthic Halosols* ≈ Gleyic Solonchaks / Salic Fluvisols
- *Aridi-Orthic Halosols* ≈ Solonchaks

<u>Histosols</u> – <u>soil order</u>. ≈ Histosols
The order is divided into two suborders:
Permagelic Histosols – soil <u>suborder</u>. ≈ Cryic Histosols
Three soil <u>groups</u> are distinguished:

- *Fibri-Permagelic Histosols* ≈ Cryic Fibric Histosols
- *Foli-Permagelic Histosols* ≈ Cryic Folic Histosols
- *Hemi-Permagelic Histosols* ≈ Cryic Hemic Histosols

Orthic Histosols – soil <u>suborder</u>. ≈ Histosols
Four soil <u>groups</u> are distinguished:

- *Fibri-Orthic Histosols* ≈ Fibric Histosols
- *Foli-Orthic Histosols* ≈ Folic Histosols
- *Hemi-Orthic Histosols* ≈ Hemic Histosols
- *Sapri-Orthic Histosols* ≈ Sapric Histosols

<u>Isohumisols</u> – <u>soil order</u>. ≈ Chernozems / Phaeozems / Rendzic Leptosols / Kastanozems
The order is divided into three <u>suborders</u>:
Lithomorphic Isohumisols – soil <u>suborder</u>. ≈ Rendzic Leptosols / Rendzic Phaeozems / Mollic Leptosols / Leptic Phaeozems
The following <u>groups</u> are distinguished within the suborder:

- *Black-Lithomorphic Isohumisols* ≈ Rendzic Leptosols / Rendzic Phaeozems / Mollic Leptosols
- *Phosphi-Litomorphic Isohumisols* ≈ Mollic Leptosols / Leptic Phaeozems

Udic Isohumisols – soil <u>suborder</u>. ≈ Chernozems / Phaeozems
The following <u>groups</u> are distinguished within the suborder:

- *Argi-Udic Isohumisols* ≈ Luvic Phaeozems
- *Hapli-Udic Isohumisols* ≈ Phaeozems
- *Stagni-Udic Isohumisols* ≈ Stagnic Phaeozems

Ustic Isohumisols – soil <u>suborder</u>. ≈ Chernozems / Kastanozems / Cryosols
The following <u>groups</u> are distinguished within the suborder:

- *Calci-Ustic Isohumisols* ≈ Calcic Chernozems
- *Cryi-Ustic Isohumisols* ≈ Mollic Cryosols / Chernozems (Gelic) / Kastanozems (Gelic)
- *Cumuli-Ustic Isohumisols* ≈ Chernozems (*Colluvic*)
- *Hapli-Ustic Isohumisols* ≈ Chernozems / Kastanozems
- *Pachi-Ustic Isohumisols* ≈ Chernic Chernozems

<u>Primosols</u> – <u>soil order</u>. ≈ Leptosols / Regosols / Arenosols / Fluvisols
The order is divided into four <u>suborders</u>:
Alluvic Primosols – soil <u>suborder</u>. ≈ Fluvisols
The following <u>groups</u> are distinguished within the suborder:

- *Aqui-Alluvic Primosols* ≈ Gleyic Fluvisols
- *Aridi-Alluvic Primosols* ≈ Fluvisols (Aridic)
- *Geli-Alluvic Primosols* ≈ Fluvisols (Gelic)
- *Udi-Alluvic Primosols* ≈ Fluvisols
- *Usti-Alluvic Primosols* ≈ Fluvisols

Anthric Primosols – soil <u>suborder</u>. ≈ Technosols
Two soil <u>groups</u> are distinguished:

- *Siltigi-Anthric Primosols* ≈ Technosols
- *Turbi-Anthric Primosols* ≈ Technosols

Orthic Primosols – soil <u>suborder</u>. ≈ Regosols
The following <u>groups</u> are distinguished within the suborder:

- *Aridi-Orthic Primosols* ≈ Regosols (Aridic)
- *Geli-Orthic Primosols* ≈ Regosols (Gelic)
- *Loessi-Orthic Primosols* ≈ Regosols (Siltic)
- *Purpli-Orthic Primosols* ≈ Regosols (*Rhodic*)
- *Rougi-Orthic Primosols* ≈ Regosols (*Chromic*)
- *Udi-Orthic Primosols* ≈ Regosols (Dystric)
- *Usti-Orthic Primosols* ≈ Regosols (Eutric)

Sandic Primosols – soil <u>suborder</u>. ≈ Arenosols
The following <u>groups</u> are distinguished within the suborder:

- *Aqui-Sandic Primosols* ≈ Endogleyic Arenosols
- *Aridi-Sandic Primosols* ≈ Arenosols (Aridic)
- *Geli-Sandic Primosols* ≈ Arenosols (Gelic)
- *Udi-Sandic Primosols* ≈ Arenosols (Dystric)
- *Usti-Sandic Primosols* ≈ Arenosols (Eutric)

Spodosols – <u>soil order</u>. ≈ Podzols
Two <u>suborders</u> are distinguished within the order:
Humic Spodosols – soil <u>suborder</u>. ≈ Carbic Podzols
There is only soil <u>group</u> in the suborder:

- *Hapli-Humic Spodosols* ≈ Carbic Podzols

Orthic Spodosols – soil <u>suborder</u>. ≈ Podzols
There is only soil <u>group</u> in the suborder:

- *Hapli-Orthic Spodosols* ≈ Podzols

Vertosols – <u>soil order</u>. ≈ Vertisols
The order is divided into three <u>suborders</u>:
Aquic Vertosols – soil <u>suborder</u>. ≈ (Gleyic) Vertisols
Two soil <u>groups</u> are distinguished:

- *Calci-Aquic Vertosols* ≈ Calcic Gleyic Vertisols
- *Hapli-Aquic Vertosols* ≈ Gleyic Vertisols

Udic Vertosols – soil <u>suborder</u>. ≈ Vertisols
The following <u>groups</u> are distinguished within the suborder:

- *Humi-Udic Vertosols* ≈ Vertisols (Humic)
- *Calci-Udic Vertosols* ≈ Calcic Vertisols
- *Hapli-Udic Vertosols* ≈ Vertisols

Ustic Vertosols – soil <u>suborder</u>. ≈ Vertisols
The following <u>groups</u> are distinguished within the suborder:

- *Calci-Ustic Vertosols* ≈ Calcic Vertisols
- *Hapli-Ustic Vertosols* ≈ Vertisols

References

CRG-CST (2001) *Chinese Soil Taxonomy*, Li Feng (ed), Science Press, Beijing and New York, 203pp

Gong Zitong (1989) 'Review of Chinese soil classifications for the past four decades', *Acta Pedologica Sinica*, vol 26, no 3, pp217–225 (in Chinese, English summary)

Gong Zitong (ed) (1994) *Chinese Soil Taxonomic Classification (First proposal)*, Institute of Soil Science, Academia Sinica, Nanjing, 93pp

Kanno, I. (1978) '"Chinese Soils" complied by the Nanking Institute of Soil Science, Academia Sinica', *Pedologist*, vol 22, no 2, pp176–181

Soil Survey Staff (1999) *Soil Taxonomy: A Basic System of Soil Classification for Making and Interpreting Soil Surveys*, USDA, Handbook 436, 2nd edition, United States Government Printing Office, Washington DC, 696pp

Spaargaren, O. (ed) (1994) *World Reference Base for Soil Resources. Draft.* World Soil Resources Reports no 84, UN Food and Agriculture Organization, Rome, 88pp

Thorp, J. (1939) *Geography of the Soils of China*, National Geological Survey of China and Institute of Geology of the National Academy of Peiping, Peking, 552pp

24
Soil Classification of Japan

Objectives and scope

Japan is a country with a high density of population and intensive agricultural production. It is not surprising that from the end of the 19th century a survey of soil resources was undertaken. Initially, the survey was completely agrogeological, then in the 1920s the concept of soil series was introduced, and, finally, a soil taxonomic classification developed starting in the 1950s (Hirai and Hamazaki, 2004). The latest version of the Japanese soil classification was published in 2002 continuing a lengthy effort of a group of pedologists (Fourth Committee for Soil Classification and Nomenclature, 2002). The classification is limited to the national borders of Japan (Table 24.1). The system includes man-made soils; however, rock outcrops and underwater sediments are not recognized as soils. Artificially flooded paddy soils are classified as soils as they represent important agricultural resources of the country. Soils transformed by agricultural practices, depending on the degree of transformation, are included in the classification at different levels. Soils completely reworked by tillage appear in the great group of *Man-made soils*, paddy soils form two groups in the great group of *Fluvic soils*, and less transformed soils are included as *Anthraquic* subgroups to a number of soil groups.

Table 24.1 *The scope of soil classification of Japan*

Superficial bodies	Representation in the system
Natural soils	National coverage
Urban soils	Partly included in the great group of man-made soils
Man-transported materials	Partly included in the great group of man-made soils
Bare rock	Not recognized as soils
Subaquatic soils	Not recognized as soils
Soils deeply transformed by agricultural activities	Included in great group, group and subgroup levels

Theoretical background

Japanese pedology was influenced by several soil science schools, mostly by the Russian, American and German. The soil survey and soil classification adopted many of the concepts and approaches of these schools. Soil classification is based on the concept of interdependence of soil genesis and soil properties. The soil survey in Japan initially used a system of soil series similar to the US (Hirai and Hamazaki, 2004); thus, the taxonomic system had to include soil series, like the US Soil Taxonomy (Soil Survey Staff, 1999). The solution was the same: the taxa had to be strictly limited, and then series had to be adjusted to their limits. The Japanese soil classification has a long history, and during its development more attention was paid to volcanic ash soils and paddy soils. These groups are classified in much more detail than other soils. The last edition of Japanese classification (Fourth Committee for Soil Classification and Nomenclature, 2002) has a number of concepts and terms borrowed from the World Reference Base for Soil Resources (WRB) (IUSS Working Group WRB, 2006).

Development

The first agrogeological classification of soils of Japan was prepared for Japan in 1882 by a German agronomist Max Fesca (Hirai and Hamazaki, 2004). In the 1920s, the American system of soil series was introduced to Japan. The most extensive soil survey using soil series was conducted in Hokkaido in those years. After World War II, American specialists worked in Japan, and developed a working classification of soils of the country. Later classification systems were developed for particular soil groups (volcanic ash soils, paddy soils), land use types (agricultural soils, forest soils) and for some islands (e.g. Hokkaido). Hokkaido Island is one of the main centres of soil survey and soil classification studies in Japan, although the agricultural and industrial development of this region is lower than in the rest of the country. In Hokkaido, there was a special committee on soil classification (Otowa, 1981). A number of classification schemes with traditional structure and traditional terminology were developed by T. Matsui (1982) beginning in the 1960s, up to the early 1980s. In the 1980s a working group to develop a uniform soil classification of Japan was created. The 1st Approximation was published in 1986 (Committee for Soil Classification and Nomenclature, 1986), and the 2nd Approximation was published in 2002 (Fourth Committee for Soil Classification and Nomenclature, 2002). The 1st Approximation was used as a legend for a soil map of Japan at a scale of 1:1,000,000.

Structure

There are five levels in the soil classification of Japan (Table 24.2). The highest level, <u>great groups</u>, is a collective taxonomic level. The great groups are divided into <u>groups.</u> Every group has a defined sequence of diagnostic horizons and is considered to be generic for this classification. Soil <u>subgroups</u> differ from the central concept of the group, or are transitional towards other groups or great groups. One of the main criteria for subgroups is the type of surface humus horizon (epipedon). Texture, mineralogical composition and temperature regime distinguish soil <u>families</u>. Soil <u>series</u> are traditional empirical entities with certain combinations of properties, mainly related to parent material. Thus, the Japanese soil classification is a hierarchical taxonomy with formal borders. At the series level it is a nominal system.

Table 24.2 *The structure of soil classification of Japan*

Level	Taxon name	Taxon characteristics	Borders between classes	Diagnostics	Terminology
0	Soils	Kingdom			
1	Great group	Collective	Formal	Chemico-morphological	Traditional
2	Group	Generic	Formal	Chemico-morphological	Traditional
3	Subgroup	Specific	Formal	Chemico-morphological	Mainly traditional
4	Family	Varietal	Formal	Regimes and physico-mineralogical	Traditional
5	Series	Generic	Formal	Chemico-morphological	Traditional

Diagnostics

The diagnostics of soil taxa in Japanese soil classification are made on the basis of the properties of the soil profile; temperature regimes are used (optionally) only at the family level. The concept of diagnostic horizons requires quantitative characterization of their properties: both morphological and chemical criteria (determined in a laboratory) are used. In the 1st Approximation, the diagnostic horizons were only partially defined quantitatively whereas in the 2nd Approximation all the horizons received exact definitions (Fourth Committee for Soil Classification and Nomenclature, 2002). The diagnostics of Japanese classification are quantitative chemico-morphological.

Terminology

The terminology of the upper levels of soil classification of Japan is mainly traditional. It is based on terms borrowed from earlier European and Russian classifications. The classification mostly uses Japanese translations of soil names rather than transcriptions or transliterations of terms. For borrowed terms the Japanese language traditionally applies one of the alphabets – *katakana*, and these soil names are listed according to the Japanese phonetic rules: for example, 'podzol' – *podozoru*, 'gley' – *gurei* (here the Japanese soil names are presented in *romaji* – Latin transcription of Japanese characters). In fact, the

English summaries of the classification do not reflect exactly the terms used in the classification. The authors, instead of translating the traditional terms being used in their classification, use the closest artificial term used by Western soil science schools; in fact, it is not translation, but a correlation of terms. For example, the Japanese soil name *kazan-hoshutsubutsu mijuku-do* means literally 'undeveloped soils of volcanic ashes', but in English summary its name is listed as *Volcanogenous Regosols*. The same is true for many other terms, including the subgroup level, where the English text uses the terms directly from Soil Taxonomy or the WRB: *terric, sapric, cumulic, andic* and so on. In the original text of the classification most of these terms are traditional. It is interesting to note that the term *ando* ('dark soil'), used as a formative element for the names of volcanic soils (*Andosols, Andisols*, etc.) is not in use in Japan. This term was proposed by the specialists who worked in Japan in the 1940s to 1950s. The Japanese use the words *kuroboku* ('ink-black') or *kurotsuchi* ('black earth').

Correlation

The correlation of the Japanese terms with the WRB groups was not very difficult; however, caution should be taken with the volcanic soils. Not all the *Kuroboku* soil group completely corresponds to the allophonic volcanic soils or to the broader *Andosols* group in the WRB (IUSS Working Group WRB, 2006). Some of these soils are not related to volcanic ash, and form in weathered igneous rocks or ancient sedimentary materials. But the same is true for the Andosols reference group: a number of new findings show that the set of properties typical for Andosols might be found in soils derived from a wider range of parent materials (Buytaert et al, 2006).

The correlation of the terms of the Japanese soil classification is listed below. The great groups are listed in alphabetical order as are the groups within the great groups. Only English variants of the terms are listed following the form proposed by the authors' English summary of their journal paper (Fourth Committee for Soil Classification and Nomenclature, 2002).

Brown Forest soils – great soil group. ≈ Cambisols / Umbrisols / Podzols
Two groups are distinguished within the great group:
Haplic Brown Forest soils – soil group. ≈ Cambisols / Podzols / Umbrisols
The following subgroups are distinguished within the group:

- *Eutric Haplic Brown Forest soils* ≈ Cambisols (Eutric)
- *Humic Haplic Brown Forest soils* ≈ Mollic Umbrisols
- *Podzolic Haplic Brown Forest soils* ≈ Enthic Podzols / Umbric Podzols
- *Pseudogleyic Haplic Brown Forest soils* ≈ Gleyic Cambisols
- *Surface-pseudogleyic Haplic Brown Forest soils* ≈ Stagnic Cambisols
- *Typic Haplic Brown Forest soils* ≈ Cambisols (Dystric)

Yellow Brown Forest soils – soil group. ≈ Cambisols (*Xanthic*)

The following subgroups are distinguished within the group:

- *Eutric Yellow Brown Forest soils* ≈ Cambisols (Eutric, *Xanthic*)
- *Pseudogleyic Yellow Brown Forest soils* ≈ Gleyic Cambisols (*Xanthic*)
- *Surface-pseudogleyic Yellow Brown Forest soils* ≈ Stagnic Cambisols (*Xanthic*)
- *Typic Yellow Brown Forest soils* ≈ Cambisols (Dystric, *Xanthic*)

Dark Red soils – great soil group. ≈ Cambisols / Luvisols / Phaeozems
Four groups are distinguished within the great group:
Dark Red Magnesian soils – soil group. ≈ Luvisols (Calcaric, *Magnesic*) / Cambisols (Calcaric, *Magnesic*)
The following subgroups are distinguished within the group:

- *Argic Dark Red Magnesian soils* ≈ Luvisols (Calcaric, *Magnesic*)
- *Typic Dark Red Magnesian soils* ≈ Cambisols (Calcaric, *Magnesic*)

Mollic Calcareous soils – soil group. ≈ Phaeozems
The following subgroups are distinguished within the group:

- *Argic Mollic Calcareous soils* ≈ Luvic Phaeozems (Calcaric)
- *Typic Mollic Calcareous soils* ≈ Phaeozems (Calcaric)

Red-brown Calcareous soils – soil group. ≈ Luvisols (Calcaric, Chromic) / Cambisols (Calcaric, Chromic)
The following subgroups are distinguished within the group:

- *Argic Red Brown Calcareous soils* ≈ Luvisols (Calcaric, Chromic)
- *Typic Red Brown Calcareous soils* ≈ Cambisols (Calcaric, Chromic)

Yellow Brown Calcareous soils – soil group. ≈ Luvisols (Calcaric) / Cambisols (Calcaric)
The following subgroups are distinguished within the group:

- *Argic Yellow Brown Calcareous soils* ≈ Luvisols (Calcaric)
- *Typic Yellow Brown Calcareous soils* ≈ Cambisols (Calcaric)

Fluvic soils – great soil group. ≈ Fluvisols
Five groups are distinguished within the great group:
Brown Fluvic soils – soil group. ≈ Fluvisols / *Fluvic* Andosols
The following subgroups are distinguished within the group:

- *Andic Brown Fluvic soils* ≈ *Andic* Fluvisols / *Fluvic* Andosols
- *Typic Brown Fluvic soils* ≈ Fluvisols
- *Wet Brown Fluvic soils* ≈ Fluvisols (Oxiaquic)

Gley Fluvic soils – soil group. ≈ Gleyic Fluvisols
The following subgroups are distinguished within the group:

- *Andic Gley Fluvic soils* ≈ *Andic* Gleyic Fluvisols
- *Peaty Gley Fluvic soils* ≈ Gleyic Histic Fluvisols
- *Sulphuric Gley Fluvic soils* ≈ Gleyic Fluvisols (Thionic)
- *Surface-gleyic Gley Fluvic soils* ≈ Stagnic Gleyic Fluvisols
- *Typic Gley Fluvic soils* ≈ Gleyic Fluvisols

Grey Fluvic soils – soil group. ≈ Mollic Fluvisols
The following subgroups are distinguished within the group:

- *Andic Grey Fluvic soils* ≈ *Andic* Mollic Fluvisols
- *Gleyic Grey Fluvic soils* ≈ Mollic Endogleyic Fluvisols
- *Peaty Grey Fluvic Soils* ≈ Mollic Histic Fluvisols
- *Sulphuric Grey Fluvic soils* ≈ Mollic Fluvisols (Thionic)
- *Surface-gleyic Grey Fluvic soils* ≈ Mollic Stagnic Fluvisols
- *Typic Grey Fluvic soils* ≈ Mollic Fluvisols

Grey Paddy soils – soil group. ≈ Hydragric Anthrosols
The following subgroups are distinguished within the group:

- *Bleached Grey Paddy soils* ≈ Hydragric Anthrosols (*Albic*)
- *Sub-aeric Grey Paddy soils* ≈ Hydragric Anthrosols
- *Typic Grey Paddy soils* ≈ Hydragric Anthrosols
- *Wet Grey Paddy soils* ≈ Hydragric Anthrosols (Oxiaquic)

Illuvial Paddy soils – soil group. ≈ Hydragric Anthrosols
The following subgroups are distinguished within the group:

- *Bleached Illuvial Paddy soils* ≈ *Luvic* Hydragric Anthrosols (*Albic*)
- *Sub-aeric Illuvial Paddy soils* ≈ *Luvic* Hydragric Anthrosols
- *Typic Illuvial Paddy soils* ≈ *Luvic* Hydragric Anthrosols
- *Wet Illuvial Paddy soils* ≈ *Luvic* Hydragric Anthrosols (Oxiaquic)

Kuroboku soils – great soil group. ≈ Andosols
Six groups are distinguished within the great group:
Allophanic Kuroboku soils – soil group. ≈ Silandic Andosols
The following subgroups are distinguished within the group:

- *Anthraquic Allophanic Kuroboku soils* ≈ Silandic Andosols (*Anthraquic*)
- *Cumulic Allophanic Kuroboku soils* ≈ Silandic Andosols (*Novic*)
- *Low-humus Allophanic Kuroboku soils* ≈ Silandic Andosols
- *Thapto-humic Allophanic Kuroboku soils* ≈ Silandic Andosols (Thaptomollic)
- *Typic Allophanic Kuroboku soils* ≈ Melanic Silandic Andosols

Brown Kuroboku soils – soil group. ≈ Fulvic Andosols
The following subgroups are distinguished within the group:

- *Cumulic Brown Kuroboku soils* ≈ Fulvic Andosols (*Novic*)
- *Non-allophanic Brown Kuroboku soils* ≈ Fulvic Aluandic Andosols
- *Thapto-humic Brown Kuroboku soils* ≈ Fulvic Andosols (Thaptomollic)
- *Typic Brown Kuroboku soils* ≈ Fulvic Andosols

Gley Kuroboku soils – soil group. ≈ Gleyic Andosols
The following subgroups are distinguished within the group:

- *Cumulic Gley Kuroboku soils* ≈ Gleyic Andosols (*Novic*)
- *Non-allophanic Gley Kuroboku soils* ≈ Gleyic Aluandic Andosols
- *Peaty Gley Kuroboku soils* ≈ Gleyic Histic Andosols
- *Typic Gley Kuroboku soils* ≈ Gleyic Andosols

Non-allophanic Kuroboku soils – soil group. ≈ Aluandic Andosols
The following subgroups are distinguished within the group:

- *Anthraquic Non-allophanic Kuroboku soils* ≈ Aluandic Andosols (*Anthraquic*)
- *Cumulic Non-allophanic Kuroboku soils* ≈ Aluandic Andosols (*Novic*)
- *Low-humus Non-allophanic Kuroboku soils* ≈ Aluandic Andosols
- *Thapto-humic Non-allophanic Kuroboku soils* ≈ Aluandic Andosols (Thaptomollic)
- *Typic Non-allophanic Kuroboku soils* ≈ Umbric Aluandic Andosols

Regosolic Kuroboku soils – soil group. ≈ Vitric Andosols
The following subgroups groups are distinguished within the group:

- *Thapto-humus Regosolic Kuroboku soils* ≈ Vitric Andosols (Thaptomollic)
- *Typic Regosolic Kuroboku soils* ≈ Vitric Andosols
- *Wet Regosolic Kuroboku soils* ≈ Vitric Andosols (Oxiaquic)

Wet Kuroboku soils – soil group. ≈ Andosols (Oxiaquic)
The following subgroups are distinguished within the group:

- *Cumulic Wet Kuroboku soils* ≈ Gleyic Andosols (Oxiaquic, *Novic*)
- *Non-allophanic Wet Kuroboku soils* ≈ Gleyic Aluandic Andosols (Oxiaquic)
- *Peaty Wet Kuroboku soils* ≈ Gleyic Histic Andosols (Oxiaquic)
- *Typic Wet Kuroboku soils* ≈ Gleyic Andosols (Oxiaquic)

Man-made soils – great soil group. ≈ Technosols / Anthrosols
Two groups are distinguished within the great group:

Landfill soils – soil group. ≈ Garbic Technosols / Terric Anthrosols
No subgroups are recognized within the group.

Reformed soils – soil group. ≈ Technosols
No subgroups are recognized within the group.

Peat soils – great soil group. ≈ Histosols
Three groups are distinguished within the great group:
High-moor Peat soils – soil group. ≈ Ombric Histosols
The following subgroups are distinguished within the group:

- *Fibric High-moor Peat soils* ≈ Ombric Fibric Histosols
- *Sapric High-moor Peat soils* ≈ Ombric Sapric Histosols
- *Terric High-moor Peat soils* ≈ *Terric* Ombric Histosols
- *Typic High-moor Peat soils* ≈ Ombric Histosols

Low-moor Peat soils – soil group. ≈ Rheic Histosols
The following subgroups are distinguished within the group:

- *Fibric Low-moor Peat soils* ≈ Rheic Fibric Histosols
- *Sapric Low-moor Peat soils* ≈ Rheic Sapric Histosols
- *Terric Low-moor Peat soils* ≈ *Terric* Rheic Histosols
- *Typic Low-moor Peat soils* ≈ Rheic Histosols

Transitional-moor Peat soils – soil group. ≈ Histosols
The following subgroups are distinguished within the group:

- *Fibric Transitional-moor Peat soils* ≈ Fibric Histosols
- *Sapric Transitional-moor Peat soils* ≈ Sapric Histosols
- *Terric Transitional-moor Peat soils* ≈ *Terric* Histosols
- *Typic Transitional-moor Peat soils* ≈ Histosols

Podzolic soils – great soil group. ≈ Podzols
One group is distinguished within the great group:
Podzolic soils – soil group. ≈ Podzols
The following subgroups are distinguished within the group:

- *Gleyic Podzolic soils* ≈ Gleyic Podzols
- *Peaty Podzolic soils* ≈ Gleyic Histic Podzols
- *Pseudogleyic Podzolic soils* ≈ Endogleyic Podzols
- *Surface-pseudogleyic Podzolic soils* ≈ Stagnic Podzols
- *Typic Podzolic soils* ≈ Podzols

Red Yellow soils – great soil group. ≈ Acrisols / Alisols / Lixisols / Cambisols / Nitisols
Two groups are distinguished within the great group:
Argic Red Yellow soils – soil group. ≈ Acrisols / Alisols / Lixisols

The following subgroups are distinguished within the group:

- *Albic Argic Red Yellow soils* ≈ Acrisols (Albic) / Alisols (Albic)
- *Anthraquic Argic Red Yellow soils* ≈ Acrisols (*Anthraquic*) / Alisols (*Anthraquic*)
- *Dark-reddish Argic Red Yellow soils* ≈ Acrisols (Rhodic) / Alisols (Rhodic)
- *Eutric Argic Red Yellow soils* ≈ Lixisols
- *Pseudogleyic Argic Red Yellow soils* ≈ Endogleyic Acrisols (Ferric) / Endogleyic Alisols (Ferric)
- *Surface-pseudogleyic Argic Red Yellow soils* ≈ Stagnic Acrisols / Stagnic Alisols
- *Typic Argic Red Yellow soils* ≈ Acrisols / Alisols

Cambic Red Yellow soils – soil group. ≈ Cambisols / Nitisols
The following subgroups are distinguished within the group:

- *Albic Cambic Red Yellow soils* ≈ Cambisols (Albic, Dystric)
- *Anthraquic Cambic Red Yellow soils* ≈ Cambisols (*Anthraquic*)
- *Dark-reddish Cambic Red Yellow soils* ≈ Nitisols
- *Eutric Cambic Red Yellow soils* ≈ Cambisols (Eutric)
- *Pseudogleyic Cambic Red Yellow soils* ≈ Endogleyic Cambisols (Ferric)
- *Surface-pseudogleyic Cambic Red Yellow soils* ≈ Stagnic Cambisols
- *Typic Cambic Red Yellow soils* ≈ Cambisols (Dystric)

Regosols – great soil group. ≈ Regosols / Arenosols / Leptosols
Four groups are distinguished within the great group:
Haplic Regosols – soil group. ≈ Regosols
The following subgroups are distinguished within the group:

- *Calcaric Haplic Regosols* ≈ Regosols (Calcaric)
- *Grumusol-like Haplic Regosols* ≈ *Vertic* Regosols
- *Rendzina-like Haplic Regosols* ≈ Regosols (Humic, Calcaric)
- *Typic Haplic Regosols* ≈ Regosols
- *Wet Haplic Regosols* ≈ Regosols (Oxiaquic)

Lithic Regosols – soil group. ≈ Leptosols
The following subgroups are distinguished within the group:

- *Calcaric Lithic Regosols* ≈ Leptosols (Calcaric)
- *Rendzina-like Lithic Regosols* ≈ Rendzic Regosols
- *Typic Lithic Regosols* ≈ Leptosols
- *Wet Lithic Regosols* ≈ Leptosols (Oxiaquic)

Sandy Regosols – soil group. ≈ Arenosols
The following subgroups are distinguished within the group:

- *Calcaric Sandy Regosols* ≈ Arenosols (Calcaric)
- *Rendzina-like Sandy Regosols* ≈ Arenosols (Calcaric, *Humic*)
- *Typic Sandy Regosols* ≈ Arenosols
- *Wet Sandy Regosols* ≈ Arenosols (*Oxiaquic*)

Volcanogeneous Regosols – soil group. ≈ Regosols (Tephric)
The following subgroups are distinguished within the group:

- *Typic Volcanogeneous Regosols* ≈ Regosols (Tephric)
- *Wet Volcanogeneous Regosols* ≈ Regosols (Tephric, Oxiaquic)

Stagnic soils – great soil group. ≈ Gleysols / Stagnosols
Two groups are distinguished within the great group:
Pseudogley soils – soil group. ≈ Gleysols
The following subgroups are distinguished within the group:

- *Aeric Pseudogley soils* ≈ Gleysols (*Oxiaquic*)
- *Gleyic Pseudogley soils* ≈ Gleysols
- *Typic Pseudogley soils* ≈ Gleysols

Stagnogley soils – soil group. ≈ Stagnosols
The following subgroups are distinguished within the group:

- *Peaty Stagnogley soils* ≈ Histic Stagnosols
- *Typic Stagnogley soils* ≈ Gleysols

References

Buytaert, W., Deckers, J. and Wyseure, G. (2006) 'Description and classification of nonallophanic Andosols in south Ecuadorian alpine grasslands (páramo)', *Geomorphology*, vol 73, nos 3–4, pp207–221

Committee for Soil Classification and Nomenclature (1986) 'Unified soil classification system of Japan (1st Approximation)', *Pedologist*, vol 30, no 2, pp123–139 (in Japanese)

Fourth Committee for Soil Classification and Nomenclature (2002) 'Unified soil classification system of Japan (2nd Approximation)', *Pedologist*, vol 46, no 1, pp36–45 (in Japanese)

Hirai, H. and Hamazaki, T. (2004) 'Historical aspects of soil classification in Japan', *Soil Science and Plant Nutrition*, vol 50, no 5, pp611–622

IUSS Working Group WRB (2006) *World Reference Base for Soil Resources*, 2nd edition, World Soil Resources Reports no 103, UN Food and Agriculture Organization, Rome

Matsui, T. (1982) 'An approximation to establish a unified comprehensive classification system for Japanese soils', *Soil Science and Plant Nutrition*, vol 28, no 2, pp235–256

Otowa, M. (1981) 'A soil classification system for soil survey of Japan', *Research Bulletin of Hokkaido Natural Agricultural Experimental Station*, vol 130, pp21–98

Soil Survey Staff (1999) *Soil Taxonomy: A Basic System of Soil Classification for Making and Interpreting Soil Surveys*, USDA, Handbook 436, 2nd edition, United States Government Printing Office, Washington DC, 696pp

25
Soil Classification of Brazil

Objectives and scope

Brazil is a dynamic developing country with a large economy. To a great extent the progress of Brazilian economy is based on its agriculture. Agricultural development required the inventory of soil resources which was started in Brazil at the end of the 1940s and continues until now (Jacomine and Camargo, 1996). Most of the national territory is covered with small- and medium-scale soil maps, including the Amazon region where biological diversity, the highest in the world, is regarded as a national heritage.

Geographically, the Brazilian classification covers only the national territory (Table 25.1). Soils transformed by humans are not included in the classification, although a few are classified as natural soils. Bare rock and subaquatic sediments are disregarded.

Table 25.1 *The scope of soil classification of Brazil*

Superficial bodies	Representation in the system
Natural soils	National coverage
Urban soils	Not included in the classification
Man-transported materials	Not included in the classification
Bare rock	Not considered as soils
Subaquatic soils	Not considered as soils
Soils deeply transformed by agricultural activities	Classified as if they are natural soils

Theoretical background

The Brazilian soil classification is similar to the US Soil Taxonomy (Soil Survey Staff, 1999). Its structure and even the names of the levels are identical with the American classification, as are many methods of diagnostics. Water and temperature regimes, however, are not included as diagnostic criteria in the upper levels of the Brazilian taxonomy. Much of the terminology is closer to that of the World Reference Base for Soil Resources (WRB) (IUSS Working Group WRB, 2006), than to that of Soil Taxonomy. In Brazil the names at lower levels are formed as a sequence of complete names from upper level taxa.

The system of soil series was borrowed from the US soil science school. However, in the US soil series appeared before taxonomic classification, and then soil taxonomy had to adjust itself to the existing empirical series. In contrast, in Brazil soil classification initially was made for higher taxonomic levels, and the series were recognized later and included in the existing taxa. In general, the US Soil Taxonomy is a main source for the Brazilian classification, but has been significantly transformed according to regional conditions and the history of the development of pedology in the country.

Development

Work on soil inventory of Brazil was started in 1947 (Jacomine and Camargo, 1996). However, before 1964 the soils were not described and classified systematically; thus it was necessary to built up a uniform system of soil classification and evaluation. Initially, the classification of Brazilian soils was very general (see, e.g. Finkl, 1982). Then the system was constructed 'from top to bottom', filling the taxa with empirical content and developing lower taxonomic levels (Beinroth, 1978). Such an approach suffered from a lack of sufficient soil survey data and it was impossible to create a classification by synthesizing field materials. On the contrary, the new classification was expected to serve as a basis for an extensive soil survey. Several options, like adopting the US Soil Taxonomy (the 7th Approximation was already published and many Latin American countries accepted it as a basic system for soil inventory) or the Food and Agriculture Organization (FAO) World Soil Map legend, had been discussed. However, the classification of tropical soils in the American taxonomy was not considered sufficient to be used for the needs of Brazil, and the FAO map was too general for practical soil survey (Costa de Lemos, 1968). Deeply weathered tropical soils were classified in detail in Brazil because these soils constitute a major proportion of the country. Since the 1960s the classification has gradually improved and become more detailed (Jacomine and Camargo, 1996). The latest available version was published in 1999 (EMBRAPA, 1999).

Structure

Soils are divided into 14 groups, called <u>orders</u>, at the first level (Table 25.2). These are groups at approximately the same level as reference groups in the WRB; some orders of Brazilian classification are broader (e.g. the NEOSSOLOS order includes *Arenosols*, *Leptosols*, *Regosols* and *Fluvisols* of the WRB, and corresponds more or less to the concept of the *Entisols* order in the US Soil Taxonomy). The orders are divided into <u>suborders</u>, based primarily on soil morphology and properties. The number of suborders is currently 44. The next level, <u>great groups</u>, is often generic; however, in some systems even suborders can be regarded as archetypes, and great groups seem to be specifications of 'central image' within these suborders (such as great groups *Distroficos* and *Eutroficos* in a number of suborders). Also, it is impossible to regard the suborder level as generic, because in many cases the suborders are too broad to be archetypes (like the NEOSSOLOS order). There are 150 great groups divided into 580 <u>subgroups</u> which are defined according to qualitative variations in soil properties. Soil <u>families</u> are distinguished within subgroups by lists of specific qualifiers for each soil (texture, mineralogical class, water and temperature regime, etc.). Soil <u>series</u> are distinguished within families as soil profiles with unique properties. The structure of the Brazilian classification is a hierarchical taxonomy with formal borders.

Table 25.2 *The structure of soil classification of Brazil*

Level	Taxon name	Taxon characteristics	Borders between classes	Diagnostics	Terminology
0	Soils				
1	Order	Collective	Formal	Chemico-morphological	Artificial
2	Suborder	Collective / Generic	Formal	Chemico-morphological	Artificial
3	Great group	Generic / Specific	Formal	Chemico-morphological	Artificial
4	Subgroup	Specific	Formal	Chemico-morphological	Artificial
5	Family	Varietal	Formal	Hydrologo-mineralogico-chemical	Mixed
6	Series	Varietal	Formal	Chemico-morphological	Traditional

Diagnostics

The diagnostics in Brazilian soil classification are selected on the basis of soil profile morphology and properties. Water and temperature regimes are taken into account only at the family level along with soil mineralogy. At all other hierarchical levels, quantitative morphological and analytical criteria are used to divide soil classes. The diagnostics can be called quantitative chemico-morphological.

Terminology

The terms used are completely artificial at the higher levels of taxonomy. Some traditional scientific terms are present at the family level, and at the series level soil names are represented by the locality name and soil texture class. The artificial terms are mainly borrowed from the WRB and, to a less extent, from the US Soil Taxonomy. The terms are written in Portuguese and are changed according to the rules of grammar of Portuguese. Capital letters are recommended for complete names of orders and suborders, and first letter capitalization for the additional great group names. Some terms are inherited from older Brazilian classifications (e.g. LATOSSOLOS), some were invented. It is interesting to note that some national classifications independently worked out almost the same terms. The changes in terminology in comparison with an older version of Brazilian classification are significant (e.g. Costa de Lemos, 1968; Beinroth, 1978), and special explanations are needed for understanding the internal correlation of different editions of the same classification.

Correlation

The correlation of the Brazilian classification with the WRB (IUSS Working Group WRB, 2006) was not a difficult task. First, the WRB borrowed from the Brazilian classification a number of concepts and approaches especially related to the tropical soils. Second, there was also a significant feedback from the WRB to the Brazilian classification, and some useful WRB concepts were accepted by the Brazilian colleagues. A usual problem for correlating tropical soils with the WRB terms is the low priority of soil colour in the WRB. Most classifications in tropical regions (Brazilian, Australian, Cuban and some others) use soil colour characteristics at the higher levels of taxonomy, while in the WRB chromatic modifiers are few in number and generally are last in the priority list of suffixes.

The correlation of Brazilian soil taxonomy terms is down to the great group level. Soil names are listed in the same order as in the original text of the classification.

ALISSOLOS – soil <u>order</u>. ≈ Alisols
Two <u>suborders</u> are distinguished within the order:
ALISSOLOS CRÔMICOS ≈ Alisols (Chromic)
Three <u>great groups</u> are distinguished within the suborder:

* *ALISSOLOS CRÔMICOS Argilúvicos* ≈ Cutanic Alisols (Chromic)
* *ALISSOLOS CRÔMICOS Húmicos* ≈ Alisols (Humic, Chromic)
* *ALISSOLOS CRÔMICOS Órticos* ≈ Alisols (Chromic)

ALISSOLOS HIPOCRÔMICOS ≈ Alisols
Two <u>great groups</u> are distinguished within the suborder:

* *ALISSOLOS HIPOCRÔMICOS Argilúvicos* ≈ Cutanic Alisols
* *ALISSOLOS HIPOCRÔMICOS Órticos* ≈ Alisols

ARGISSOLOS – soil <u>order</u>. ≈ Acrisols / Lixisols
Four <u>suborders</u> are distinguished within the order:
ARGISSOLOS ACINZENTADOS ≈ Acrisols / Lixisols
Two <u>great groups</u> are distinguished within the suborder:

- *ARGISSOLOS ACINZENTADOS Distróficos* ≈ Acrisols
- *ARGISSOLOS ACINZENTADOS Eutróficos* ≈ Lixisols

ARGISSOLOS AMARELOS ≈ Acrisols (*Xanthic*) / Lixisols (*Xanthic*)
Two <u>great groups</u> are distinguished within the suborder:

- *ARGISSOLOS AMARELOS Distróficos* ≈ Acrisols (*Xanthic*)
- *ARGISSOLOS AMARELOS Eutroficos* ≈ Lixisols (*Xanthic*)

ARGISSOLOS VERMELHO-AMARELOS ≈ Acrisols (Chromic) / Lixisols (Chromic)
Three <u>great groups</u> are distinguished within the suborder:

- *ARGISSOLOS VERMELHO-AMARELOS Alumínicos* ≈ Acrisols (Alumic, Chromic)
- *ARGISSOLOS VERMELHO-AMARELOS Distróficos* ≈ Acrisols (Chromic)
- *ARGISSOLOS VERMELHO-AMARELOS Eutróficos* ≈ Lixisols (Chromic)

ARGISSOLOS VERMELHOS ≈ Acrisols (Rhodic) / Lixisols (Rhodic)
Three <u>great groups</u> are distinguished within the suborder:

- *ARGISSOLOS VERMELHOS Distróficos* ≈ Acrisols (Rhodic)
- *ARGISSOLOS VERMELHOS Eutroférricos* ≈ Acrisols (Ferric, Rhodic) / Lixisols (Ferric, Rhodic)
- *ARGISSOLOS VERMELHOS Eutróficos* ≈ Lixisols (Rhodic)

CAMBISSOLOS – soil <u>order</u>. ≈ Cambisols
Three <u>suborders</u> are distinguished within the order:
CAMBISSOLOS HÍSTICOS ≈ Folic Cambisols
Two <u>great groups</u> are distinguished within the suborder:

- *CAMBISSOLOS HÍSTICOS Alumínicos* ≈ Folic Cambisols (Alumic)
- *CAMBISSOLOS HÍSTICOS Distróficos* ≈ Folic Cambisols (Dystric)

CAMBISSOLOS HAPLICOS ≈ Cambisols
Eleven <u>great groups</u> are distinguished within the suborder:

- *CAMBISSOLOS HAPLICOS Alumínicos* ≈ Cambisols (Alumic)
- *CAMBISSOLOS HAPLICOS Carbonáticos* ≈ Cambisols (Calcaric)
- *CAMBISSOLOS HAPLICOS Distroférricos* ≈ Cambisols (Dystric)
- *CAMBISSOLOS HAPLICOS Eutroférricos* ≈ Cambisols (Eutric)
- *CAMBISSOLOS HAPLICOS Petroférricos* ≈ Plinthic Cambisols
- *CAMBISSOLOS HAPLICOS Sálicos* ≈ Endosalic Cambisols
- *CAMBISSOLOS HAPLICOS Sódicos* ≈ Cambisols (Sodic)

- *CAMBISSOLOS HAPLICOS Ta Distróficos*[1] ≈ Cambisols (Dystric)
- *CAMBISSOLOS HAPLICOS Ta Eutróficos* ≈ Cambisols (Eutric)
- *CAMBISSOLOS HAPLICOS Tb Eutróficos* ≈ Cambisols (Eutric)
- *CAMBISSOLOS HAPLICOS Tb Distróficos* ≈ Cambisols (Dystric)

CAMBISSOLOS HÚMICOS ≈ Cambisols (Humic)
Four <u>great groups</u> are distinguished within the suborder:

- *CAMBISSOLOS HÚMICOS Alumínicos* ≈ Cambisols (Alumic, Humic)
- *CAMBISSOLOS HÚMICOS Alumnoférricos* ≈ Cambisols (Alumic, Humic)
- *CAMBISSOLOS HÚMICOS Distroférricos* ≈ Cambisols (Humic, Dystric)
- *CAMBISSOLOS HÚMICOS Distróficos* ≈ Cambisols (Humic, Dystric)

CHERNOSSOLOS – soil <u>order</u>. ≈ Chernozems / Phaeozems / Rendzic Leptosols
Four <u>suborders</u> are distinguished within the order:
CHERNOSSOLOS ARGILÚVICOS ≈ Luvic Chernozems / Luvic Phaeozems
Three <u>great groups</u> are distinguished within the suborder:

- *CHERNOSSOLOS ARGILÚVICOS Carbonáticos* ≈ Luvic Calcic Chernozems
- *CHERNOSSOLOS ARGILÚVICOS Férricos* ≈ Luvic Phaeozems (Chromicos)
- *CHERNOSSOLOS ARGILÚVICOS Órticos* ≈ Luvic Chernozems / Luvic Phaeozems

CHERNOSSOLOS EBÂNICOS ≈ Chernic Chernozems
Two <u>great groups</u> are distinguished within the suborder:

- *CHERNOSSOLOS EBÂNICOS Carbonáticos* ≈ Calcic Chernic Chernozems
- *CHERNOSSOLOS EBÂNICOS Órticos* ≈ Chernic Chernozems

CHERNOSSOLOS HÁPLICOS ≈ Chernozems / Phaeozems
Three <u>great groups</u> are distinguished within the suborder:

- *CHERNOSSOLOS HÁPLICOS Carbonáticos* ≈ Calcic Chernozems
- *CHERNOSSOLOS HÁPLICOS Férricos* ≈ Phaeozems (Chromic)
- *CHERNOSSOLOS HÁPLICOS Órticos* ≈ Chernozems

---

[1]Symbol Ta at the great groups level means soils with high activity clays, and Tb, soils with low activity clays

CHERNOSSOLOS RÊNDZICOS ≈ Rendzic Leptosols / Rendzic Phaeozems
Two <u>great groups</u> are distinguished within the suborder:

- *CHERNOSSOLOS RÊNDZICOS Líticos* ≈ Rendzic Leptosols
- *CHERNOSSOLOS RÊNDZICOS Saprolíticos* ≈ Rendzic Phaeozems

ESPODOSSOLOS – soil <u>order</u>. ≈ Podzols
Two <u>suborders</u> are distinguished within the order:
ESPODOSSOLOS CÁRBICOS ≈ Carbic Podzols
Three <u>great groups</u> are distinguished within the suborder:

- *ESPODOSSOLOS CÁRBICOS Hidromórficos* ≈ Gleyic Albic Carbic Podzols
- *ESPODOSSOLOS CÁRBICOS Hiperespressos* ≈ Albic Carbic Podzols
- *ESPODOSSOLOS CÁRBICOS Órticos* ≈ Albic Carbic Podzols

ESPODOSSOLOS FERROCÁRBICOS ≈ Podzols
Three great groups are distinguished within the suborder:

- *ESPODOSSOLOS FERROCÁRBICOS Hidromórficos* ≈ Gleyic Albic Podzols
- *ESPODOSSOLOS FERROCÁRBICOS Hiperespressos* ≈ Albic Podzols
- *ESPODOSSOLOS FERROCÁRBICOS Órticos* ≈ Albic Podzols

GLEISSOLOS – soil <u>order</u>. ≈ Gleysols
Four <u>suborders</u> are distinguished within the order:
GLEISSOLOS HÁPLICOS ≈ Gleysols
Six <u>great groups</u> are distinguished within the suborder:

- *GLEISSOLOS HÁPLICOS Alumínicos* ≈ Gleysols (Alumic)
- *GLEISSOLOS HÁPLICOS Ta Distróficos* ≈ Gleysols (Dystric)
- *GLEISSOLOS HÁPLICOS Ta Carbonáticos* ≈ Calcic Gleysols
- *GLEISSOLOS HÁPLICOS Ta Eutróficos* ≈ Gleysols (Eutric)
- *GLEISSOLOS HÁPLICOS Tb Distróficos* ≈ Gleysols (Dystric)
- *GLEISSOLOS HÁPLICOS Tb Eutróficos* ≈ Gleysols (Eutric)

GLEISSOLOS MELÂNICOS ≈ Umbric Gleysols / Mollic Gleysols
Four <u>great groups</u> are distinguished within the suborder:

- *GLEISSOLOS MELÂNICOS Alumínicos* ≈ Umbric Gleysols (Alumic)
- *GLEISSOLOS MELÂNICOS Carbonáticos* ≈ Calcic Mollic Gleysols
- *GLEISSOLOS MELÂNICOS Distróficos* ≈ Umbric Gleysols
- *GLEISSOLOS MELÂNICOS Eutróficos* ≈ Mollic Gleysols

GLEISSOLOS SÁLICOS ≈ Endosalic Gleysols / Gleysols (Sodic) / Gleyic Solonchaks

Two <u>great groups</u> are distinguished within the suborder:

- *GLEISSOLOS SÁLICOS Órticos* ≈ Endosalic Gleysols / Gleyic Solonchaks
- *GLEISSOLOS SÁLICOS Sódicos* ≈ Gleysols (Sodic)

GLEISSOLOS TIOMÓRFICOS ≈ Gleysols (Thionic)
Three <u>great groups</u> are distinguished within the suborder:

- *GLEISSOLOS TIOMÓRFICOS Hísticos* ≈ Histic Gleysols (Thionic)
- *GLEISSOLOS TIOMÓRFICOS Húmicos* ≈ Gleysols (Thionic, Humic)
- *GLEISSOLOS TIOMÓRFICOS Órticos* ≈ Gleysols (Thionic)

LATOSSOLOS – soil <u>order</u>. ≈ Ferralsols
Four <u>suborders</u> are distinguished within the order:
LATOSSOLOS AMARELOS ≈ Ferralsols (Xanthic)
Six great groups are distinguished within the suborder:

- *LATOSSOLOS AMARELOS Ácricos* ≈ Acric Ferralsols (Xanthic)
- *LATOSSOLOS AMARELOS Acriférricos* ≈ Acric Ferralsols (Xanthic)
- *LATOSSOLOS AMARELOS Coesos* ≈ Ferralsols (Xanthic)
- *LATOSSOLOS AMARELOS Distroférricos* ≈ Ferralsols (Dystric, Xanthic)
- *LATOSSOLOS AMARELOS Distróficos* ≈ Ferralsols (Dystric, Xanthic)
- *LATOSSOLOS AMARELOS Eutróficos* ≈ Ferralsols (Eutric, Xanthic)

LATOSSOLOS BRUNOS ≈ Ferralsols (Humic)
Three <u>great groups</u> are distinguished within the suborder:

- *LATOSSOLOS BRUNOS Ácricos* ≈ Acric Ferralsols (Humic)
- *LATOSSOLOS BRUNOS Alumínicos* ≈ Ferralsols (Alumic, Humic)
- *LATOSSOLOS BRUNOS Distróficos* ≈ Ferralsols (Humic, Dystric)

LATOSSOLOS VERMELHO-AMARELOS ≈ Ferralsols (Chromic)
Five <u>great groups</u> are distinguished within the suborder:

- *LATOSSOLOS VERMELHO-AMARELOS Ácricos* ≈ Acric Ferralsols (Chromic)
- *LATOSSOLOS VERMELHO-AMARELOS Acriférricos* ≈ Acric Ferralsols (Chromic)
- *LATOSSOLOS VERMELHO-AMARELOS Distroférricos* ≈ Ferralsols (Dystric, Chromic)
- *LATOSSOLOS VERMELHO-AMARELOS Distróficos* ≈ Geric Ferralsols (Dystric, Chromic)
- *LATOSSOLOS VERMELHO-AMARELOS Eutróficos* ≈ Ferralsols (Eutric, Chromic)

LATOSSOLOS VERMELHOS ≈ Ferralsols (Rhodic)
Eight <u>great groups</u> are distinguished within the suborder:

- *LATOSSOLOS VERMELHOS Ácricos* ≈ Acric Ferralsols (Rhodic)
- *LATOSSOLOS VERMELHOS Acriférricos* ≈ Acric Ferralsols (Rhodic)
- *LATOSSOLOS VERMELHOS Alumnoférricos* ≈ Ferralsols (Alumic, Rhodic)
- *LATOSSOLOS VERMELHOS Distroférricos* ≈ Ferralsols (Dystric, Rhodic)
- *LATOSSOLOS VERMELHOS Distróficos* ≈ Ferralsols (Dystric, Rhodic)
- *LATOSSOLOS VERMELHOS Eutroférricos* ≈ Ferralsols (Eutric, Rhodic)
- *LATOSSOLOS VERMELHOS Eutróficos* ≈ Ferralsols (Eutric, Rhodic)
- *LATOSSOLOS VERMELHOS Petroférricos* ≈ Plinthic Ferralsols (Rhodic)

LUVISSOLOS – soil <u>order</u>. ≈ Luvisols
Two <u>suborders</u> are distinguished within the order:
LUVISSOLOS CRÔMICOS ≈ Luvisols (Chromic)
Three <u>great groups</u> are distinguished within the suborder:

- *LUVISSOLOS CRÔMICOS Carbonáticos* ≈ Calcic Luvisols (Chromic)
- *LUVISSOLOS CRÔMICOS Órticos* ≈ Luvisols (Chromic)
- *LUVISSOLOS CRÔMICOS Pálicos* ≈ Luvisols (Profondic, Chromic)

LUVISSOLOS HIPOCRÔMICOS ≈ Luvisols
Two <u>great groups</u> are distinguished within the suborder:

- *LUVISSOLOS HIPOCRÔMICOS Carbonáticos* ≈ Calcic Luvisols
- *LUVISSOLOS HIPOCRÔMICOS Órticos* ≈ Luvisols

NEOSSOLOS – soil <u>order</u>. ≈ Fluvisols / Leptosols / Regosols / Arenosols
Four <u>suborders</u> are distinguished within the order:
NEOSSOLOS FLÚVICOS (Solos Aluviais) ≈ Fluvisols
Seven <u>great groups</u> are distinguished within the suborder:

- *NEOSSOLOS FLÚVICOS Carbonáticos* ≈ Calcic Fluvisols
- *NEOSSOLOS FLÚVICOS Psamíticos* ≈ Fluvisols (Arenic)
- *NEOSSOLOS FLÚVICOS Sálicos* ≈ Salic Fluvisols
- *NEOSSOLOS FLÚVICOS Sódicos* ≈ Fluvisols (Sodic)
- *NEOSSOLOS FLÚVICOS Ta Eutróficos* ≈ Fluvisols (Eutric)
- *NEOSSOLOS FLÚVICOS Tb Distróficos* ≈ Fluvisols (Dystric)
- *NEOSSOLOS FLÚVICOS Tb Eutróficos* ≈ Fluvisols (Eutric)

NEOSSOLOS LITÓLICOS (Solos Litólicos) ≈ Leptosols / Leptic Histosols
Six <u>great groups</u> are distinguished within the suborder:

- *NEOSSOLOS LITÓLICOS Carbonáticos* ≈ Calcic Leptosols
- *NEOSSOLOS LITÓLICOS Distróficos* ≈ Leptosols (Dystric) / Epileptic Regosols (Dystric)
- *NEOSSOLOS LITÓLICOS Eutróficos* ≈ Leptosols (Eutric) / Epileptic Regosols (Eutric)
- *NEOSSOLOS LITÓLICOS Hísticos* ≈ Histic Leptosols / Epileptic Histosols
- *NEOSSOLOS LITÓLICOS Húmicos* ≈ Umbric Leptosols / Mollic Leptosols
- *NEOSSOLOS LITÓLICOS Psamíticos* ≈ Leptosols (*Arenic*) / Epileptic Regosols (Arenic)

NEOSSOLOS REGOLÍTICOS (Regossolos) ≈ Regosols
Three great groups are distinguished within the suborder:

- *NEOSSOLOS REGOLÍTICOS Distróficos* ≈ Regosols (Dystric)
- *NEOSSOLOS REGOLÍTICOS Eutróficos* ≈ Regosols (Eutric)
- *NEOSSOLOS REGOLÍTICOS Psamíticos* ≈ Regosols (Arenic)

NEOSSOLOS QUARZARÊNICOS (Areias Quartzosas) ≈ Arenosols
Two great groups are distinguished within the suborder:

- *NEOSSOLOS QUARZARÊNICOS Hidromórficos* ≈ Gleyic Arenosols
- *NEOSSOLOS QUARZARÊNICOS Órticos* ≈ Arenosols

NITOSSOLOS – soil order. ≈ Nitisols
Two suborders are distinguished within the order:
NITOSSOLOS HÁPLICOS ≈ Nitisols
Three great groups are distinguished within the suborder:

- *NITOSSOLOS HÁPLICOS Alumínicos* ≈ Nitisols (Alumic)
- *NITOSSOLOS HÁPLICOS Distróficos* ≈ Nitisols (Dystric)
- *NITOSSOLOS HÁPLICOS Eutróficos* ≈ Nitisols (Eutric)

NITOSSOLOS VERMELHOS ≈ Nitisols (Rhodic)
Four great groups are distinguished within the suborder:

- *NITOSSOLOS VERMELHOS Distroférricos* ≈ Nitisols (Dystric, Rhodic)
- *NITOSSOLOS VERMELHOS Distróficos* ≈ Nitisols (Dystric, Rhodic)
- *NITOSSOLOS VERMELHOS Eutroférricos* ≈ Nitisols (Eutric, Rhodic)
- *NITOSSOLOS VERMELHOS Eutróficos* ≈ Nitisols (Eutric, Rhodic)

ORGANOSSOLOS – soil order. ≈ Histosols
Four suborders are distinguished within the order:
ORGANOSSOLOS FÓLICOS ≈ Folic Histosols
Three great groups are distinguished within the suborder:

- *ORGANOSSOLOS FÓLICOS Fíbricos* ≈ Fibric Folic Histosols
- *ORGANOSSOLOS FÓLICOS Hêmicos* ≈ Hemic Folic Histosols
- *ORGANOSSOLOS FÓLICOS Sápricos* ≈ Sapric Folic Histosols

ORGANOSSOLOS HÁPLICOS ≈ Histosols
Three <u>great groups</u> are distinguished within the suborder:

- *ORGANOSSOLOS HÁPLICOS Fíbricos* ≈ Fibric Histosols
- *ORGANOSSOLOS HÁPLICOS Hêmicos* ≈ Hemic Histosols
- *ORGANOSSOLOS HÁPLICOS Sápricos* ≈ Sapric Histosols

ORGANOSSOLOS MÉSICOS ≈ Histosols
Three <u>great groups</u> are distinguished within the suborder:

- *ORGANOSSOLOS MÉSICOS Fíbricos* ≈ Fibric Histosols
- *ORGANOSSOLOS MÉSICOS Hêmicos* ≈ Hemic Histosols
- *ORGANOSSOLOS MÉSICOS Sápricos* ≈ Sapric Histosols

ORGANOSSOLOS TIOMÓRFICOS ≈ Histosols (Thionic)
Three <u>great groups</u> are distinguished within the suborder:

- *ORGANOSSOLOS TIOMÓRFICOS Fíbricos* ≈ Fibric Histosols (Thionic)
- *ORGANOSSOLOS TIOMÓRFICOS Hêmicos* ≈ Hemic Histosols (Thionic)
- *ORGANOSSOLOS TIOMÓRFICOS Sápricos* ≈ Sapric Histosols (Thionic)

PLANOSSOLOS – soil <u>order</u>. ≈ Planosols
Three <u>suborders</u> are distinguished within the order:
PLANOSSOLOS HÁPLICOS ≈ Planosols
Three <u>great groups</u> are distinguished within the suborder:

- *PLANOSSOLOS HÁPLICOS Distróficos* ≈ Planosols (Dystric)
- *PLANOSSOLOS HÁPLICOS Eutróficos* ≈ Planosols (Eutric)
- *PLANOSSOLOS HÁPLICOS Sálicos* ≈ Endosalic Planosols

PLANOSSOLOS HIDROMÓRFICOS ≈ Gleyic Planosols
Three <u>great groups</u> are distinguished within the suborder:

- *PLANOSSOLOS HIDROMÓRFICOS Distróficos* ≈ Gleyic Planosols (Dystric)
- *PLANOSSOLOS HIDROMÓRFICOS Eutróficos* ≈ Gleyic Planosols (Eutric)
- *PLANOSSOLOS HIDROMÓRFICOS Sálicos* ≈ Endosalic Gleyic Planosols

PLANOSSOLOS NÁTRICOS ≈ Solodic Planosols (Sodic)
Three <u>great groups</u> are distinguished within the suborder:

- *PLANOSSOLOS NÁTRICOS Carbonáticos* ≈ Calcic Solodic Planosols (Sodic)
- *PLANOSSOLOS NÁTRICOS Órticos* ≈ Solodic Planosols (Sodic)
- *PLANOSSOLOS NÁTRICOS Sálicos* ≈ Endosalic Solodic Planosols (Sodic)

PLINTOSSOLOS – soil <u>order</u>. ≈ Plinthosols
Three <u>suborders</u> are distinguished within the order:
PLINTOSSOLOS ARGILÚVICOS ≈ Acric Plinthosols / Lixic Plinthosols
Three <u>great groups</u> are distinguished within the suborder:

- *PLINTOSSOLOS ARGILÚVICOS Alumínicos* ≈ Acric Plinthosols (Alumic)
- *PLINTOSSOLOS ARGILÚVICOS Distróficos* ≈ Acric Plinthosols
- *PLINTOSSOLOS ARGILÚVICOS Eutróficos* ≈ Lixic Plinthosols

PLINTOSSOLOS HÁPLICOS ≈ Plinthosols
Two <u>great groups</u> are distinguished within the suborder:

- *PLINTOSSOLOS HÁPLICOS Distróficos* ≈ Plinthosols (Dystric)
- *PLINTOSSOLOS HÁPLICOS Eutróficos* ≈ Plinthosols (Eutric)

PLINTOSSOLOS PÉTRICOS ≈ Petric Plinthosols / Pisolithic Plinthosols
Three <u>great groups</u> are distinguished within the suborder:

- *PLINTOSSOLOS PÉTRICOS Concrecionários Dystróficos* ≈ Pisolithic Plinthosols (Dystric)
- *PLINTOSSOLOS PÉTRICOS Concrecionários Eutróficos* ≈ Pisolithic Plinthosols (Eutric)
- *PLINTOSSOLOS PÉTRICOS Litoplínticos* ≈ Petric Plinthosols

VERTISSOLOS – soil <u>order</u>. ≈ Vertisols
Three <u>suborders</u> are distinguished within the order:
VERTISSOLOS CROMADOS ≈ Vertisols (Chromic)
Four great groups are distinguished within the suborder:

- *VERTISSOLOS CROMADOS Carbonáticos* ≈ Calcic Vertisols (Chromic)
- *VERTISSOLOS CROMADOS Órticos* ≈ Vertisols (Chromic)
- *VERTISSOLOS CROMADOS Sálicos* ≈ Salic Vertisols (Chromic)
- *VERTISSOLOS CROMADOS Sódicos* ≈ Sodic Vertisols (Chromic)

VERTISSOLOS EBÂNICOS ≈ Pellic Vertisols
Three <u>great groups</u> are distinguished within the suborder:

- *VERTISSOLOS EBÂNICOS Carbonáticos* ≈ Calcic Vertisols (Pellic)
- *VERTISSOLOS EBÂNICOS Órticos* ≈ Vertisols (Pellic)
- *VERTISSOLOS EBÂNICOS Sódicos* ≈ Vertisols (Sodic, Pellic)

VERTISSOLOS HIDROMÓRFICOS ≈ Gleyic Vertisols
Four <u>great groups</u> are distinguished within the suborder:

- *VERTISSOLOS HIDROMÓRFICOS Carbonáticos* ≈ Calcic Gleyic Vertisols
- *VERTISSOLOS HIDROMÓRFICOS Órticos* ≈ Gleyic Vertisols
- *VERTISSOLOS HIDROMÓRFICOS Sálicos* ≈ Gleyic Salic Vertisols
- *VERTISSOLOS HIDROMÓRFICOS Sódicos* ≈ Gleyic Vertisols (Sodic)

References

Beinroth, F. H. (1978) 'Relationship between U.S. Soil Taxonomy, the Brasilian soil classification system and FAO/UNESCO soil units', in E. Bornemisza and A. Alvarado (eds) *Soil Management in Tropical America*, Proceedings of a Seminar held at CIAT, Cali, Colombia, 10–14 February 1974, Gordon Press, Raleigh, NC, pp92–108

Costa de Lemos, R. (1968) 'The main tropical soils of Brasil', *Approaches to Soil Classification*, FAO World Soil Resources Report no 32, UN Food and Agriculture Organization, Rome, pp95–106

EMBRAPA (1999) *Sistema Brasileiro de Clasificação de solos*, Embrapa Produção de Informação, Brasília – Embrapa Solos, Rio de Janeiro, 412pp

Finkl, C. W. Jr. (ed) (1982) *Soil Classification*, Benchmark Papers in Soil Science, vol 1, Hutchinson Ross Publishing Company, Stroudsburg, PA, 391pp

IUSS Working Group WRB (2006) *World Reference Base for Soil Resources*, 2nd edition, World Soil Resources Reports no 103, UN Food and Agriculture Organization, Rome, 128pp

Jacomine, P. T. K. and Camargo, M. N. (1996) 'Classificaçao pedologica nacional em vigor', in V. H. Alvarez, L. E. F. Fontes and M. P. F. Fontes (eds) *O solo nos grandes dominios morfoclimaticos do Brasil o e desenvolvimento sustentado*, SBCS-UFV, Viçosa, Brazil, pp675–689

Soil Survey Staff (1999) *Soil Taxonomy: A Basic System of Soil Classification for Making and Interpreting Soil Surveys*, USDA, Handbook 436, 2nd edition, United States Government Printing Office, Washington DC, 696pp

26
Soil Classification of Cuba

written together with Alberto Hernández Jiménez

Objectives and scope

Cuba is one of the few Latin American countries that uses its own soil classification. Apart from certain political reasons, the importance of soil classification for Cuba is determined by the high level of development of agriculture in the country. From the beginning of the Spanish colonization the island was recognized for its warm tropical climate, and a number of tropical crops, especially sugar cane, were cultivated there.

The country has very little natural vegetation, and much of the land is used for agriculture, thus knowledge about soil resources has been critical for the development of agriculture. The classification of soils is intended for soil inventory, interpretation and mapping at any scale in Cuba (Table 26.1). The system does not include superficial substrates of urban and industrial areas. Some transported ground may be included in the type *Recultivado Antrópico* ('Recultivated anthropogenic soils'), and most soils deeply transformed by agriculture are included in *Antrosoles*. Underwater sediments and bare rock are not regarded as soils.

Theoretical background

Soil classification of Cuba is based on the concepts of soil genesis, insofar as all properties of soils result from certain processes.

The soils classification of Cuba was influenced by four main schools of pedology: Soviet (Russian), French, American and Chinese. The Soviet and Chinese schools gave special attention to soil-forming processes. The French school gave the principles of soil evolution based on the mineralogical and geochemical approach to the study and classification of tropical soils. The American school was a source of overall structure of classification and the

Table 26.1 *The scope of soil classification of Cuba*

Superficial bodies	Representation in the system
Natural soils	National coverage
Urban soils	Not included in the classification
Transported materials	Not generally included in the classification; partly included in *Recultivado Antrópico*
Bare rock	Not considered as soils
Subaquatic soils	Not considered as soils
Soils deeply transformed by agricultural activities	Included in *Antrosoles*

use of diagnostic horizons. The Cuban system is not a mere compilation of the classifications of the countries mentioned above. The Cuban school has produced an original soil classification.

Development

In the beginning of the 20th century the soils of Cuba were actively studied by US soil scientists, and the book *The Soils of Cuba* (Bennett and Allison, 1928) was highly regarded by Cuban soil science. Later soil scientists actively collaborated with specialists from the Soviet Union and China who proposed a number of general classification schemes for Cuban soils. Important contributions were made by French specialists with vast experience in studying soils of the tropics. The first draft of genetic soil classification was elaborated for a basic soil map (scale 1:250,000) in collaboration with specialists from the Soil Science Institute of Nanjing (Hernández Jiménez et al, 1971; Instituto de Suelos, 1971). An increasing demand for agricultural production in the country required a more detailed soil classification for soil maps at 1:25,000 (Dirección Nacional de Suelos y Fertilizantes, 1990). This classification built on results of the first version, and added new results obtained in collaboration with Soviet and French specialists (Instituto de Suelos, 1975). This classification was used in Cuba for 20 years before it was decided to revise it (Instituto de Suelos, 1999). The second version was slightly expanded, and its diagnostics were adjusted to current concepts. A detailed history of the development of soil classification in Cuba is presented in a recent monograph (Hernández Jiménez et al, 2006) that is recommended to everybody interested in the history of soil science.

Structure

The upper, collective level of soil taxonomy of Cuba is a group of types (Table 26.2). These groups bring together soils according to the main pedogenetic processes resulting in certain diagnostic horizons. Soil types are the basic, generic level of classification and represent 'central images' of existing soils. Soil subtypes are distinguished within types according to qualitative differences in their properties, or as transitions between soil types. Soil <u>genera</u> are distinguished within subtypes according to quantitative or semiquantitative characteristics of chemical, mineralogical and other properties of the soil or their modifications. The <u>species</u> within the same genus are grouped by the thickness and quantity of organic matter of the A horizon. Differentiation of <u>varieties</u> within a species is made by the mechanical composition of the soil's A horizon. The classification is a hierarchical taxonomy with formal borders.

Table 26.2 *The structure of soil classification of Cuba*

Level	Taxon name	Taxon characteristics	Borders between classes	Diagnostics	Terminology
0	Soils	Kingdom			
1	Group	Collective	Mostly formal	Chemico-morphological	Traditional / artificial
2	Type	Generic	Mostly formal	Chemico-morphological	Traditional / artificial
3	Subtype	Specific	Mostly formal	Chemico-morphological	Traditional / artificial
4	Genus	Specific 2	Formal	Chemico-morphological	Mainly traditional
5	Species	Varietal 1	Formal	Chemico-morphological	Traditional
6	Variety	Varietal 2	Formal	Texture	Traditional

Diagnostics

Soil diagnostics are made on the basis of the morphology, chemical properties and mineralogical composition of a given profile. The diagnostics of some soils require extensive chemical analyses, but they are not obligatory. If soil morphology permits making definite diagnostics in the field, that will suffice, and chemical analyses are only applied in doubtful cases. No moisture or temperature regimes criteria are used. The soil diagnostics in Cuba can be called quantitative chemico-morphological.

Terminology

The classification uses mainly traditional scientific terms (*Alítico, Ferrítico, Halomórficos*, etc.) for soil groups, with some terms borrowed from the World Reference Base for Soil Resources (WRB) (*Antrosol, Fluvisol, Vertisol*). For soil types and lower levels of the taxonomy a mixture of artificial terms is used similar to those of the WRB, along with traditional soil names. The terminology may be called mixed traditional and artificial.

Correlation

A correlation of the terms of soil classification of Cuba (Instituto de Suelos, 1999) with the terms of the WRB is provided down to the type level (Hernández Jiménez et al, 2004). Soil groups and their included soil types are listed alphabetically.

Alítico – soil group. ≈ Alisols / Acrisols
There are five soil types within the group:

- *Alta Actividad Arcillosa Amarillento* ≈ Alisols (*Xanthic*)
- *Alta Actividad Arcillosa Rojo Amarillento* ≈ Cutanic Alisols (Chromic)
- *Baja Actividad Arcillosa Amarillento* ≈ Cutanic Acrisols (Alumic, Xanthic)
- *Baja Actividad Arcillosa Rojo* ≈ Cutanic Acrisols (Alumic, Rhodic)
- *Baja Actividad Arcillosa Rojo Amarillento* ≈ Cutanic Acrisols (Alumic, Chromic)

Antrosol – soil group. ≈ Anthrosols / Technosols
There are three soil types within the group:

- *Hidromórfico Antrópico* ≈ Hydragric Anthrosols / Endogleyic Anthrosols
- *Recultivado Antrópico* ≈ Terric Anthrosols / Garbic Technosols / Histic Technosols
- *Salino Antrópico* ≈ Salic Anthrosols

Ferrálico – soil group. ≈ Ferralic Cambisols / Nitisols
There are two soil types within the group:

- *Ferrálico Amarillento* ≈ Ferralic Cambisols (*Xanthic*) / Lixisols (*Xanthic*)
- *Ferrálico Rojo* ≈ Ferralic Cambisols (Rhodic) / Nitisols (Rhodic)

Ferralítico – soil group. ≈ Ferralsols / Nitisols / Acrisols
There are three soil types within the group:

- *Ferralítico Amarillento Lixivado* ≈ Acric Ferralsols (Xanthic) / Acric Nitisols (*Xanthic*)
- *Ferralítico Rojo* ≈ Ferralsols (Rhodic) / Nitisols (Rhodic)
- *Ferralítico Rojo Lixivado* ≈ Acric Ferralsols (Rhodic) / Acric Nitisols (Rhodic)

Ferrítico – soil group. ≈ Ferralsols / Nitisols
There are two soil types within the group:

- *Ferrítico Amarillo* ≈ Ferralsols (Rhodic) / Nitisols (*Xanthic*)
- *Ferrítico Rojo Oscuro* ≈ Ferralsols (Rhodic) / Nitisols (Rhodic)

Fersialítico – soil <u>group</u>. ≈ Cambisols / Luvisols
There are three soil <u>types</u> within the group:

- *Fersialítico Amarillento* ≈ Cambisols (*Xanthic*)
- *Fersialítico Pardo Rojizo* ≈ Cambisols (Chromic) / Luvisols (Chromic)
- *Fersialítico Rojo* ≈ Cambisols (Chromic) / Luvisols (Chromic)

Fluvisol – soil <u>group</u>. ≈ Fluvisols
Only one soil <u>type</u> is in this group:

- *Fluvisol* ≈ Fluvisols

Halomórficos – soil <u>group</u>. ≈ Solonchak / Solonetz
There are two soil <u>types</u> within the group:

- *Salíno* ≈ Solonchaks
- *Sódico* ≈ Solonetz

Hidromórficos – soil <u>group</u>. ≈ Gleysols / Plinthisols
There are three soil <u>types</u> within the group:

- *Gley Húmico* ≈ Histic Gleysols / Mollic Gleysols / Umbric Gleysols / Gleysols (Humic)
- *Gley Nodular Ferruginoso* ≈ Gleysols (*Ferric*) / Pisolithic Plinthisols
- *Gley Vértico* ≈ Gleyic Vertisols

Histosol – soil <u>group</u>. ≈ Histosols / Gleysols
There are three soil <u>types</u> within the group:

- *Histosol Fíbrico* ≈ Fibric Histosols / Histic Gleysols
- *Histosol Mésico* ≈ Hemic Histosols / Histic Gleysols
- *Histosol Sáprico* ≈ Sapric Histosols / Histic Gleysols

Húmico sialítico – soil <u>group</u>. ≈ Rendzic Leptosols / Phaeozems
There are two soil <u>types</u> within the group:

- *Húmico Calcimórfico* ≈ Phaeozems (Calcaric)
- *Rendzina* ≈ Rendzic Leptosols / Leptosols (Calcaric, *Rhodic*) / Rendzic Phaeozems

Pardo Sialítico – soil <u>group</u>. ≈ Cambisols
There are two soil <u>types</u> within the group:

- *Pardo* ≈ Cambisols (Eutric)
- *Pardo Grisáceo* ≈ Cambisols

Poco evolucionado – soil <u>group</u>. ≈ Arenosols / Leptosols
There are three soil <u>types</u> within the group:

- *Arenosol* ≈ Arenosols
- *Lithosol* ≈ Lithic Leptosols
- *Protorendzina* ≈ Rendzi-Lithic Leptosols / Lithic Leptosols (Calcaric, *Rhodic*)

Vertisol – soil <u>group</u>. ≈ Vertisols
There are two soil <u>types</u> within the group:

- *Vertisol Pélico* ≈ Vertisols (Pellic)
- *Vertisol Crómico* ≈ Vertisols (Chromic)

References

Bennett, H. H. and Allison, R. V. (1928) *The Soils of Cuba*, Tropical Plant Research Foundation, Washington DC, 410pp

Dirección Nacional de Suelos y Fertilizantes (1990) *El Mapa Genético, Escala 1:25 000 de los Suelos de Cuba*, X Congreso Latinoamericano de la Ciencia del Suelo, Ministerio de Agricultura, La Habana

Hernández Jiménez, A., Ascanio García, M. O and Pérez Jiménez, J. M. (1971) 'Informe sobre el mapa genético de los suelos de Cuba en escala 1:250 000', La Habana, *Revista de Agricultura*, no 4, pp1–20.

Hernández Jiménez, A., Ascanio García, M. O., Cabrera Rodríguez, A., Moralez Díaz, M. and Medina Basso, N. (2004) 'Correlación de la nueva versión de clasificación genética de los suelos de Cuba con World Reference Base', in A. Hernández Jiménez and M. O. Ascanio García (eds) *Problemas Actuales de Clasificación de Suelos: Énfasis en Cuba*, Universidad Veracruzana, Xalapa, Ver., México, pp203–221

Hernández Jiménez, A., Ascanio García, M. O., Moralez Díaz, M. and León Valido, A. (2006) *La Historia de la Clasificación de Suelos en Cuba*, editorial Félix Varela, Instituto Nacional de Ciencias Agricolas, La Habana, 98pp.

Instituto de Suelos (1971) 'Mapa genético de los suelos de Cuba en escala 1:250 000, 19 hojas a color', Instituto Cubano de Geodesia y Cartografía, La Habana

Instituto de Suelos (1975) *Segunda Clasificación Genética de los Suelos de Cuba*, Serie Suelos 23, Academia de Ciencias de Cuba, Ciudad de La Habana, 36pp

Instituto de Suelos (1999) *Nueva Versión de Clasificación genética de los Suelos de Cuba*, AGRINFOR, Ministerio de la Agricultura, Ciudad de La Habana, 64pp

27
Australian Soil Classification

Objectives and scope

Australia has a developed and extensive agricultural sector; grazing is especially important. Processes of land degradation and desertification are widespread in Australia, and soil science provides a necessary scientific basis for
land conservation, improvement and reclamation. It is quite natural that soil
classifications were developed in this country as early as the beginning of the
20th century. The objectives of the Australian classification were to serve as a
basis for soil inventory and mapping. The recent version of soil classification
of Australia (Isbell, 2002) has the same aim and has national coverage (Table
27.1).

The soil cover of the continent is unique and is not readily compared with
other parts of the world. An attempt to use the same system of soil classification
in Australia and New Zealand failed (Hewitt, 1992). Soils transformed by
humans are included in the classification, both agricultural lands and industrial
and urban substrates. The classification even covers unusual areas like airports,
golf grounds and 'scalped' soils (where the soil has been removed down to
parent material), which are not recognized as soils in most other classifications.
Underwater soils are partially included in the classification to the suborder
Intertidal Hydrosols. Bare rock is not regarded as soil.

Table 27.1 *The scope of the Australian soil classification*

Superficial bodies	Representation in the system
Natural soils	National coverage
Urban soils	Included as several suborders in the order *Anthroposols*
Man-transported materials	Included in the suborder *Dredgic Anthroposols*
Bare rock	Not recognized as soil
Subaquatic soils	Included in the order *Hydrosols*
Soils deeply transformed by agricultural activities	Included as several suborders in the order *Anthroposols*

Theoretical background

The soil classification of Australia (Isbell, 2002) is similar in principle to the US Soil Taxonomy (Soil Survey Staff, 1999). In fact, the US classification might be called a source system for the Australian system; however, there are important differences between these two classifications. The main difference is the list of soil archetypes reflecting in part the spatial distribution of different soil orders in the respective countries. For example, Vertisols cover less than 1 per cent of the total area of the US, but more than 15 per cent in Australia (Isbell, 1996), thus it is natural that this order is described in more detail in the Australian classification. The diversity of soils with clay-enriched B horizons is also greater in Australia, and some of these soils have unique properties due to their polygenetic origin. These unique soils have no analogies in the US Soil Taxonomy. Soils formed on volcanic ash and pumice are not distinguished as a separate order in Australia because they cover insignificant areas. The presence of an organic-matter-enriched upper horizon, diagnostic for *Mollisols* in Soil Taxonomy and for *Chernozems*, *Phaeozems* and *Kastanozems* in the World Reference Base for Soil Resources (WRB), is used only at the second level of the Australian classification. Anthropogenically transformed soils are distinguished in a separate order (like in the Chinese, French and Russian classifications, as well as in the WRB).

Development

The first Australian soil classification, proposed by J. Prescott, was published in the 1920s; it was a general scheme based mainly on a theoretical model rather than on empirical data (Moore et al, 1983). It was based on the ideals of the Russian school of pedology, and used mainly factors and concepts of soil genesis for soil diagnostics. Later research has led the pedologists of Australia to conclude that climatic criteria were not very useful for predicting soil distribution, as soil formation occurs there mainly on ancient surfaces which have passed through several cycles of weathering and pedogenesis, and a soil profile seldom reflects actual factors of soil formation. At the end of the 1940s C. Stephens proposed a new morphogenetic classification. This classification used only soil properties for diagnostics but appeared not to be very successful mainly due to its weakly developed structure. The highest levels of the taxonomy were not informative and were not used in practice, and the lowest taxa had no strict definitions (Isbell, 1992). To a great extent the classification suffered from a conflict between the upper levels of the taxonomy and a system of soil series imported from the US. The series served practically as archetypes, and no 'central images' at a higher level were proposed. Stephens (1956) pointed out that 'no level of the classification is basic, only the soil profile itself is basic'. In the 1950s the Australian pedologists hoped that the new US classification would serve as an international system, and it would also be applicable for Australia. After publication of the 6th Approximation of the American classification, Guy D. Smith visited Australia and described more than 200 soil profiles. However, the 7th Approximation was a disappointment as Australian soils were still not evident in the US classification, with many referred to as *Alfisols* or *Vertisols*, and the details of the taxonomy were not sufficient to reflect the peculiarity of the Australian soil cover.

During this time K. Northcote worked out a system of classification based on quantitative criteria for soil taxa. He grouped the soils of Australia according to the most important features at the scale of the country. The classification was made analytically, from top to bottom, and allowed one to divide a number of soil series that had been grouped together by other classification systems but which differed significantly in their properties (Northcote, 1971). This classification had five levels: division, subdivision, section, class and principal soil form. Every level was divided on the basis of mutually exclusive properties. The classification was called a 'Factual Key' (Northcote, 1971). The terminology used in the Factual Key had no equivalents in the world. Special codes of letters and figures were used instead of soil names, though some soil taxa had parallel English names. Four divisions were called O (organic soils), U (soils with a uniform profile), G (soil with a gradual change of texture with depth) and D (soils with an abrupt texture boundary. The Factual Key appeared to be effective for soil survey and served as a basis for soil mapping for more than 25 years in Australia. In practice, the soil scientists in this country until recently used two 'languages': traditional terms proposed by Stephens (1956) were used for farmers and the Factual Key (Northcote, 1971) was used by pedologists. Finally, in the mid-1990s a new classification was proposed by Isbell (1996); minor corrections were made for the second edition (Isbell, 2002). This classification is detailed enough to provide soil survey at any scale within Australia.

Structure

The structure of the Australian soil classification is similar to that of the US Soil Taxonomy (Table 27.2). The highest level of the taxonomy is soil <u>order</u>, which is defined according to a similar construction of a soil profile, or by a single diagnostic horizon or property, or by water regime (e.g. for *Hydrosols*). The orders are divided into <u>suborders</u> by colour, texture, carbonates content or the presence of additional diagnostic horizons and properties. <u>Great groups</u>, as well as <u>subgroups</u>, are distinguished by additional diagnostic characteristics. Though the sequence of the names of the levels of taxonomy is the same in the two classifications, the meanings of these levels are completely different. For example, in most orders the separation into suborders is made just by the colour of the soil. Within subgroups soils are divided into <u>families</u> by quantitative characteristics, such as texture, soil depth, carbonates content and so forth. Within families there are <u>series</u>; or, more strictly, existing soil series are referred to a certain family.

According to its structure the Australian soil classification is a hierarchical taxonomy with partially formalized borders. At the series level it is a nominal system.

Table 27.2 *The structure of the Australian soil classification*

Level	Taxon name	Taxon characteristics	Borders between classes	Diagnostics	Terminology
0	Soils	Kingdom			
1	Order	Generic / Collective	Fuzzy	Morphological	Artificial
2	Suborder	Generic / Specific	Fuzzy	Morphological	Mixed
3	Great group	Specific	Formalized	Morphological	Mixed
4	Subgroup	Specific	Formalized	Morphological	Mixed
5	Family	Varietal	Formal	Morphological	Traditional
6	Series	Generic	Formal	Morphological	Traditional

Diagnostics

The object of classification of the Australian classification is a soil profile itself without any reference to temperature or water regime. The only exception is the *Hydrosol* order where the definition of the order includes the hydrological regime of soils, and some of its suborders are classified according to the period of soil flooding. An important distinction of the Australian system is that it is based mainly on field diagnostics. Very few analytical procedures are required and most of them can be performed in the field (e.g. pH determination with a field pH-meter). However, carbonate content is important for many great groups and even suborders (in *Calcarosols* order) and should be determined in a laboratory. With few exceptions the diagnostics of Australian soil classification can be called morphological.

The limits of taxa are said to be determined by quantitative criteria. Diagnostic horizons are used for defining soil orders and suborders yet most of the criteria for these horizons are qualitative. For example, some orders are distinguished only by the structure of the B horizon; the *Dermosols* are defined as 'soils having more than weakly expressed structure of B horizon'. At the suborders level colour characteristics are mainly used. For example, the *Vertosol* order is divided into suborders of *Black, Brown, Grey, Red* and *Yellow Vertosols*.

Terminology

Soil classification of Australia rejects traditional terminology. New artificial soil names were constructed as the taxa are different from all the other classifications (e.g. the *Organosols* order is narrower than the order of *Histosols* in Soil Taxonomy or the *Histosols* group in the WRB because tidal peatlands in Australia are excluded from this order and included in the *Hydrosols* order. The names of the orders are based mainly on Latin roots. Some order names are similar to the terms of Soil Taxonomy or the WRB with modifications in one or two letters to avoid mistaking them as identical groups of soils (*Ferrosols, Vertosols, Podosols*). The attempt to retain original meanings of terms has not always been successful; for example the Australian, Chinese and Romanian schemes have *Vertosols* in order not to change the meaning of *Vertisols* in other classifications.

The origin of some terms in the Australian classification is obvious (*Anthroposols, Dermosols, Calcarosols, Organosols, Sodosols*). Others need explanation, such as: *Kandisols*, originated from Latin *kandis* – 'white clay' (the same root is used for low activity clay-enriched *kandic* horizons in the US Soil Taxonomy); *Tenosols*, from Latin *tenos* – 'young'; and *Kurosols*, from Japanese *kuroi* – 'black'.

At lower levels of the classification artificial terms are mixed with ordinary English words, denoting colour, mineralogical composition and miscellaneous soil properties. Adding terms from higher levels allows the construction of the names of lower levels. For example, in the *Chromosol* order there is a *Black Chromosols* suborder, and in this suborder there is a *Magnesial Black Chromosol* great group, and so on.

Correlation

The Australian classification is one of the most difficult to correlate with the WRB. The main problem is that many soils in Australia have no analogies in other parts of the world and are not represented in other classifications. Another problem is that the latest version of the Australian classification disregards many traditional archetypes by recognizing quite different taxonomic limits. For example, the order of Podosols in the Australian classification is much broader than Spodosols in the Soil Taxonomy or Podzols in the WRB and includes 'giant podzols' with extremely thick (several metres) albic horizons which are often classified as undeveloped sandy soils elsewhere. Many soil properties are used in the Australian classification at taxonomic levels different from that in the WRB, for example colour. Some criteria, such as a period of soil flooding, have no analogies in other classifications.

We attempt to correlate Australian soil taxa (Isbell, 2002) down to the great group level with the terms of the WRB (IUSS Working Group WRB, 2006). Soil orders, suborders within the orders, and great groups within the suborders are each listed alphabetically.

Anthroposols – soil order. ≈ Anthrosols / Technosols
The following seven suborders are distinguished within the order:
Cumulic Anthroposols ≈ Irragric Anthrosols / Terric Anthrosols
No great groups are distinguished within the suborder.
Dredgic Anthroposols ≈ Technosols / Terric Anthrosols
No great groups are distinguished within the suborder.
Garbic Anthroposols ≈ Garbic Technosols
No great groups are distinguished within the suborder.
Hortic Anthroposols ≈ Hortic Anthrosols
No great groups are distinguished within the suborder.
Scalpic Anthroposols ≈ Anthropic Regosols
No great groups are distinguished within the suborder.
Spolic Anthroposols ≈ Spolic Technosols

No <u>great groups</u> are distinguished within the suborder.
Urbic Anthroposols ≈ Urbic Technosols
No <u>great groups</u> are distinguished within the suborder.

<u>Calcarosols</u> – soil <u>order</u>. ≈ Calcisols / Gypsisols / Durisols
The following seven <u>suborders</u> are distinguished within the order:
Calcic Calcarosols ≈ Calcisols / Calcic Durisols / *Calcic* Leptosols
The suborder is divided into the following <u>great groups</u>:

- *Argic Calcic Calcarosols* ≈ Luvic Calcisols
- *Duric Calcic Calcarosols* ≈ Calcic Durisols
- *Lithic Calcic Calcarosols* ≈ *Leptic* Calcisols / *Calcic* Leptosols
- *Marly Calcic Calcarosols* ≈ Calcisols (*Calcaric*)
- *Paralithic Calcic Calcarosols* ≈ Calcisols (Skeletic) / Leptosols (Calcaric)
- *Pedal Calcic Calcarosols* ≈ *Cambic* Calcisols
- *Petrocalcic Calcic Calcarosols* ≈ Petric Calcisols
- *Regolithic Calcic Calcarosols* ≈ Skeletic Calcisols

Hypercalcic Calcarosols ≈ Hypercalcic Calcisols / Calcic Durisols / *Calcic* Leptosols
The suborder is divided into the following <u>great groups</u>:

- *Argic Hypercalcic Calcarosols* ≈ Luvic Hypercalcic Calcisols
- *Duric Hypercalcic Calcarosols* ≈ Hypercalcic Durisols
- *Lithic Hypercalcic Calcarosols* ≈ *Leptic* Hypercalcic Calcisols / *Calcic* Leptosols
- *Marly Hypercalcic Calcarosols* ≈ Hypercalcic Calcisols (*Calcaric*)
- *Paralithic Hypercalcic Calcarosols* ≈ Hypercalcic Calcisols (Skeletic) / *Hypercalcic* Leptosols
- *Pedal Hypercalcic Calcarosols* ≈ *Cambic* Hypercalcic Calcisols
- *Petrocalcic Hypercalcic Calcarosols* ≈ Hypercalcic Petric Calcisols
- *Regolithic Hypercalcic Calcarosols* ≈ Hypercalcic Calcisols (Skeletic)
- *Rendic Hypercalcic Calcarosols* ≈ Epihypercalcic Calcisols

Hypergypsic Calcarosols ≈ Calcic Gypsisols
No <u>great groups</u> are distinguished within the suborder.
Hypocalcic Calcarosols ≈ Hypocalcic Calcisols
The suborder is divided into the following <u>great groups</u>:

- *Argic Hypocalcic Calcarosols* ≈ Luvic Hypocalcic Calcisols / Lixic Hypocalcic Calcisols
- *Duric Hypocalcic Calcarosols* ≈ Hypocalcic Durisols
- *Lithic Hypocalcic Calcarosols* ≈ *Leptic* Hypocalcic Calcisols / *Calcic* Leptosols
- *Marly Hypocalcic Calcarosols* ≈ Hypocalcic Calcisols (*Calcaric*)
- *Paralithic Hypocalcic Calcarosols* ≈ Hypocalcic Calcisols (Skeletic)
- *Pedal Hypocalcic Calcarosols* ≈ *Cambic* Hypocalcic Calcisols

- *Petrocalcic Hypocalcic Calcarosols* ≈ Endopetric Hypocalcic Calcisols
- *Regolithic Hypocalcic Calcarosols* ≈ Leptic Hypocalcic Calcisols

Lithocalcic Calcarosols ≈ Petric Calcisols
The suborder is divided into the following <u>great groups</u>:

- *Argic Lithocalcic Calcarosols* ≈ Luvic Petric Calcisols
- *Duric Lithocalcic Calcarosols* ≈ Petrocalcic Durisols
- *Lithic Lithocalcic Calcarosols* ≈ *Leptic* Petric Calcisols
- *Marly Lithocalcic Calcarosols* ≈ Petric Calcisols (*Calcaric*)
- *Paralithic Lithocalcic Calcarosols* ≈ Petric Calcisols (Skeletic)
- *Pedal Lithocalcic Calcarosols* ≈ *Cambic* Petric Calcisols
- *Regolithic Lithocalcic Calcarosols* ≈ Petric Calcisols

Shelly Calcarosols ≈ Haplic Calcisols
No <u>great groups</u> are distinguished within the suborder.
Supracalcic Calcarosols ≈ Calcisols
The suborder is divided into the following <u>great groups</u>:

- *Argic Supracalcic Calcarosols* ≈ Luvic Petric Calcisols (*Fractipetric*)
- *Duric Supracalcic Calcarosols* ≈ Calcic Durisols
- *Lithic Supracalcic Calcarosols* ≈ *Leptic* Calcisols / *Calcic* Leptosols
- *Marly Supracalcic Calcarosols* ≈ Calcisols (*Calcaric*)
- *Paralithic Supracalcic Calcarosols* ≈ Calcisols (Skeletic) / *Calcic* Leptosols
- *Pedal Supracalcic Calcarosols* ≈ *Cambic* Calcisols
- *Petrocalcic Supracalcic Calcarosols* ≈ Petric Calcisols
- *Regolithic Supracalcic Calcarosols* ≈ Calcisols (Skeletic)

<u>Chromosols</u> – soil <u>order</u>. ≈ Luvisols / Lixisols / Luvic Durisols / Lixic Plinthisols
The following five <u>suborders</u> are distinguished within the order:
Black Chromosols ≈ Luvisols / Lixisols / Luvic Durisols / Lixic Plinthisols
The suborder is divided into the following <u>great groups</u>:

- *Calcic Black Chromosols* ≈ Calcic Luvisols / Calcic Lixisols
- *Duric Black Chromosols* ≈ Luvic Durisols / Lixic Durisols / Luvic Calcisols
- *Dystrophic Black Chromosols* ≈ Luvisols (Epidystric) / Lixisols (Epidystric)
- *Eutrophic Black Chromosols* ≈ Luvisols (Hypereutric) / Lixisols (Hypereutric)
- *Hypercalcic Black Chromosols* ≈ Luvic Calcisols
- *Hypocalcic Black Chromosols* ≈ Calcic Luvisols / Calcic Lixisols
- *Lithocalcic Black Chromosols* ≈ Luvisols (*Calcaric*) / Lixisols (*Calcaric*)
- *Magnesic Black Chromosols* ≈ Luvisols (Magnesic) / Lixisols (Magnesic)
- *Mesotrophic Black Chromosols* ≈ Luvisols
- *Pedaric Black Chromosols* ≈ Luvisols / Lixisols
- *Petrocalcic Black Chromosols* ≈ Luvic Petric Calcisols
- *Petroferric Black Chromosols* ≈ Lixic Plinthisols
- *Subplastic Black Chromosols* ≈ Luvisols / Lixisols
- *Supracalcic Black Chromosols* ≈ Calcic Luvisols / Calcic Lixisols

Brown Chromosols ≈ Luvisols (Chromic) / Lixisols (Chromic) / Luvic Durisols (Chromic) / Lixic Plinthisols (Chromic)
Great groups in the suborder and their correlation are identical to those of *Black Chromosols*, except for the name *Brown Chromosols*.

Grey Chromosols ≈ Luvisols (Hyperochric) / Lixisols (Hyperochric) / Luvic Durisols (Hyperochric) / Lixic Plinthisols (Hyperochric)
Great groups in the suborder are identical to those of *Black Chromosols*, except for the name *Grey Chromosols*.

Red Chromosols ≈ Luvisols (Rhodic) / Lixisols (Rhodic) / Luvic Durisols (Rhodic) / Lixic Plinthisols (Rhodic)
Great groups in the suborder are identical to those of *Black Chromosols*, except for the name *Red Chromosols*.

Yellow Chromosols ≈ Luvisols (*Xanthic*) / Lixisols (*Xanthic*) / Luvic Durisols (*Xanthic*) / Lixic Plinthisols (*Xanthic*)
Great groups in the suborder are identical to those of *Black Chromosols*, except for the name *Yellow Chromosols*.

Dermosols – soil order. ≈ Cambisols / Chernozems / Kastanozems / Phaeozems / Umbrisols
The following five suborders are distinguished within the order:
Black Dermosols ≈ Mollic Umbrisols / Chernozems / Phaeozems / Umbrisols / Nitisols
The suborder is divided into the following great groups:

- *Calcic Black Dermosols* ≈ Chernozems
- *Duric Black Dermosols* ≈ Duric Phaeozems / Duric Chernozems / (*Umbric*) Durisols
- *Dystrophic Black Dermosols* ≈ Umbrisols
- *Eutrophic Black Dermosols* ≈ Mollic Umbrisols / Chernozems / Phaeozems
- *Hypercalcic Black Dermosols* ≈ Calcic Chernozems
- *Hypocalcic Black Dermosols* ≈ Chernozems
- *Lithocalcic Black Dermosols* ≈ Hypercalcic Chernozems
- *Magnesic Black Dermosols* ≈ Umbrisols (*Magnesic*) / Chernozems (*Magnesic*) / Phaeozems (*Magnesic*)
- *Mesotrophic Black Dermosols* ≈ Mollic Umbrisols / Phaeozems
- *Pedaric Black Dermosols* ≈ Mollic Nitisols / Umbric Nitisols
- *Petrocalcic Black Dermosols* ≈ Petrocalcic Chernozems
- *Petroferric Black Dermosols* ≈ *Mollic* Plinthisols / Umbric Plinthisols
- *Subplastic Black Dermosols* ≈ Umbrisols / Chernozems / Phaeozems
- *Supracalcic Black Dermosols* ≈ Calcic Chernozems

Brown Dermosols ≈ Phaeozems / Kastanozems / Umbrisols
Great groups in the suborder are identical to those of *Black Dermosols*, except for the name *Brown Dermosols*.

Grey Dermosols ≈ Cambisols / Phaeozems
<u>Great groups</u> in the suborder are identical to those of *Black Dermosols*, except for the name *Grey Dermosols*.

Red Dermosols ≈ Cambisols (Chromic) / Nitisols (Chromic)
<u>Great groups</u> in the suborder are identical to those of *Black Dermosols*, except for the name *Red Dermosols*.

Yellow Dermosols ≈ Cambisols (*Xanthic*) / Nitisols (*Xanthic*)
<u>Great groups</u> in the suborder are identical to those of *Black Dermosols*, except for the name *Yellow Dermosols*.

<u>Ferrosols</u> – soil <u>order</u>. ≈ Ferralsols / Nitisols
The following five <u>suborders</u> are distinguished within the order:
Black Ferrosols ≈ Ferralsols (Humic)
The suborder is divided into the following <u>great groups</u>:

- *Calcareous Black Ferrosols* ≈ *Calcic* Ferralsols (Humic)
- *Dystrophic Black Ferrosols* ≈ Ferralsols (Humic, Dystric)
- *Eutrophic Black Ferrosols* ≈ Ferralsols (Humic, Eutric)
- *Magnesic Black Ferrosols* ≈ Ferralsols (Humic, *Magnesic*)
- *Mesotrophic Black Ferrosols* ≈ Ferralsols (Humic)

Brown Ferrosols ≈ Ferralsols (Chromic)
<u>Great groups</u> in the suborder are identical to those of *Black Ferrosols*, except for the name *Brown Ferrosols*.

Grey Ferrosols ≈ Ferralsols
<u>Great groups</u> in the suborder are identical to those of *Black Ferrosols*, except for the name *Grey Ferrosols*.

Red Ferrosols ≈ Ferralsols (Rhodic)
<u>Great groups</u> in the suborder are identical to those of *Black Ferrosol*, except for the name *Red Ferrosols*.

Yellow Ferrosols ≈ Ferralsols (Xanthic)
<u>Great groups</u> in the suborder are identical to those of *Black Ferrosols*, except for the name *Yellow Ferrosols*.

<u>Hydrosols</u> – soil <u>order</u>. ≈ Gleysols / Fluvisols / Salic Histosols / Gleyic Leptosols / Gleyic Solonchaks
The following seven <u>suborders</u> are distinguished within the order:
Extratidal Hydrosols ≈ Gleysols / Gleyic Leptosols
The suborder is divided into the following <u>great groups</u>:

- *Calcarosolic Extratidal Hydrosols* ≈ Calcic Gleysols
- *Chromosolic Extratidal Hydrosols* ≈ Luvic Gleysols

- *Dermosolic Extratidal Hydrosols* ≈ *Cambic* Gleysols
- *Kandosolic Extratidal Hydrosols* ≈ Gleysols
- *Kurosolic Extratidal Hydrosols* ≈ Alic Gleysols / Acric Gleysols
- *Petroferric Extratidal Hydrosols* ≈ Plinthic Gleysols
- *Rudosolic Extratidal Hydrosols* ≈ Gleysols
- *Sodosolic Extratidal Hydrosols* ≈ Gleysols (Sodic)
- *Sulphidic Extratidal Hydrosols* ≈ Gleysols (Protothionic)
- *Sulphuric Extratidal Hydrosols* ≈ Gleysols (Orthothionic)
- *Tenosolic Extratidal Hydrosols* ≈ *Leptic* Gleysols / Gleyic Leptosols

Hypersalic Hydrosols ≈ Gleyic Hypersalic Solonchaks
The suborder is divided into the following great groups:

- *Epicalcareous Hypersalic Hydrosols* ≈ Calcic Gleyic Hypersalic Solonchaks
- *Gypsic Hypersalic Hydrosols* ≈ Gypsic Gleyic Hypersalic Solonchaks
- *Halic Hypersalic Hydrosols* ≈ Gleyic Hypersalic Solonchaks (Chloridic)
- *Haplic Hypersalic Hydrosols* ≈ Gleyic Hypersalic Solonchaks
- *Mottled Hypersalic Hydrosols* ≈ Gleyic Hypersalic Solonchaks (Oxiaquic)
- *Sulphidic Hypersalic Hydrosols* ≈ Gleyic Hypersalic Solonchaks (*Protothionic*)

Intertidal Hydrosols ≈ Gleyic Tidalic Fluvisols / Histosols
The suborder is divided into the following great groups:

- *Arenaceous Intertidal Hydrosols* ≈ Gleyic Tidalic Fluvisols (Arenic)
- *Argillaceous Intertidal Hydrosols* ≈ Gleyic Tidalic Fluvisols (Clayic)
- *Epicalcareous Intertidal Hydrosols* ≈ Calcic Gleyic Tidalic Fluvisols
- *Faunic Intertidal Hydrosols* ≈ Gleyic Tidalic Fluvisols
- *Histic Intertidal Hydrosols* ≈ *Tidalic* Histosols
- *Histic-Sulphidic Intertidal Hydrosols* ≈ *Tidalic* Histosols (Protothionic)
- *Lutaceous Intertidal Hydrosols* ≈ Gleyic Fluvisols (Siltic)
- *Sulfidic Intertidal Hydrosols* ≈ Gleyic Tidalic Fluvisols (Protothionic)

Oxyaquic Hydrosols ≈ Gleyic Calcisols / Gleyic Luvisols / Gleyic Lixisols / Gleyic Acrisols / Gleyic Alisols / Gleyic Cambisols / Gleyic Regosols / Plinthic Gleysols / Gleyic Solonetz / Thionic Gleysols / Gleyic Leptosols
The suborder is divided into the following great groups:

- *Calcarosolic Oxyaquic Hydrosols* ≈ Gleyic Calcisols (Oxyaquic)
- *Chromosolic Oxyaquic Hydrosols* ≈ Gleyic Luvisols (Oxyaquic) / Gleyic Lixisols (Oxyaquic)
- *Dermosolic Oxyaquic Hydrosols* ≈ Gleyic Cambisols (Oxyaquic)
- *Kandosolic Oxyaquic Hydrosols* ≈ Gleyic Cambisols (Oxyaquic) / Gleyic Regosols (Oxyaquic)
- *Kurosolic Oxyaquic Hydrosols* ≈ Gleyic Acrisols (Oxyaquic) / Gleyic Alisols (Oxyaquic)
- *Petroferric Oxyaquic Hydrosols* ≈ Plinthic Gleysols (Oxyaquic)

- *Rudosolic Oxyaquic Hydrosols* ≈ Gleyic Regosols (Oxyaquic)
- *Sodosolic Oxyaquic Hydrosols* ≈ Gleyic Solonetz (Oxyaquic)
- *Sulphidic Oxyaquic Hydrosols* ≈ Protothionic Gleysols (Oxyaquic)
- *Sulphuric Oxyaquic Hydrosols* ≈ Orthothionic Gleysols (Oxyaquic)
- *Tenosolic Oxyaquic Hydrosols* ≈ Gleyic Leptosols (Oxyaquic)

Redoxic Hydrosols ≈ Gleyic Calcisols / Gleyic Luvisols / Gleyic Lixisols / Gleyic Acrisols / Gleyic Alisols / Gleyic Cambisols / Gleyic Regosols / Plinthic Gleysols / Gleyic Solonetz / Thionic Gleysols / Gleyic Leptosols
Great groups in the suborder are identical to those of *Oxyaquic Hydrosols*, except for the name Redozic Hydrosols.

Salic Hydrosols ≈ Gleyic Solonchaks
The suborder is divided into the following great groups:

- *Calcarosolic Salic Hydrosols* ≈ Calcic Gleyic Solonchaks
- *Chromosolic Salic Hydrosols* ≈ Luvic Gleyic Solonchaks
- *Dermosolic Salic Hydrosols* ≈ *Cambic* Gleyic Solonchaks
- *Kandosolic Salic Hydrosols* ≈ Gleyic Solonchaks
- *Kurosolic Salic Hydrosols* ≈ Luvic Gleyic Solonchaks
- *Petroferric Salic Hydrosols* ≈ *Plinthic* Gleyic Solonchaks
- *Rudosolic Salic Hydrosols* ≈ Gleyic Solonchaks
- *Sodosolic Salic Hydrosols* ≈ Gleyic Solonchaks (Sodic)
- *Sulphidic Salic Hydrosols* ≈ Gleyic Solonchaks (*Protothionic*)
- *Sulphuric Salic Hydrosols* ≈ Gleyic Solonchaks (*Orthothionic*)
- *Tenosolic Salic Hydrosols* ≈ *Leptic* Gleyic Solonchaks

Supratidal Hydrosols ≈ Gleyic Fluvisols
The suborder is divided into the following great groups:

- *Epicalcareous Supratidal Hydrosols* ≈ Calcaric Gleyic Fluvisols
- *Gypsic Supratidal Hydrosols* ≈ Gypsiric Gleyic Fluvisols
- *Haplic Supratidal Hydrosols* ≈ Gleyic Fluvisols
- *Mottled Supratidal Hydrosols* ≈ Gleyic Fluvisols (Oxiaquic)
- *Sulphidic Supratidal Hydrosols* ≈ Gleyic Fluvisols (Protothionic)

Kandosols – soil order. ≈ Cambisols / Calcisols / Plinthisols
The following five suborders are distinguished within the order:
Black Kandosols ≈ Cambisols (Humic) / Calcisols (*Humic*) / Plinthisols (Humic)
The suborder is divided into the following great groups:

- *Calcic Black Kandosols* ≈ *Calcic* Cambisols (Humic)
- *Dystrophic Black Kandosols* ≈ Cambisols (Humic, Dystric)
- *Eutrophic Black Kandosols* ≈ Cambisols (Humic, Eutric)
- *Hypercalcic Black Kandosols* ≈ Calcisols (*Humic*)
- *Hypocalcic Black Kandosols* ≈ *Calcic* Cambisols (Humic)
- *Lithocalcic Black Kandosols* ≈ Cambisols (Calcaric, *Humic*)

- *Magnesic Black Kandosols* ≈ Cambisols (Humic, *Magnesic*)
- *Mesotrophic Black Kandosols* ≈ Cambisols (Humic)
- *Petrocalcic Black Kandosols* ≈ Petric Calcisols (*Humic*)
- *Petroferric Black Kandosols* ≈ Plinthisols (Humic)
- *Placic Black Kandosols* ≈ Cambisols (Humic, *Placic*)
- *Supracalcic Black Kandosols* ≈ Calcisols (*Humic*)

Brown Kandosols ≈ Cambisols (Chromic) / Durisols (Chromic) / Calcisols (Chromic) / Plinthisols (Chromic)
The suborder is divided into the following <u>great groups</u>:

- *Calcic Brown Kandosols* ≈ *Calcic* Cambisols (Chromic)
- *Duric Brown Kandosols* ≈ Durisols (Chromic)
- *Dystrophic Brown Kandosols* ≈ Cambisols (Dystric, Chromic)
- *Eutrophic Brown Kandosols* ≈ Cambisols (Eutric, Chromic)
- *Hypercalcic Brown Kandosols* ≈ Calcisols (Chromic)
- *Hypocalcic Brown Kandosols* ≈ *Calcic* Cambisols (Chromic)
- *Lithocalcic Brown Kandosols* ≈ Calcisols (Chromic, *Calcaric*)
- *Magnesic Brown Kandosols* ≈ Cambisols (Chromic, *Magnesic*)
- *Mellic Brown Kandosols* ≈ Cambisols (Chromic)
- *Mesotrophic Brown Kandosols* ≈ Cambisols (Chromic)
- *Petrocalcic Brown Kandosols* ≈ Petric Calcisols (Chromic)
- *Petroferric Brown Kandosols* ≈ Plinthisols (Chromic)
- *Placic Brown Kandosols* ≈ Cambisols (Chromic, Placic)
- *Supracalcic Brown Kandosols* ≈ Calcisols (Chromic)

Grey Kandosols ≈ Cambisols / Calcisols / Plinthisols
<u>Great groups</u> in the suborder are identical to those of *Black Kandosols*, except for the name *Grey Kandosols*.

Red Kandosols ≈ Cambisols (Rhodic) / Durisols (*Rhodic*) / Calcisols (*Rhodic*) / Plinthisols (*Rhodic*)
<u>Great groups</u> in the suborder are identical to those of *Brown Kandosols*, except for the name *Red Kandosols*.

Yellow Kandosols ≈ Cambisols (*Xanthic*) / Calcisols (*Xanthic*) / Plinthisols (*Xanthic*)
<u>Great groups</u> in the suborder are identical to those of *Black Kandosols*, except for the name *Yellow Kandosols*.

<u>Kurosols</u> – soil <u>order</u>. ≈ Acrisols / Alisols / Lixisols
The following five <u>suborders</u> are distinguished within the order:
Black Kurosols ≈ Acrisols (Humic) / Alisols (Humic) / Lixisols (Humic)
The suborder is divided into the following <u>great groups</u>:

- *Dystrophic Black Kurosols* ≈ Acrisols (Humic) / Alisols (Humic)
- *Eutrophic Black Kurosols* ≈ Lixisols (Humic)

- *Magnesic Black Kurosols* ≈ Lixisols (*Magnesic*)
- *Magnesic-Natric Black Kurosols* ≈ Alisols (*Sodic, Magnesic*)
- *Mesotrophic Black Kurosols* ≈ Acrisols (Humic) / Alisols (Humic)
- *Natric Black Kurosols* ≈ Alisols (*Sodic*)
- *Petroferric Black Kurosols* ≈ Plinthic Acrisols (Humic) / Plinthic Alisols (Humic)

Brown Kurosols ≈ Acrisols (Chromic) / Alisols (Chromic) / Lixisols (Chromic)
<u>Great groups</u> in the suborder are identical to those of *Black Kurosols*, except for the name *Brown Kurosols*.

Grey Kurosols ≈ Acrisols / Alisols / Lixisols
<u>Great groups</u> in the suborder are identical to those of *Black Kurosols*, except for the name *Grey Kurosols*.

Red Kurosols ≈ Acrisols (Rhodic) / Alisols (Rhodic) / Lixisols (Rhodic)
<u>Great groups</u> in the suborder are identical to those of *Black Kurosols*, except for the name *Red Kurosols*.

Yellow Kurosols ≈ Acrisols (*Xanthic*) / Alisols (*Xanthic*) / Lixisols (*Xanthic*)
<u>Great groups</u> in the suborder are identical to those of *Black Kurosols*, except for the name *Yellow Kurosols*.

<u>Organosols</u> – soil <u>order</u>. ≈ Histosols
The following three <u>suborders</u> are distinguished within the order:
Fibric Organosols ≈ Fibric Histosols
The suborder is divided into the following <u>great groups</u>:

- *Acidic Fibric Organosols* ≈ Fibric Histosols (Dystric)
- *Basic Fibric Organosols* ≈ Fibric Histosols (Eutric)
- *Calcareous Fibric Organosols* ≈ Calcic Fibric Histosols
- *Folic Fibric Organosols* ≈ Fibric Folic Histosols
- *Sulphidic Fibric Organosols* ≈ Fibric Histosols (Protothionic)
- *Sulphuric Fibric Organosols* ≈ Fibric Histosols (Orthothionic)

Hemic Organosols ≈ Hemic Histosols
<u>Great groups</u> in the suborder are identical to those of *Fibric Organosols*, except for the name *Hemic Organosols*.

Sapric Organosols ≈ Sapric Histosols
<u>Great groups</u> in the suborder are identical to those of *Fibric Organosols*, except for the name *Sapric Organosols*.

<u>Podosols</u> – soil <u>order</u>. ≈ Podzols / Arenosols
The following three <u>suborders</u> are distinguished within the order:
Aeric Podosols ≈ Podzols / Hyperalbic Arenosols

The suborder is divided into the following <u>great groups</u>:

- *Humosesquic Aeric Podosols* ≈ Podzols
- *Humosesquic / Sesquic Aeric Podosols* ≈ Podzols
- *Pipey Aeric Podosols* ≈ Podzols
- *Sesquic Aeric Podosols* ≈ Rustic Podzols

Aquic Podosols ≈ Gleyic Podzols
The suborder is divided into the following <u>great groups</u>:

- *Humic / Alsilic Aquic Podosols* ≈ Gleyic Carbic Podzols
- *Humic Aquic Podosols* ≈ Gleyic Carbic Podzols

Semiaquic Podosols ≈ Endogleyic Podzols
The suborder is divided into the following <u>great groups</u>:

- *Humic / Alsilic Semiaquic Podosols* ≈ Endogleyic Carbic Podzols
- *Humic / Humosesquic Semiaquic Podosols* ≈ Endogleyic Carbic Podzols
- *Humic Semiaquic Podosols* ≈ Endogleyic Carbic Podzols
- *Humic / Sesquic Semiaquic Podosols* ≈ Endogleyic Carbic Podzols
- *Humosesquic Semiaquic Podosols* ≈ Endogleyic Podzols
- *Pipey Semiaquic Podosols* ≈ Endogleyic Podzols
- *Sesquic Semiaquic Podosols* ≈ Endogleyic Rustic Podzols

<u>Rudosols</u> – soil <u>order</u>. ≈ Regosols / Arenosols / Fluvisols / Solonchaks / Gypsisols / Calcisols
The following nine <u>suborders</u> are distinguished within the order:
Arenic Rudosols ≈ Arenosols
No <u>great groups</u> are distinguished within the suborder.

Carbic Rudosols ≈ Humic Regosols
No <u>great groups</u> are distinguished within the suborder.

Clastic Rudosols ≈ Regosols (Skeletic) / Fluvisols (Skeletic) / Regosols (Tephric, Skeletic)
The suborder is divided into the following <u>great groups</u>:

- *Bauxitic Clastic Rudosols* ≈ Regosols (*Gibbsic*)
- *Colluvic Clastic Rudosols* ≈ Regosols (Skeletic)
- *Ferric Clastic Rudosols* ≈ Regosols (Skeletic)
- *Fluvic Clastic Rudosols* ≈ Fluvisols (Skeletic)
- *Lithosolic Clastic Rudosols* ≈ Leptic Regosols (Skeletic)
- *Tephric Clastic Rudosols* ≈ Regosols (Tephric, Skeletic)

Hypergypsic Rudosols ≈ Hypergypsic Gypsisols
No <u>great groups</u> are distinguished within the suborder.

Hypersalic Rudosols ≈ Solonchaks
The suborder is divided into the following great groups:

- *Gypsic Hypersalic Rudosols* ≈ Gypsic Solonchaks
- *Halic Hypersalic Rudosols* ≈ Solonchaks (Chloridic)
- *Sulphidic Hypersalic Rudosols* ≈ Solonchaks (*Protothionic*)

Leptic Rudosols ≈ Leptic Regosols / Petric Durisols / Petric Plinthisols / Petric Calcisols
The suborder is divided into the following great groups:

- *Duric Leptic Rudosols* ≈ Petric Durisols
- *Ferric-Petroferric Leptic Rudosols* ≈ Petric Plinthisols
- *Lithic Leptic Rudosols* ≈ Leptic Regosols
- *Paralithic Leptic Rudosols* ≈ Leptic Regosols
- *Petrocalcic Leptic Rudosols* ≈ Petric Calcisols
- *Petroferric Leptic Rudosols* ≈ Petric Plinthisols

Lutic Rudosols ≈ Regosols (Siltic)
No great groups are distinguished within the suborder.

Shelly Rudosols ≈ Regosols (Calcaric, Skeletic)
No great groups are distinguished within the suborder.

Stratic Rudosols ≈ Haplic Fluvisols
No great groups are distinguished within the suborder.

Sodosols – soil order. ≈ Solonetz
The following five suborders are distinguished within the order:
Black Sodosols ≈ Humic Solonetz
The suborder is divided into the following great groups:

- *Duric Black Sodosols* ≈ Duric Solonetz (Humic)
- *Effervescent Black Sodosols* ≈ Calcic Solonetz (Humic)
- *Hypernatric Black Sodosols* ≈ Solonetz (Humic)
- *Mesonatric Black Sodosols* ≈ Solonetz (Humic)
- *Mottled-Hypernatric Black Sodosols* ≈ Stagnic Solonetz (Humic)
- *Mottled-Mesonatric Black Sodosols* ≈ Stagnic Solonetz (Humic)
- *Mottled-Subnatric Black Sodosols* ≈ Stagnic Solonetz (Humic) / Solodic Planosols (Humic)
- *Pedaric Black Sodosols* ≈ Solonetz (Humic)
- *Petrocalcic Black Sodosols* ≈ Petrocalcic Solonetz (Humic)
- *Petroferric Black Sodosols* ≈ *Plinthic* Solonetz (Humic)
- *Subnatric Black Sodosols* ≈ Solonetz (Humic) / Solodic Planosols (Humic)

Brown Sodosols ≈ Solonetz (*Chromic*)
Great groups in the suborder are identical to those of *Black Sodosols*, except for the name *Brown Sodosols*.

Grey Sodosols ≈ Solonetz
<u>Great groups</u> in the suborder are identical to those of *Black Sodosols*, except for the name *Grey Sodosols*.

Red Sodosols ≈ Solonetz (*Rhodic*)
<u>Great groups</u> in the suborder are identical to those of *Black Sodosols*, except for the name *Red Sodosols*.

Yellow Sodosols ≈ Solonetz (*Xanthic*)
<u>Great groups</u> in the suborder are identical to those of *Black Sodosols*, except for the name *Yellow Sodosols*.
<u>Tenosols</u> – soil <u>order</u>. ≈ Leptosols / Chernozems / Phaeozems / Durisols / Plinthosols / Calcisols / Cambisols / Andosols / Albeluvisols
The following 12 <u>suborders</u> are distinguished within the order:
Bleached-Leptic Tenosols ≈ Lithic Leptosols (*Albic*) / Epipetric Durisols (*Albic*) / Petric Plinthosols (Albic) / Epipetric Calcisols (*Albic*)
The suborder is divided into the following <u>great groups</u>:

- *Ferric Bleached-Leptic Tenosols* ≈ Plinthosols (Albic)
- *Ferric-Petroferric Bleached-Leptic Tenosols* ≈ Epipetric Plinthosols (Albic)
- *Lithic Bleached-Leptic Tenosols* ≈ Lithic Leptosols (*Albic*)
- *Paralithic Bleached-Leptic Tenosols* ≈ Lithic Leptosols (*Albic*)
- *Petrocalcic Bleached-Leptic Tenosols* ≈ Epipetric Calcisols (*Albic*)
- *Petroferric Bleached-Leptic Tenosols* ≈ Petric Plinthosols (Albic)
- *Regolithic Bleached-Leptic Tenosols* ≈ Leptosols (*Albic*)
- *Silpanic Bleached-Leptic Tenosols* ≈ Epipetric Durisols (*Albic*)

Black-Orthic Tenosols ≈ Leptosols (Humic) / Cambisols (Humic) / Andosols (Humic) / Durisols (Humic) / Plinthosols (Humic) / Andosols (Humic)
The suborder is divided into the following <u>great groups</u>:

- *Andic Black-Orthic Tenosols* ≈ Andosols (Humic)
- *Arenic Black-Orthic Tenosols* ≈ Arenosols (*Humic*)
- *Argic Black-Orthic Tenosols* ≈ Luvisols (Humic)
- *Bauxitic Black-Orthic Tenosols* ≈ *Gibbsic* Cambisols (Humic)
- *Duric Black-Orthic Tenosols* ≈ Durisols (*Humic*)
- *Ferric Black-Orthic Tenosols* ≈ Plinthosols (Humic)
- *Ferric-Duric Black-Orthic Tenosols* ≈ *Duric* Plinthosols (Humic)
- *Ferric-Petroferric Black-Orthic Tenosols* ≈ Petric Plinthosols (Humic)
- *Inceptic Black-Orthic Tenosols* ≈ Cambisols (Humic, Skeletic)
- *Lithic Black-Orthic Tenosols* ≈ Leptosols (Humic)
- *Paralithic Black-Orthic Tenosols* ≈ Leptosols (Humic)
- *Petrocalcic Black-Orthic Tenosols* ≈ Petric Calcisols (Humic)
- *Petroferric Black-Orthic Tenosols* ≈ Petric Plinthosols (Humic)
- *Regolithic Black-Orthic Tenosols* ≈ Leptosols (Humic)
- *Shelly Black-Orthic Tenosols* ≈ Leptosols (Calcaric, Humic)

- *Silpanic Black-Orthic Tenosols* ≈ Petric Durisols (*Humic*)
- *Tephric Black-Orthic Tenosols* ≈ Cambisols (Tephric, Humic)

Bleached-Orthic Tenosols ≈ Leptosols (*Albic*) / Cambisols (*Albic*) / Durisols (*Albic*) / Plinthosols (Albic) / Albic Luvisols / Andosols (*Albic*)
<u>Great groups</u> in the suborder are identical to those of *Black-Orthic Tenosols*, except for the name *Bleached Orthic Tenosols*.

Brown-Orthic Tenosols ≈ Leptosols (*Chromic*) / Cambisols (Chromic) / Durisols (Chromic) / Plinthosols (Chromic)
<u>Great groups</u> in the suborder are identical to those of *Bleached-Orthic Tenosols*, except for the name *Brown-Orthic Tenosols*.
Calcenic Tenosols ≈ Leptosols (Calcaric) / Cambisols (Calcaric) / Durisols (Calcaric) / Plinthosols (Calcaric) / Luvisols (*Calcaric*) / Andosols (Calcaric)
The suborder is divided into the following <u>great groups</u>:

- *Andic Calcenic Tenosols* ≈ Andosols (Calcaric)
- *Arenic Calcenic Tenosols* ≈ Arenosols (Calcaric)
- *Argic Calcenic Tenosols* ≈ Luvisols (*Calcaric*)
- *Duric Calcenic Tenosols* ≈ Durisols (Calcaric)
- *Ferric Calcenic Tenosols* ≈ Plinthosols (Calcaric)
- *Lithic Calcenic Tenosols* ≈ Leptosols (Calcaric)
- *Paralithic Calcenic Tenosols* ≈ Leptosols (Calcaric)
- *Petrocalcic Calcenic Tenosols* ≈ Petric Calcisols (Calcaric)
- *Regolithic Calcenic Tenosols* ≈ Leptosols (Calcaric)
- *Silpanic Calcenic Tenosols* ≈ Petric Durisols (Calcaric)
- *Tephric Calcenic Tenosols* ≈ Cambisols (Calcaric, Tephric)

Chernic Tenosols ≈ Phaeozems / Chernozems / Melanic Andosols / Umbric Plinthosols
The suborder is divided into the following <u>great groups</u>:

- *Andic Chernic Tenosols* ≈ Melanic Andosols
- *Bauxitic Chernic Tenosols* ≈ Phaeozems (*Gibbsic*)
- *Ferric Chernic Tenosols* ≈ Umbric Plinthosols
- *Inceptic Chernic Tenosols* ≈ Phaeozems (Skeletic)
- *Lithic Chernic Tenosols* ≈ Leptic Phaeozems
- *Marly Chernic Tenosols* ≈ Rendzic Phaeozems
- *Paralithic Chernic Tenosols* ≈ Leptic Phaeozems
- *Petrocalcic Chernic Tenosols* ≈ Petrocalcic Chernozems
- *Petroferric Chernic Tenosols* ≈ Umbric Petric Plinthosols
- *Placic Chernic Tenosols* ≈ (*Placic*) Phaeozems
- *Regolithic Chernic Tenosols* ≈ Leptic Phaeozems
- *Shelly Chernic Tenosols* ≈ Chernozems (Calcaric)
- *Silpanic Chernic Tenosols* ≈ Duric Phaeozems
- *Tephric Chernic Tenosols* ≈ Phaeozems (Tephric)

Chernic-Leptic Tenosols ≈ Mollic Leptosols / Rendzic Leptosols / Leptic Melanic Andosols / Leptic Umbric Plinthosols / Chernozems
The suborder is divided into the following <u>great groups</u>:

- *Duric Chernic-Leptic Tenosols* ≈ Epipetric Durisols
- *Ferric-Petroferric Chernic-Leptic Tenosols* ≈ Umbric Epipetric Plinthosols
- *Lithic Chernic-Leptic Tenosols* ≈ Mollic Leptosols
- *Paralithic Chernic-Leptic Tenosols* ≈ Mollic Leptosols
- *Petrocalcic Chernic-Leptic Tenosols* ≈ Rendzic Leptosols
- *Petroferric Chernic-Leptic Tenosols* ≈ Umbric Epipetric Plinthosols

Grey-Orthic Tenosols ≈ Leptosols / Cambisols / Andosols / Durisols / Plinthosols
<u>Great groups</u> in the suborder are identical to those of *Bleached-Orthic Tenosols,* except for the name *Grey-Orthic Tenosols.*

Leptic Tenosols ≈ Leptosols / Leptic Andosols / Leptic Plinthosols
<u>Great groups</u> in the suborder are identical to those of *Chernic-Leptic Tenosols,* except for the name *Leptic Tenosols.*

Red-Orthic Tenosols ≈ Leptosols / Cambisols / Andosols / Durisols / Plinthosols
<u>Great groups</u> in the suborder are identical to those of *Bleached-Orthic Tenosols,* except for the name *Red Orthic Tenosols.*

Sesqui-Nodular Tenosols ≈ Leptosols (*Ferric*) / Cambisols (*Ferric*) / Durisols (*Ferric*) / Plinthosols (*Ferric*) / Luvisols (Ferric)
The suborder is divided into the following <u>great groups</u>:

- *Argic Sesqui-Nodular Tenosols* ≈ Luvisols (Ferric)
- *Duric Sesqui-Nodular Tenosols* ≈ Durisols (*Ferric*)
- *Inceptic Sesqui-Nodular Tenosols* ≈ Cambisols (Skeletic, *Ferric*)
- *Lithic Sesqui-Nodular Tenosols* ≈ Leptosols (*Ferric*)
- *Paralithic Sesqui-Nodular Tenosols* ≈ Leptosols (*Ferric*)
- *Petrocalcic Sesqui-Nodular Tenosols* ≈ Petric Calcisols (*Ferric*)
- *Petroferric Sesqui-Nodular Tenosols* ≈ Petric Plinthosols (*Ferric*)
- *Regolithic Sesqui-Nodular Tenosols* ≈ Leptosols (*Ferric*)
- *Silpanic Sesqui-Nodular Tenosols* ≈ Durisols (*Ferric*)

Yellow-Orthic Tenosols ≈ Leptosols / Cambisols / Andosols / Durisols / Plinthosols
<u>Great groups</u> in the suborder are identical to those of *Bleached-Orthic Tenosols,* except for the name *Yellow Orthic Tenosols.*

<u>Vertosols</u> – soil <u>order</u>. ≈ Vertisols
The following six <u>suborders</u> are distinguished within the order:
Aquic Vertosols ≈ Stagnic Vertisols
No <u>great groups</u> are distinguished within the suborder due to the lack of information.

Black Vertosols ≈ Vertisols (Pellic)
The suborder is divided into the following <u>great groups</u>:

- *Crusty Black Vertosols* ≈ Vertisols (Pellic)
- *Epipedal Black Vertosols* ≈ Vertisols (Pellic)
- *Massive Black Vertosols* ≈ Mazic Vertisols (Pellic)
- *Self-mulching Black Vertosols* ≈ Grumic Vertisols (Pellic)

Brown Vertosols ≈ Vertisols (Chromic)
<u>Great groups</u> in the suborder are identical to those of *Black Vertosols*, except for the name *Brown Vertosols*.

Grey Vertosols ≈ Vertisols
<u>Great groups</u> in the suborder are identical to those of *Black Vertosols*, except for the name *Grey Vertosols*.

Red Vertosols ≈ Vertisols (*Rhodic*)
<u>Great groups</u> in the suborder are identical to those of *Black Vertosols*, except for the name *Red Vertosols*.

Yellow Vertosols ≈ Vertisols (*Xanthic*)
<u>Great groups</u> in the suborder are identical to those of *Black Vertosols*, except for the name *Yellow Vertosols*.

References

Hewitt, A. E. (1992) 'Soil classification in New Zealand: Legacy and lessons', *Australian Journal of Soil Research*, vol 30, no 5, pp843–854

Isbell, R. F. (1992) 'A brief history of national soil classification in Australia since the 1920s', *Australian Journal of Soil Research*, vol 30, no 5, pp825–842

Isbell, R. F. (1996) *Australian Soil Classification*, CSIRO Land & Water, Canberra, 160pp

Isbell, R. F. (2002) *Australian Soil Classification*, revised edition, CSIRO Land & Water, Canberra, 144pp

IUSS Working Group WRB (2006) *World Reference Base for Soil Resources*, 2nd edition, World Soil Resources Reports no 103, UN Food and Agriculture Organization, Rome, 128pp

Moore, A. W., Isbell, R. F. and Northcote, K. H. (1983) 'Classification of Australian soils', in A. W. Moore (ed) *Soils: An Australian Viewpoint*, CSIRO, Melbourne and Academic Press, London, pp253–266

Northcote, K. H. (1971) *A Factual Key for the Recognition of Australian Soils*, Rellim Technical Publication no VII, Kuratta Park, 122pp

Soil Survey Staff (1999) *Soil Taxonomy: A Basic System of Soil Classification for Making and Interpreting Soil Surveys*, USDA, Handbook 436, 2nd edition, United States Government Printing Office, Washington DC, 696pp

Stephens, C. G. (1956) *A Manual of Australian Soils*, Communications of the Scientific and Industrial Research Organisation, Melbourne, 54pp

28
Soil Classification of New Zealand

Objectives and scope

From the beginning of colonization New Zealand was mainly an agricultural country; thus interest in soil resources was quite natural. Soil survey, which had been using soil series as mapping units, soon needed a uniform system for the inventory of soil resources. The latest classification (Hewitt, 1998) is designed for large- and medium-scale soil surveys. The classification has national coverage (Table 28.1). The focus on agricultural lands determined the limits of the scope of the classification. Urban and industrial areas are not included, nor are bare rock or underwater sediments. Soils deeply transformed by agriculture are included in a special order of *Anthropic soils*.

Table 28.1 *The scope of soil classification of New Zealand*

Superficial bodies	Representation in the system
Natural soils	National coverage
Urban soils	Not included in the classification
Man-transported materials	Not included in the classification
Bare rock	Not considered as soil
Subaquatic soils	Not considered as soil
Soils deeply transformed by agricultural activities	Included in the special order of *Anthropic soils*

Theoretical background

The soil classification of New Zealand is constructed as a hierarchical taxonomy to make generalization of maps possible; soil groupings are based on measurable soil properties and not subjective opinions on soil genesis. Soil classes are constructed in such a way that it is possible to determine a soil's main properties from its placement in the system (Hewitt, 1992). The borders of classes are defined by strict quantitative criteria like in the US Soil Taxonomy (Soil Survey Staff, 1999) which served as a source system. Pedologists of New

Zealand rejected or reworked some of important concepts of their American colleagues. Soil regimes are not included in diagnostics of soil taxa. The main mapping units for large-scale soil survey are soil series; however they are regarded more as landscape units than as soil individuals.

Development

Soil classification existed in New Zealand even before the appearance of the Europeans (Hewitt, 1992). The native inhabitants of the islands, the Maori, created a detailed system of soil names, yet this indigenous classification had little influence on the development of soil classification and mapping in the country. Taylor presented the first systematic classification of New Zealand soils as a legend for a national soil map in 1948 (Hewitt, 1992). In 1953 the soil scientists of New Zealand reported completion of the soil survey of the country; however, it was mainly small-scale mapping. As a detailed soil survey was needed, so was a detailed soil classification. Taylor and Pohlen proposed such a classification in 1962 consisting of seven hierarchical levels. The terminology of this classification was complex: both traditional and 'technical' names were used. Hewitt (1992) notes that technical names obscured the genetic understanding of taxa rather than helping to interpret them, and were soon forgotten.

The publication of the US Soil Taxonomy with its pragmatic basis greatly affected the development of soil classification in New Zealand. The possibility of using Soil Taxonomy as an official classification in New Zealand was seriously discussed but was rejected in 1983 (Hewitt, 1992) when it was considered to be a national classification and did not reflect the peculiarity of New Zealand soils. The classification actually in use in New Zealand was published in the early 1990s (Hewitt, 1992) and a later version includes only minor modifications and corrections (Hewitt, 1998).

Structure

The New Zealand system has four hierarchical levels (Table 28.2). Soil orders are mainly generic taxa that reflect the development of the profile and/or the main soil-forming process as determined by the presence of a diagnostic horizon. Several orders are too general for the generic level, especially those related to poorly developed soils. Some soil archetypes are found at the soil group level. Most groups and subgroups specify soil characteristics. Within subgroups are soil series, determined by three main parameters: soil-forming material, texture and soil depth. The concept of soil series is broader than in the US and is approximately at the level of soil families in the Soil Taxonomy (Soil Survey Staff, 1999). The soil classification of New Zealand is a hierarchical taxonomy with formal borders.

Table 28.2 *The structure of soil classification of New Zealand*

Level	Taxon name	Taxon characteristics	Borders between classes	Diagnostics	Terminology
0	Soils				
1	Order	Generic	Formal	Chemico-morphological	Mixed
2	Group	Specific	Formal	Chemico-morphological	Mixed
3	Subgroup	Specific	Formal	Chemico-morphological	Mixed
6	Series	Varietal	Formal	Chemico-morphological	Traditional

Diagnostics

The source of diagnostics in New Zealand classification is the soil profile. Soil-forming conditions are not taken into account. Some of the order names may be confusing (e.g. *Semiarid soils*), but diagnostics are selected based on soil properties, not moisture regimes. Diagnostic horizons and properties are characterized mainly by their morphology and field chemical tests. The diagnostics in this classification are designed for field work with a minimum of laboratory analyses. They may be called quantitative chemico-morphological.

Terminology

One of the requirements noted by Hewitt (1992) for the national classification was a desire to use understandable English words. Most of the terms in the classification are meaningful words, such as *Organic soils*, *Allophanic soils*, *Pumice soils* and *Brown soils*, and some are traditional soil science terms of Russian origin (*Podzols*, *Gley soils*). It was impossible to completely avoid artificial terminology, and there are words at the order (e.g. *Ultic soils*, *Melanic soils*, etc.) and lower levels based on Latin roots.

Correlation

As noted above, the design of the New Zealand soil classification is similar to that of the US Soil Taxonomy (Soil Survey Staff, 1999). However, the correlation of the terms between the New Zealand classification and the World Reference Base for Soil Resources (WRB) (IUSS Working Group WRB, 2006) was much easier that that between the US Soil Taxonomy and the WRB. The main difficulty for the latter correlation was the use of moisture and temperature regimes as diagnostic criteria at higher levels of taxonomy in the US classification, which could not be reflected in the terms of the WRB. In New Zealand, those regimes are not used for soil diagnostics, and the correlation is easier. The other factor that facilitates the correlation is that the classification of soils derived from volcanic materials, which are widespread in New Zealand, was to a great extent imported to the WRB from the New Zealand taxonomy. However, some of soil groups in New Zealand have no direct analogies in the

WRB – for example, *Mafic* groups (soils developed in the regoliths of basic ferromagnesian rocks), to which the closest modifier in the WRB is *Magnesic* (soils with high exchangeable Mg concentration); the correspondence is not complete, because high content of iron oxides and hydroxides is not reflected by WRB modifiers.

The list of correlation of terms of the soil classification of New Zealand with the terms of the WRB is provided down to the level of groups. The names of orders are listed alphabetically and the names of soil groups are alphabetical within the orders.

ALLOPHANIC SOILS ≈ Andosols
The order is divided into the following groups:

- *Gley Allophanic Soils* ≈ Gleyic Andosols
- *Impeded Allophanic Soils* ≈ Petroduric Andosols / Duric Andosols
- *Orthic Allophanic Soils* ≈ Andosols
- *Perch-gley Allophanic soils* ≈ *Stagnic* Andosols

ANTHROPIC SOILS ≈ Anthrosols / Regosols
The order is divided into the following groups:

- *Fill Anthropic Soils* ≈ Hortic Anthrosols
- *Mixed Anthropic Soils* ≈ Plaggic Anthrosols / Hortic Anthrosols
- *Refuse Anthropic Soils* ≈ Terric Anthrosols
- *Truncated Anthropic Soils* ≈ Regosols

BROWN SOILS ≈ Cambisols / Arenosols
The order is divided into the following groups:

- *Acid Brown Soils* ≈ Cambisols (Dystric)
- *Allophanic Brown Soils* ≈ Andic Cambisols
- *Firm Brown Soils* ≈ Fragic Cambisols
- *Mafic Brown Soils* ≈ Cambisols (*Magnesic*)
- *Orthic Brown Soils* ≈ Cambisols
- *Oxidic Brown Soils* ≈ Ferralic Cambisols
- *Sandy Brown Soils* ≈ Brunic Arenosols

GLEY SOILS ≈ Gleysols / Fluvisols
The order is divided into the following groups:

- *Acid Gley Soils* ≈ Gleysols (Dystric)
- *Orthic Gley Soils* ≈ Gleysols
- *Oxidic Gley Soils* ≈ Plinthic Gleysols
- *Recent Gley Soils* ≈ Gleyic Fluvisols
- *Sandy Gley Soils* ≈ Gleysols (Arenic)
- *Sulphuric Gley Soils* ≈ Gleysols (Thionic) / Fluvisols (Thionic)

GRANULAR SOILS ≈ Luvisols / Lixisols
The order is divided into the following groups:

- *Melanic Granular Soils* ≈ Umbric Luvisols / Umbric Lixisols
- *Orthic Granular Soils* ≈ Luvisols / Lixisols
- *Oxidic Granular Soils* ≈ Luvisols (Chromic) / Lixisols (Chromic)
- *Perch-gley Granular Soils* ≈ Stagnic Luvisols / Stagnic Lixisols

MELANIC SOILS ≈ Chernozems / Phaeozems / Kastanozems / Vertisols / Leptosols
The order is divided into the following groups:

- *Mafic Melanic Soils* ≈ Chernozems (*Magnesic*) / Phaeozems (*Magnesic*)
- *Orthic Melanic Soils* ≈ Chernozems / Phaeozems
- *Perch-gley Melanic Soils* ≈ Stagnic Chernozems / Stagnic Phaeozems
- *Rendzic Melanic Soils* ≈ Rendzic Leptosols / Rendzic Phaeozems
- *Vertic Melanic Soils* ≈ Vertisols

ORGANIC SOILS ≈ Histosols / Histic Gleysols
The order is divided into the following groups:

- *Fibric Organic Soils* ≈ Fibric Histosols / Histic Gleysols
- *Humic Organic Soils* ≈ Sapric Histosols / Histic Gleysols
- *Litter Organic Soils* ≈ Folic Histosols
- *Mesic Organic Soils* ≈ Hemic Histosols / Histic Gleysols

OXIDIC SOILS ≈ Ferralsols / Plinthisols
The order is divided into the following groups:

- *Nodular Oxidic Soils* ≈ Ferralsols (Ferric)
- *Orthic Oxidic Soils* ≈ Ferralsols
- *Perch-gley Oxidic Soils* ≈ Stagnic Plinthisols

PALLIC SOILS ≈ Stagnosols / Planososls
The order is divided into the following groups:

- *Argillic Pallic Soils* ≈ Luvic Planosols / Lixic Planosols
- *Duric Pallic Soils* ≈ Duric Planosols
- *Fragic Pallic Soils* ≈ Fragic Planosols
- *Immature Pallic Soils* ≈ Ruptic Planosols
- *Laminar Pallic Soils* ≈ *Lamellic* Planosols
- *Perch-gley Pallic Soils* ≈ Stagnic Planosols / Stagnosols

PODZOLS ≈ Podzols
The order is divided into the following groups:

- *Densipan Podzols* ≈ Albic Podzols (*Densic*)
- *Groundwater-gley Podzols* ≈ Gleyic Albic Podzols
- *Orthic Podzols* ≈ Albic Podzols
- *Pan Podzols* ≈ Albic Ortstenic Podzols
- *Perch-gley Podzols* ≈ Stagnic Albic Podzols

PUMICE SOILS ≈ Vitric Andosols

- *Impeded Pumice Soils* ≈ Duric Vitric Andosols / Fragic Vitric Andosols
- *Orthic Pumice Soils* ≈ Vitric Andosols
- *Perch-gley Pumice Soils* ≈ *Stagnic* Vitric Andosols

RAW SOILS ≈ Fluvisols / Gleysols / Arenosols / Regosols
The order is divided into the following groups:

- *Fluvial Raw Soils* ≈ *Protic* Fluvisols
- *Gley Raw Soils* ≈ Gleysols / Endogleyic Regosols
- *Hydrothermal Raw Soils* ≈ Regosols
- *Orthic Raw Soils* ≈ Regosols
- *Rocky Raw Soils* ≈ Leptosols
- *Sandy Raw Soils* ≈ Protic Arenosols
- *Tephric Raw Soils* ≈ Regosols (Tephric)

RECENT SOILS ≈ Fluvisols / Regosols / Arenosols / Leptosols
The order is divided into the following groups:

- *Fluvial Recent Soils* ≈ Fluvisols
- *Hydrothermal Recent Soils* ≈ Regosols
- *Orthic Recent Soils* ≈ Regosols
- *Rocky Recent Soils* ≈ Leptosols
- *Sandy Recent Soils* ≈ Arenosols
- *Tephric Recent Soils* ≈ Regsols (Tephric)

SEMIARID SOILS ≈ Luvisols / Lixisols / Solonetz / Cambisols
The order is divided into the following groups:

- *Aged-argillic Semiarid Soils* ≈ Luvisols (Profondic, Aridic) / Lixisols (Profondic, Aridic)
- *Argillic Semiarid Soils* ≈ Luvisols (Aridic) / Lixisols (Aridic)
- *Immature Semiarid Soils* ≈ Cambisols (Aridic)
- *Solonetzic Semiarid Soils* ≈ Solonetz

ULTIC SOILS ≈ Acrisols / Nitisols
The order is divided into the following groups:

- *Albic Ultic Soils* ≈ Acrisols (Albic)
- *Densipan Ultic Soils* ≈ Acrisols (*Densic*) / Nitisols (*Densic*)
- *Perch-gley Ultic Soils* ≈ Stagnic Acrisols / *Stagnic* Nitisols
- *Sandy Ultic Soils* ≈ Acrisols (Arenic)
- *Yellow Ultic Soils* ≈ Acrisols (*Xanthic*) / Nitisols (*Xanthic*)

References

Hewitt, A. E. (1992) 'Soil classification in New Zealand: Legacy and lessons', *Australian Journal of Soil Research*, vol 30, no 5, pp843–854

Hewitt, A. E. (1998) *New Zealand Soil Classification*, 2nd edition, Maanaki Whenua –Landcare New Zealand Ltd, Dunedin, Landcare Research Science Series No 1, Lincoln, Canterbury, New Zealand, 122pp

IUSS Working Group WRB (2006) *World Reference Base for Soil Resources*, 2nd edition, World Soil Resources Reports no 103, UN Food and Agriculture Organization, Rome, 128pp

Soil Survey Staff (1999) *Soil Taxonomy: A Basic System of Soil Classification for Making and Interpreting Soil Surveys*, USDA, Handbook 436, 2nd edition, United States Government Printing Office, Washington DC, 696pp

29
Soil Classification of Ghana

Objectives and scope

The Ghana Interim Soil Classification was made as a provisional tool for soil mapping in Ghana, but was also used for soil inventory in a number of British Commonwealth countries (Adjei-Gyapong and Asiamah, 2002). The classification is focused on natural soils (Table 29.1). The soils of urban and industrial areas, not very widespread in East Africa, are not present in the classification. Exposed rock is partially included in several great groups (various Lithosols) in the Lithopeds suborder: these soils are distinguished in landscapes with common outcrops of consolidated rock or ironpan. The soils under shallow water are included in a special suborder *Hydrosols* with two suborders for fresh and brackish water. Cultivated soils are classified in the same manner as the natural ones; the soils deeply transformed by agriculture are classed with natural soils.

Table 29.1 *The scope of soil classification of Ghana*

Superficial bodies	Representation in the system
Natural soils	Used for Ghana and a number of other African countries
Urban soils	Not included in the classification
Man-transported materials	Not included in the classification
Bare rock	Partially included in the *Lithosols* groups
Subaquatic soils	Included in a special suborder of *Hydropeds*
Soils deeply transformed by agricultural activities	Classified as if they are natural soils

Theoretical background

The soil classification of Ghana was influenced by the early classification works of the British school of pedology (e.g. Robinson, 1936) and by the data provided by numerous soil survey campaigns in the African countries of the

Commonwealth. Few of these reports are still available outside the countries where the survey was done. The soil classification of Ghana is a combination of two different approaches: one, a soil series empirical classification and the other, a general geographical and environmental approach to soil grouping. The description and classification of soil series was completely focused on soil morphology and properties, while the upper level taxa were characterized by the environmental conditions and overall soil morphology rather than by quantitative characteristics of soil (Adjei-Gyapong and Asiamah, 2002). The structure and the names of the taxonomic levels (orders, suborders and great soil groups) are the same as in the US Soil Taxonomy (Soil Survey Staff, 1999).

Development

The soil classification of Ghana was developed by the Director of the West African Cocoa Research Institute, C. F. Charter, and his assistant, H. Brammer, in the 1950s (Adjei-Gyapong and Asiamah, 2002). They successfully applied their system both in Ghana and in other African countries. Unfortunately, most of their publications were in 'grey' literature, which is almost unavailable now. The only publication now used is a short review of soils of Ghana made by Brammer (1962). The classification is called 'interim' because it was not completely finished, and a number of discrepancies still exist in the system. However, it is still used in Ghana for soil mapping and in scientific publications.

Structure

The soil taxonomy of Ghana has six hierarchical levels (Table 29.2). The upper level of the <u>orders</u> is conceptual, and shows which factors (climatic, topographic/hydrologic or lithologic/evolutionary ones) are the most important. The second level of <u>suborders</u> specifies what kind of climate or topoforms determines pedogenesis. The level of <u>families</u> gives soil names based on their own properties but in a general way. The level of <u>great groups</u> is generic; the level of <u>subgroups</u> specifies soils characteristics. The latter level was developed only for the most abundant soils in Ghana. Soil <u>series</u> with corresponding phases are at the lowest level of taxonomy: like in the US and the UK, the series form a special empirical generic level. The classification of Ghana is a hierarchical taxonomy with fuzzy borders between the classes.

Diagnostics

The diagnostics in the soil classification of Ghana are qualitative. The definition of classes is given by the description of a 'central image' rather than by strict limits between different units. At the highest collective levels mainly external factors are used for diagnostics: the climate, topoforms and specific parent materials. At the level of great soil groups the qualitative characteristics of soils are used together with some environmental criteria such as forest or

Table 29.2 *The structure of soil classification of Ghana*

Level	Taxon name	Taxon characteristics	Borders between classes	Diagnostics	Terminology
0	Soils	Kingdom			
1	Order	Collective	Fuzzy	Conceptual	Artificial
2	Suborder	Collective	Fuzzy	Climatic-landscape	Artificial
3	Family	Collective	Fuzzy	Chemico-morphological	Artificial / traditional
4	Great group	Generic 1	Fuzzy	Landscape-chemico-morphological	Artificial / traditional
5	Subgroup	Specific	Fuzzy	Morphological	Artificial / traditional
6	Series	Generic 2	Fuzzy	Chemico-morphological	Localities' names

savanna vegetation. The diagnostics of soil series are mostly morphological with supplementary use of simple laboratory analyses.

Terminology

The names of the highest levels are artificial, and reflect the pedogenetic concepts of the authors. At the level of families and lower the terminology is more traditional, although the names invented by using a Latin root 'sol', such as Ochrosol, Oxysol, Rubrisol, Hydrosol and so on, are used in the classification. This terminology is artificial, but the constructed soil names are accompanied by normal English modifiers, like 'acid', 'red', 'dune-sand' and others; thus, the terminology is characterized as a mixed one. Some soil names, later used in the Food and Agriculture Organization (FAO) soil map legend, are present in this classification: Regosols, Planosols and Gleisols (a slightly modified version of Gleysols). It helps to understand the origin of the terms of the FAO legend. A few names are confusing: for example, in the Ghanaian classification Planosols are not soils with an abrupt textural change, as in the World Reference Base for Soil Resources (WRB) (IUSS Working Group WRB, 2006), but are soils of plain areas with near-surface water tables which include analogues of the Podzols, Plinthisols and Vertisols WRB reference groups.

Correlation

A limited correlation of the soil classification of Ghana with the terms of the WRB (IUSS Working Group WRB, 2006) is available in the paper of Adjei-Gyapong and Asiamah (2002). Thus we present our version. Since the hierarchical structure of the classification is intricate, we list the soils as they are listed in the scheme presented by Brammer (1962).

CLIMATOPHYTIC EARTHS – soil order. This order includes soils developed mainly under the influence of the climate and vegetation rather than by their position in the relief or by specific parent materials; it corresponds more or less to the concept of 'zonal' soils. There are two suborders within the order:

Hydropeds – soil suborder. These are freely drained soils, where percolating water reaches the groundwater level. There are two soil group families in the suborder:

Latosols ≈ Acrisols / Ferralsols / Nitisols / Lixisols / Plinthisols
Within the soil group family there are three great soil groups:

- *Forest Ochrosols* ≈ Acrisols / Nitisols / Plinthisols
- *Savanna Ochrosols* ≈ Lixisols / Plinthisols
- *Forest Oxysols* ≈ Acrisols / Ferralsols

Basisols ≈ Luvisols / Alisols / Cambisols / Umbrisols / Phaeozems
Within the soil group family there are four great soil groups:

- *Forest Rubrisols* ≈ Alisols / Cambisols
- *Savanna Rubrisols* ≈ Luvisols / Cambisols
- *Forest Brunosols* ≈ Cambisols / Umbrisols
- *Savanna Brunosols* ≈ Cambisols / Phaeozems

Xeropeds – soil suborder. These are soils, where the percolating water front does not reach the groundwater level. For Ghana the existence of this suborder was not confirmed, and it has no units of lower level.

TOPOHYDRIC EARTHS – soil order. This order includes soils formed mainly under the influence of the relief and close groundwater table. The concept more or less corresponds to the 'intrazonal' soils.

Planopeds – soil suborder. There are four soil group families in the suborder:

Very Acid Planosols ≈ Gleyic Podzols
Within the soil group family there is one great soil group:

- *Groundwater Podsols* ≈ Gleyic Podzols

Acid Planosols ≈ Plinthosols
Within the soil group family there is one great soil group:

- *Groundwater Laterites* ≈ Plinthosols

Calcium Planosols ≈ Calcic Vertisols
Within the soil group family there are two great soil groups:

- *Tropical Black Earths* ≈ Calcic Vertisols (Pellic)
- *Tropical Brown Earths* ≈ Calcic Vertisols (Chromic)

Sodium Planosols ≈ Solonetz / Planosols (Sodic)
Within the soil group family there is one great soil group:

- *Tropical Grey Earths* ≈ Solonetz / Planosols (Sodic)

Depressiopeds – soil suborder. There are five soil group families in the suborder:
Very Acid Gleisols ≈ Gleysols (Hyperdystric) / Gleysols (Thionic)
Within the soil group family there is one great soil group:

- *Savanna Grey Very Acid Gleisols* ≈ Gleysols (Hyperdystric) / Gleysols (Thionic)

Acid Gleisols ≈ Gleysols (Dystric) / Endogleyic Cambisols (Dystric)
Within the soil group family there are four great soil groups:

- *Savanna Black Acid Gleisols* ≈ Mollic Gleysols (Dystric)
- *Savanna Brown Acid Gleisols* ≈ Endogleyic Cambisols (Dystric)
- *Forest Grey Acid Gleisols* ≈ Gleysols (Dystric)
- *Savanna Grey Acid Gleisols* ≈ Gleysols (Dystric)

Neutral Gleisols ≈ Gleysols (Eutric) / Endogleyic Cambisols (Eutric)
Within the soil group family there are four great soil groups:

- *Forest Black Neutral Gleisols* ≈ Mollic Gleysols (Eutric)
- *Savanna Brown Neutral Gleisols* ≈ Endogleyic Cambisols (Eutric)
- *Forest Grey Neutral Gleisols* ≈ Gleysols (Eutric)
- *Savanna Grey Neutral Gleisols* ≈ Gleysols (Eutric)

Calcium Vleisols ≈ Calcic Solonchaks
Within the soil group family there are three great soil groups:

- *Black Vleisols* ≈ Calcic Mollic Solonchaks
- *Brown Vleisols* ≈ Calcic Solonchaks
- *Grey Vleisols* ≈ Calcic Gleyic Solonchaks

Sodium Vleisols ≈ Solonetz / Solonchaks
Within the soil group family there are two great soil groups:

- *Solonetz* ≈ Solonetz
- *Solonchaks* ≈ Solonchaks (Sodic)

Cumulopeds – soil suborder. There is one soil group family in the suborder:
Cumulosols ≈ Histosols
Within the soil group family there are three great soil groups:

- *Very Acid Bog* ≈ Histosols (Hyperdystric) / Histosols (Thionic)
- *Acid Bog* ≈ Histosols (Dystric)
- *Saline Bog* ≈ Salic Histosols

<u>Hydropeds</u> – soil suborder. There is one soil group family in the suborder:
Hydrosols ≈ Subaquatic Fluvisols
Within the soil group family there are two great soil groups:

- *Neutral Hydrosols* ≈ Subaquatic Fluvisols
- *Saline Hydrosols* ≈ Salic Subaquatic Fluvisols

LITHOCHRONIC EARTHS – soil order. The order includes immature soils, either eroded or formed on young sediments. The concept is similar to that of 'azonal' soils.
Lithopeds – soil suborder. There are two soil group families in the suborder:
Basiomorphic Lithosols ≈ Leptosols
Within the soil group family there are four great soil groups:

- *Black Basiomorphic Lithosols* ≈ Mollic Leptosols / Umbric Leptosols
- *Brown Basiomorphic Lithosols* ≈ Leptosols (*Chromic*)
- *Red Basiomorphic Lithosols* ≈ Leptosols (*Rhodic*)
- *Yellow Basiomorphic Lithosols* ≈ Leptosols (*Xanthic*)

Non-Basimorphic Lithosols ≈ Leptosols (Dystric) / Epipetric Plinthisols
Within the soil group family there is one great soil group:

- *Non-Basiomorphic Lithosols* ≈ Leptosols (Dystric) / Epipetric Plinthisols

Regopeds – soil suborder. There is one soil group family in the suborder:
Regosols ≈ Arenosols / Regosols / Calcisols
Within the soil group family there are three great soil groups:

- *Dune-sand Regosols with calcareous pan* ≈ Petric Calcisols (Arenic)
- *Dune-sand Regosols without calcareous pan* ≈ Arenosols
- *Regosols* ≈ Regosols / Arenosols

Alluviopeds – soil suborder. There is one soil group family in the suborder:
Alluviosols ≈ Fluvisols
Within the soil group family there are three great soil groups:

- *Black Alluviosols* ≈ Mollic Fluvisols
- *Brown Alluviosols* ≈ Fluvisols
- *Gley Alluviosols* ≈ Gleyic Fluvisols

References

Adjei-Gyapong, T. and Asiamah, R. D. (2002) 'The interim Ghana soil classification system and its relation with the World Reference Base for Soil Resources', Quatorzième reunion du Sous-Comité ouest et centre africain de corrélation des sols pour la mise en valeur des terres, Abomey, Bénin, 9–13 October 2000, World Soil Resources Reports no 98, UN Food and Agriculture Organization, Rome, pp51–76

Brammer, H. (1962) 'Soils', in J. B. Wills (ed) *Agriculture and Landuse in Ghana*, Oxford University Press, London, Accra and New York, pp88–126

IUSS Working Group WRB (2006) *World Reference Base for Soil Resources*, 2nd edition, World Soil Resources Reports no 103, UN Food and Agriculture Organization, Rome, 128pp

Robinson, G. W. (1936) *Soils, their Origin, Constitution and Classification. An Introduction to Pedology*, Thos Murby & Co, London, 442pp

Soil Survey Staff (1999) *Soil Taxonomy: A Basic System of Soil Classification for Making and Interpreting Soil Surveys*, USDA, Handbook 436, 2nd edition, United States Government Printing Office, Washington DC, 696pp

30
Soil Classification of the Republic of South Africa

Objectives and scope

The soil classification of South Africa is a regional system (Soil Classification Working Group, 1991). It is aimed at soil mapping at various scales. Also, like most soil classifications, it gives the state-of-the-art of research on soil genesis and properties in South Africa. Though this classification has much in common with many other classifications, it has a unique structure (dimeric table) and specific terminology, which make it difficult to use outside the country. Since the classification is agriculture-oriented, it does not include 'exotic' bodies like urban soils, underwater soils, mine waste materials and so forth (Table 30.1).

Table 30.1 *The scope of the South African soil classification*

Superficial bodies	Representation in the system
Natural soils	National coverage
Urban soils	Not recognized as soils
Man-transported materials	One soil form, *Witbank*, includes man-made deposits
Bare rock	Not recognized as soils
Subaquatic soils	Not recognized as soils
Soils deeply transformed by agricultural activities	Classified as if they are natural soils

Theoretical background

The source system for part of the South African classification is the US Soil Taxonomy (Soil Survey Staff, 1999). The main ideas borrowed from Soil Taxonomy are polypedon, soil series and diagnostic horizons concepts. In the field, South African pedologists divide soil into so-called 'master horizons': O, A, E, B, C, R and G. These horizons are then studied to classify them as diagnostic horizons and materials. A combination of diagnostic horizons results in certain 'soil forms'. These forms are combinations of topsoil and subsoil horizons similar to the format proposed by E. A. Fitzpatrick (1988). Though the principle of distinguishing soil classes on the basis of sequences of horizons can be regarded as formal, the horizons themselves reflect soil genesis. This principle of South African classification is clearly outlined in a preface to the text of the classification. In general, the theory of soil classification of South Africa is similar to most international classifications.

Development

The development of soil science research and soil survey in South Africa resulted in the need for soil classification in the 1970s. Initially, mostly so-called 'simplified soil series' were used for mapping: these included only topsoil texture, since no sufficient information was available to characterize 'classical soil series'. Though only a few soil groups had enough information to be defined as soil series in the classical sense, the South African pedologists decided to establish a working group on soil classification and arrange the existing knowledge into a classification system. The first edition of South African soil classification (Soil Classification Working Group, 1977) included mainly soil forms (first taxonomic level classes), with few families (second-level taxa). The second edition (Soil Classification Working Group, 1991) includes many more families. Actually, the information available would permit the publishing of a new revised edition of the South African classification, but the dates for its preparation and publication have not been indicated.

Structure

At first glance, the South African classification has a traditional hierarchical structure with three levels (Table 30.2). The first level (forms) is a combination of diagnostic topsoil and subsoil horizons. The second level (families) is defined by the presence of additional characteristics in diagnostic horizons or by specific features of parent material, such as hardness or calcareous nature. The texture classes are defined by the texture of topsoil horizons.

As mentioned above, the South African classification looks like an ordinary hierarchical taxonomy. However, the definition of soil forms is unique in soil science. In the first edition (Soil Classification Working Group, 1977) all the forms were arranged in a table, resembling the periodic table of chemical elements. In the rows there were topsoil diagnostic horizons; in the columns, subsoil horizons. Forms appeared in cells of the table as combinations of topsoil and subsoil horizons. Like a periodic table, the scheme permitted the developing of a number of genetic and geographic interpretations. For example, one could find spatial links between soils in the same column or in the same row. Some cells were empty, because no soils with such a combination of horizons existed in South Africa. For example, organic topsoil was found only together with a gleyic horizon. It is unfortunate that in the second edition

Table 30.2 *The structure of the South African soil classification*

Level	Taxon name	Taxon characteristics	Borders between classes	Diagnostics	Terminology
0	Soils	Kingdom			
1	Form	Generic	Formal	Chemico-morphological	Proper names (localities' names)
2	Family	Varietal		Chemico-morphological	Proper names (localities' names)
3	Texture class	Specific	Formal	Physical	Scientific names

(Soil Classification Working Group, 1991) the structure is hidden – and the soil forms are presented as a list, not as a table.

Diagnostics

The source of diagnostics in South African classification is the soil profile. No factors of soil formation or soil regimes are taken into account. Soil horizons are identified using mainly field morphological criteria. However, a number of chemical analyses are required for most diagnostic horizons both at the form and family levels, such as organic matter content, cation exchange capacity and base saturation, iron extracted by sodium dithionate-citrate-bicarbonate and pyrophosphate extracts. Texture is to be determined using the pipette method both for families diagnostics and for soil texture classes. Generally, the diagnostics in this classification can be called quantitative chemico-morphological.

Terminology

The names of both series and soil forms in South Africa are proper names, originating from the place of first description. It is the same system as is used in many countries having soil series systems. Usually proper names are typical for nominal systems; for example, the system of soil series in the US is a nominal one, if used without higher taxonomic levels. However, in South Africa the situation is different. Every proper name in soil classification serves as a scientific term. First, the borders of soil forms are defined formally. Second, nominal systems, or 'lists', are potentially infinite; soil forms in South African classification are included into strict frames of the table. The terminology of this classification illustrates that soil names in some classifications are not very important. It is possible to give traditional names to soils, to invent artificial terms, to use proper names or just number the taxa – everything depends on the tradition and concepts of the designers. South Africa is a relatively small country, and the names of localities give additional information not only on soils, but also inferences about the landscape where they formed. However, it is not very convenient for use outside the country.

Correlation

The correlation of soil form names of the South African classification with the terms of the World Reference Base for Soil Resources (WRB) is attempted. Partially the correlation is based on terms of FAO-UNESCO (1974) provided in the text of the first edition of the South African classification (Soil Classification Working Group, 1977). However, both classifications have changed significantly since then, so this correlation is a fresh start. The correspondence of these two classifications is not very good because their main concepts of taxa and diagnostic horizons are different, and some qualitative criteria and analytical procedures also differ. For example, an important concept in the WRB (IUSS Working Group WRB, 2006) is textural differentiation (the presence of the *argic* horizon) but is considered in the South African classification only at the second level of taxonomy. The main diagnostic criteria at the first taxonomic level for B horizons are their colour and structure.

Only the first level of the soil classification of South Africa is correlated. Soil forms are listed below in alphabetic order.

- *Addo* ≈ Calcaric Calcisols
- *Arcadia* ≈ Vertisols
- *Askham* ≈ Petric Calcisols / Petric Luvic Calcisols
- *Augrabies* ≈ Haplic Calcisols / Luvic Calcisols
- *Avalon* ≈ Plinthosols
- *Bainsvley* ≈ Plinthosols / Plinthic Ferralsols / Plinthic Acrisols / Plinthic Lixisols
- *Bloemdal* ≈ Ferralsols / Acrisols / Lixisols
- *Bonheim* ≈ Luvic Chernozems / Luvic Phaeozems / Luvic Kastanozems
- *Brandvlei* ≈ Calcaric Cambisols
- *Cartref* ≈ Albic Leptic Luvisols
- *Champagne* ≈ Histic Gleysols / Histosols
- *Clovelly* ≈ Cambisols / Luvisols
- *Coega* ≈ Petric Calcisols
- *Concordia* ≈ Albic Podzols
- *Constantia* ≈ Rustic Podzols / Albic Acrisols / Albic Alisols
- *Dresden* ≈ Petric Plinthosols
- *Dundee* ≈ Fluvisols
- *Estcourt* ≈ Albic Solonetz / Stagnic Solonetz / Solodic Planosols
- *Etosha* ≈ Calcaric Cambisols
- *Fernwood* ≈ Hyperalbic Arenosols
- *Gamoep* ≈ Calcaric Leptic Cambisols / Calcaric Leptic Luvisols
- *Garies* ≈ Durisols (Chromic)
- *Glencoe* ≈ Haplic Plinthosols / Plinthic Cambisols

- *Glenrosa* ≈ Hyperskeletic Leptosols / Leptic Luvisols
- *Griffin* ≈ Cambisols (Chromic) / Luvisols (Chromic)
- *Groenkop* ≈ Leptic Podzols
- *Houwhoek* ≈ Leptic Albic Podzols
- *Hutton* ≈ Ferralsols (Rhodic) / Acrisols (Chromic) / Lixisols (Chromic)
- *Immerpan* ≈ Petrocalcic Kastanozems / Petrocalcic Phaeozems
- *Inanda* ≈ Umbric Ferralsols (Rhodic) / Haplic Ferralsols (Rhodic)
- *Inhoek* ≈ Chernozems / Haplic Phaeozems / Haplic Kastanozems
- *Jonkersberg* ≈ Entic Placic Podzols
- *Katspruit* ≈ Haplic Gleysols / Calcic Gleysols
- *Kimberley* ≈ Ferralsol (Calcaric)
- *Kinkelbos* ≈ Calcic Albic Luvisols
- *Klapmuts* ≈ Albic Luvisols
- *Knersvlakte* ≈ Petric Durisols
- *Kranskop* ≈ Cambisols (Chromic) / Umbrisols (Chromic) / Acrisols
- *Kroonstad* ≈ Gleyic Planosols / Gleyic Stagnosols
- *Lamotte* ≈ Endogleyic Albic Podzols
- *Longlands* ≈ Albic Plinthosols
- *Lusiki* ≈ Luvisols
- *Magwa* ≈ Cambisols (Chromic) / Umbrisols (Chromic) / Acrisols
- *Mayo* ≈ Leptic Phaeozems / Leptic Kastanozems / Rendzic Hyperskeletic Leptosols / Rendzic Phaeozems
- *Milkwood* ≈ Leptic Chernozems / Leptic Phaeozems / Leptic Kastanozems / Mollic Leptosols
- *Mispah* ≈ Leptosols
- *Molopo* ≈ Cambisols (Calcaric) / Luvisols (Calcaric)
- *Montagu* ≈ Endogleyic Calcic Cambisols / Endoleptic Calcic Luvisols
- *Namib* ≈ Arenosols
- *Nomanci* ≈ Umbric Hyperskeletic Leptosols / Leptic Umbrisols
- *Oakleaf* ≈ Cambisols / Luvisols
- *Oudshoorn* ≈ Petric Durisols
- *Pinedene* ≈ Endogleyic Cambisols / Endogleyic Alisols / Endogleyic Luvisols
- *Pinegrove* ≈ Haplic Podzols
- *Plooysburg* ≈ Leptic Ferralsols (Calcaric)
- *Prieska* ≈ Calcaric Leptic Calcisols
- *Rensburg* ≈ Gleyic Vertisols
- *Sepane* ≈ Endogleyic Luvisols
- *Shepstone* ≈ Rhodi-Albic Acrisols / Rhodi-Albic Alisols
- *Shortlands* ≈ Nitisols / Lixisols / Acrisols
- *Sterkspuit* ≈ Solonetz
- *Steendal* ≈ Phaeozems (Calcaric) / Kastanozems (Calcaric)
- *Swartland* ≈ Chromic Luvisols / Hyperochric Luvisols
- *Sweetwater* ≈ Cambisols / Umbrisols
- *Trawal* ≈ Calcic Petric Durisol
- *Tsitsikamma* ≈ Albic Placic Podzols

- *Tukulu* ≈ Endogleyic Cambisols
- *Valsrivier* ≈ Luvisols
- *Vilafontes* ≈ Albic Luvisols / Albic Lixisols
- *Wasbank* ≈ Albic Petric Plinthosols
- *Westleigh* ≈ Haplic Plinthosols / Plinthic Acrisols / Plinthic Lixisols
- *Willowbrook* ≈ Mollic Gleysols / Gleyic Phaeozems
- *Witbank* ≈ Technosols
- *Witfontain* ≈ Endogleyic Podzols

References

FAO-UNESCO (1974) *Legend of the Soil Map of the World*, UN Food and Agriculture Organization, Rome

Fitzpatrick, E. A. (1988) *Soil Horizons Designation and Classification. A Coordinate System for Defining Soil Horizons and their Use as the Basic Elements in Soil Classification for Different Purposes*, ISRIC Technical Paper no 17, Wageningen, The Netherlands, 142pp

IUSS Working Group WRB (2006) *World Reference Base for Soil Resources*, 2nd edition World Soil Resources Reports no 103, UN Food and Agriculture Organization, Rome, 128pp

Soil Classification Working Group (1977) *Soil Classification: A Binomial System for South Africa*, Science Bulletin no 390, Department of Agriculture Technical Survey, Pretoria, 150pp

Soil Classification Working Group (1991) *Soil Classification: A Taxonomic System for South Africa*, 2nd (revised) edition, Memoirs on the Agricultural Natural Resources of South Africa no 15, Department of Agricultural Development, Pretoria, 257pp

Soil Survey Staff (1999) *Soil Taxonomy: A Basic System of Soil Classification for Making and Interpreting Soil Surveys*, USDA, Handbook 436, 2nd edition, United States Government Printing Office, Washington DC, 696pp

31
Outdated, Extinct and Underdeveloped Classifications

The development of classifications

This book deals mainly with classifications currently in use. It means that these classifications are used for communication by the scientific community, for soil mapping and inventory, and are commonly taught in university courses.

Not all the classifications we have presented fit all of the criteria listed above.

Soil classification develops together with our knowledge about the soil, and therefore represents a status at a given point in time. Some developments are gradual, and are regarded as different versions of the same system. For example, the second edition of the US Soil Taxonomy (Soil Survey Staff, 1999) is an updated version of the first edition (Soil Survey Staff, 1975), that was, in its turn, a developed version of the 7th Approximation. However, in the history of soil classifications there are stages where an old scheme is rejected almost completely, and replaced by a new construction. Such occurred with the 1938 US system (Baldwin et al, 1938) that was abandoned and superseded by the 7th Approximation. The process resembles somewhat that described by Kuhn (1970) as a shift of a paradigm; however, on closer inspection soil is still soil but the organization of the available knowledge has changed. Several soil classification systems, including the Russian (Shishov et al, 2004), French (AFES, 1998), Chinese (CRG-CST, 2001) and Australian (Isbell, 2002), recently passed thresholds resulting in changes in concepts, structure, diagnostics and terminology. The acceptance of a new different classification system would normally lead to the extinction of a prior one. In some cases the replacement of the old classification by the new one is not very fast, and the systems coexist for a long period. Usually the institutions responsible for soil mapping have a decisive role. If such an institution decides to change the classification, the scientific community soon accepts it. If there are several such institutions in disagreement in a country, the process may last much longer.

We consider classifications that are already replaced by new systems but are still used somewhere for mapping, scientific communication or in education, as <u>outdated</u> classifications. The systems that are completely out of use we consider as <u>extinct</u> classifications. It is important to note that both kinds are useful because many scientific publications, cartographic materials and databases were made using these systems.

Apart from the classifications replaced by more recent systems, there are schemes that did not develop, either intentionally or by chance, to the level of a normal classification. These marginal constructions may consist of schematic classifications, map legends, soil inventories, authorial classifications or classification proposals. A few examples are provided below.

Outdated classifications

There are a number of outdated classifications that are officially 'closed' but are still in use among specialists. A striking example is the French classification prepared by the Commission for Soil Cartography (CPCS, 1967). It was used both in France and in the former French colonies which still had close scientific links. This classification had well-developed systematics of tropical soils, and was successfully applied in many African countries. After the older classification was changed into a completely novel Reference Base system (AFES, 1998), its use in France officially finished. However, it was still used for soil survey in many French-speaking countries in Africa. In Senegal, for example, large-scale soil maps are produced based on the old French system.

In China, the system currently in use is the Chinese Soil Taxonomy prepared by the Nanjing Institute of Soil Science (CRG-CST, 2001). However, a number of older specialists continue to use the older genetic soil classification (Gong Zitong, 1989) and until recently the older classification was even included in some university courses.

A more complex situation exists with the Soviet classification which is still used in Russia and some post-Soviet countries (Egorov et al, 1977). Much of its pedogenetic basis is out of date, which is recognized even by the opponents of the new Russian system (Shishov et al, 2004). Because the Soviet classification is not officially replaced by the new system, it is not yet an outdated one.

Extinct classifications

Some classifications are being phased out as a new national system is accepted. Gradually all the documents, maps, textbooks and scientific publications are adjusted to the new system. Perhaps the greatest challenge in the history of soil classification was related to the change of the older soil classification of the US (Baldwin et al, 1938; Thorp and Smith, 1949, see Table 31.1) to the system we know now as Soil Taxonomy. The older classification, which was good for its time, was completely replaced by the new taxonomic system and now is only of historical interest. Initially the new taxonomy was met negatively (Simonson, 1989) both inside the country and abroad. Long-term efforts of

Table 31.1 *The upper levels of the old soil classification of the USA (Baldwin and Smith, 1949)*

Orders	Suborders	Great groups
Zonal soils	1. Soils of the cold zone	Tundra soils
	2. Light-coloured soils of arid regions	Desert soils Red desert soils Sierozem Brown soil Reddish brown soil
	3. Dark-coloured soils of semiarid, subhumid and humid grasslands	Chestnut soils Reddish chestnut soils Chernozem soil Prairie soils Reddish prairie soils
	4. Soils of the forest–grassland transition	Degraded chernozem Non-calcic brown, or Shantung brown soils
	5. Light-coloured podzolized soils of timbered regions	Podzol soils Grey wooded or grey wooded podzolized soils Brown podzolic soils Grey-brown podzolic soils Red-yellow podzolic soils
	6. Lateritic soils of forested warm-temperate and tropical regions	Reddish-brown lateritic soils Yellowish-brown lateritic soils Lateritic soils
Intrazonal soils	1. Halomorphic (saline and alkaline) soils of imperfectly drained arid regions and littoral deposits	Solonchak, or saline soils Solonetz soils Soloth
	2. Hydromorphic soils of marshes, swamps, seep areas and flats	Humic-glei soils Alpine-meadow soils Bog soils Half-bog soils Low-humus glei soils Planosols Groundwater podzol soils Groundwater laterite soils
	3. Calcimorphic soils	Brown forest soils (Braunerde) Rendzina soils
Azonal soils		Lithosols Regosols (including Dry Sands) Alluvial soils

the National Cooperative Soil Survey promoting the new classification resulted in its complete acceptance within the US and in much of the world. Another example of the extinction of an older classification is the classification of Australian soils by Stephens (1956).

In most cases the competition between the old and new classifications finished with the establishment of the more recent version. Also, there are examples when a new classification fails to win, leading to the abandonment of both old and new versions. For example, in Finland a genetic classification was in use from the beginning of the 20th century (Frosterus, 1923). Then leading Finnish pedologists proposed replacing it with an agrogeological classification mainly using the terms of Quaternary geology (Aaltonen et al, 1949). Since the system was somewhat duplicating the classification used for geological maps, it was ignored, and for long time Quaternary sediments maps were produced in Finland instead of soil maps. Unfortunately, the older system was also abandoned. Current soil mapping is done to a very limited extent using the US Soil Taxonomy or the World Reference Base for Soil Resources (WRB) as a legend.

Not all classifications need to be replaced by new versions to disappear. There are a number of systems in the world that are in danger of extinction, like, for example, the soil classification of The Netherlands. Soil classification there is not used any more for mapping, nor is it included in university programmes. It is used only by scientists, mainly of the older generation.

Practically never in the history of soil science has a developed national classification system disappeared due to its replacement by a foreign or international classification. It was more common that underdeveloped classifications could not compete with the international ones. We do observe in the European Union a process of harmonization of national soil classifications with the umbrella system – WRB (IUSS Working Group WRB, 2006). Some of the classifications correspond well at higher levels of taxonomy with the WRB units, and even use the same terminology. It may indicate a process of peaceful aggregation of at least some of the European soil classification systems.

Should the extinction of national soil classifications be regarded as a dangerous process? The answer is positive if we speak about the extinction of classifications due to the loss of interest in continuing soil research, because then there is no new information or knowledge. If they disappear due to their replacement by newer ones, it is a normal process of the progress of science. Harmonization and consolidation of soil classifications is definitely a positive process which gives hope for the development of a universal system of soil classification sometime in the future.

Underdeveloped classifications

An extensive and diverse group of underdeveloped, 'almost classifications', will be mentioned only briefly. There are hundreds of soil inventories and soil map legends that cannot be regarded as classifications because they have no diagnostics and no defined structure. They are more or less schematic conceptual soil groupings which might have been developed to the level of a classification but were not.

By our definition of soil classification systems nearly all early schemes, including those of Hilgard, Dokuchaev and even Kubiëna (Sibirtsev, 1951), would be excluded. Of special interest are grouping systems that were almost developed into soil classifications but for some reason did not reach that level. For example, in Italy a soil inventory was made and a provisional soil classification was proposed (Principi, 1943) but it was never elaborated as a normal soil taxonomy.

A detailed soil inventory was started with a provisional soil classification in Peru (Oficina Nacional de Evaluación de Recursos Naturales, 1969) but the work was stopped and the classification was forgotten. Now few students in Peru know of any soil classification other than the US Soil Taxonomy. However, the underdeveloped national classification was not replaced by Soil Taxonomy, rather it did not develop because of the difficult socioeconomic situation in the country and lack of public demand for soil inventory. The US Soil Taxonomy is used only for educational purposes; currently no soil mapping is done in Peru.

An attempt to develop a world soil classification was done by Papadakis (1969), an Argentinian who worked in the Food and Agricultural Organization (FAO) for many years. His classification was never widely considered or discussed because the scientific community was focused on preparation of the FAO-UNESCO Soil Map of the World.

There are many schematic classifications that never developed into national systems. Mainly these schemes were created for scientific and educational purposes and often had no lower taxonomic levels and criteria of diagnostics. Such schematic classifications were especially widespread in the Soviet Union whereby textbooks and many scientific monographs proposed their own soil classifications (e.g. Glazovskaya, 1972; Volobuyev, 1973; Rozanov, 1977 and many others). In part it may have reflected a common antipathy among scientists towards any 'official system' of soil classification.

Maybe it is fortunate that all these systems did not develop further into competing regional or national ones. Undeveloped soil classifications, however, have provided many interesting and thought-provoking ideas about obtaining soil information and organizing it in novel ways.

References

Aaltonen, V. T., Aarnio, B. and Hyyppa, E. (1949) 'A critical review of soil terminology and soil classification in Finland in the year 1949', *Journal of Scientific Agricultural Society of Finland*, vol 21, pp55–66

AFES (1998) *A Sound Reference Base for Soils* (The 'Referentiel pedologique': text in English), INRA, Paris, 322pp

Baldwin, M., Kellog, C. E. and Thorp, J. (1938) 'Soil classification', in United States Department of Agriculture, *Soils and Men: Yearbook of Agriculture 1938*, US Government Printing Office, Washington DC, pp979–1001

CPCS (1967) *Classification des Sols*, Ecole nationale supérieure agronomique, Grignon, France, 87pp (in French)

CRG-CST (2001) *Chinese Soil Taxonomy*, Li Feng (ed), Science Press, Beijing and New York, 203pp

Egorov, V. V., Fridland, V. M., Ivanova, E. N., Rozov, N. N., Nosin, V. A. and Fraev, T. A. (1977) *Classification and Diagnostics of Soils of USSR*, Kolos Press, Moscow, 221pp (in Russian)

Frosterus, B. (1923) 'Die Klassifikation der Boden und Bodenarten Finlands', *Comite Internacional de Pedologie*, IV Commission, no 9, pp141–175 (in German)

Glazovskaya, M. A. (1972) *Soils of the World, vol I: The Main Soil Families and Types*, Moscow State University Publishers, Moscow, 231pp (in Russian)

Gong Zitong (1989) 'Review of Chinese soil classifications for the past four decades', *Acta Pedologica Sinica*, vol 26, no 3, pp217–225 (in Chinese, English summary)

Isbell, R. F. (2002) *Australian Soil Classification*, revised edition, CSIRO Land & Water, Canberra, 144pp

IUSS Working Group WRB (2006) *World Reference Base for Soil Resources*, 2nd edition, World Soil Resources Reports no 103, UN Food and Agriculture Organization, Rome, 128pp

Kuhn, T. S. (1970) *The Structure of Scientific Revolution,* University of Chicago Press, Chicago

Oficina Nacional de Evaluación de Recursos Naturales (1969) *Inventario de Estudios de Suelos del Perú*, ONERN, Lima, 446pp (in Spanish)

Papadakis, J. (1969) *Soils of the World*, Elsevier Publishers, Amsterdam, 422pp

Principi, P. (1943) *I Terreni d'Italia*, Sociedade Anonima Editrice Dante Alighieri, Albrighi, Segati XXI, Genoa, Rome, Naples and Gitta di Castello, 242pp (in Italian)

Rozanov, B. G. (1977) *Soil Cover of the Globe*, Moscow State University Publishers, Moscow, 248pp (in Russian)

Shishov, L. L., Tonkonogov, V. D., Lebedeva, I. I. and Gerasimova, M. I. (2004) *Classification and Diagnostics of Soils of Russia*, Oykumena, Smolensk, 342pp (in Russian)

Sibirtsev, N. M. (1951) 'Soil Science', in *Selected Works*, Agricultural Literature Press, Moscow, vol 1, pp19–465 (in Russian)

Simonson, R. W. (1989) *Historical Highlights of Soil Survey and Soil Classification with Emphasis on the United States, 1899–1970*, International Soil Reference and Information Centre, Wageningen, Technical Paper no 18, 83pp

Soil Survey Staff (1975) *Soil Taxonomy: A Basic System of Soil Classification for Making and Interpreting Soil Surveys*, USDA, Handbook 436, United States Government Printing Office, Washington DC, 754pp

Soil Survey Staff (1999) *Soil Taxonomy: A Basic System of Soil Classification for Making and Interpreting Soil Surveys*, USDA, Handbook 436, 2nd edition, United States Government Printing Office, Washington DC, 696pp

Stephens, C. G. (1956) *A Manual of Australian Soils*, Communications of the Scientific and Industrial Research Organisation, Melbourne, 54pp

Thorp, J. and Smith, G. D. (1949) 'Higher categories of soil classification', *Soil Science*, vol 67, no 2, pp117–126

Volobuyev, V. R. (1973) *System of World Soils*, Elm Press, Baku, Azerbaijan, 308pp (in Russian)

32
Classifications of Paleosols

The scope of paleosols classification

The classification of paleosols is a particular task for paleopedological, paleogeographical and archaeological studies. Though it is limited to a relatively narrow part of a scientific community, and never used for soil survey and inventory, we include a short list of existing paleosols classifications which are actually in use. All the paleosol classifications are based on the existing classifications, such as the US Soil Taxonomy (Soil Survey Staff, 1999) and the World Reference Base for Soil Resources (WRB) (IUSS Working Group WRB, 2006), but are significantly modified for the reasons mentioned below.

The scope of the term 'paleosol' is very extensive. The age of paleosols, according to general agreement, should be more than Holocene (Nettleton and Olson, 1999), and the upper limit is not defined, noting that Pre-Cambrian paleosols have been described (Retallack, 1990). Thus, paleosols can be represented both by almost unaltered soil profiles and by completely lithified sedimentary rock which only hypothetically may be regarded as soil. Paleosols include not only buried soils older than Holocene, but also superficial soils that have undergone significant climatic change in the past (surface paleosols), and soils that have been buried in the past but currently are exposed on the surface (exhumed paleosols) (Bronger and Cutt, 1989). The surface and exhumed paleosols are generally classified as normal superficial soils, in places causing problems for the classifiers. For example, the German soil classification includes special taxa for deeply weathered surface paleosols that do not correspond to the actual environment (Ad-hoc-AG Boden, 2005).

Difficulties for paleosol classification

For a long time, buried paleosols were classified in the same manner as superficial soils (e.g. Retallack, 1990). However, this approach had certain limitations. Practically all buried soils, especially the most ancient ones, have undergone diagenetic alteration and their properties are different from the initial ones. Thus, a number of the soil characteristics generally important for classification should be extrapolated or omitted. According to Dahms and Holiday (1998), using the same classification of surface soils for paleosols includes a kind of 'circular logic': we name a soil, in part, according to our interpretation of paleoenvironment and then confirm the interpretation by the presence of a certain soil unit. For example, it is difficult to estimate pre-existing base saturation for a buried soil with an *argillic* horizon saturated with diagenetic carbonates, thus it is impossible to distinguish Alfisols from Ultisols, and these two orders indicate different environments. Some diagnostic features have depth requirements that are impossible to estimate in some buried soils because the majority of them are truncated. It is especially difficult to classify paleosols in terms of the US Soil Taxonomy (Soil Survey Staff, 1999) because current moisture and temperature regimes are used at the highest levels of the taxonomy. Consequently, many paleopedologists prefer using provisional names for their objects, like Unit1, Red Unit and so on, and then discuss them in the terms of particular properties and processes (e.g. Sedov et al, 2001). However, this approach makes it difficult to correlate soils both geographically and stratigraphically as no standardized terminology is used.

Paleosols classifications based on the US Soil Taxonomy

A special classification for paleosols was proposed by Mack et al (1993). The general idea was to characterize buried soils according to their surface analogies, using for differentiating characteristics only stable soil properties which are not altered by diagenesis. The labile properties, such as salinity, base saturation and acidity, were not taken into account; the authors used mainly morphological and mineralogical criteria in their system. Consequently, the whole classification system had to be modified. The following soil orders were distinguished:

- Argillisols – soil with a horizon of clay illuviation (*argillic*, *kandic* or *natric*);
- Calcisols – soils with a cemented horizon of calcium carbonate accumulation;
- Gleysols – soils with morphological evidence of iron and manganese depletion;
- Gypsisols – soils with a cemented horizon of gypsum accumulation;
- Histosols – soils with an organic layer;
- Oxisols – deeply weathered soils;
- Protosols – weakly developed soils;
- Spodosols – soils with a horizon of iron, aluminium and humus accumulation;
- Vertisols – clayey soils with morphological evidence of swelling and shrinking (slickensides and cracks).

In the cases in which the definition of a paleosol was similar to that of a surface analogy in the Soil Taxonomy (Soil Survey Staff, 1999), the soil names were left the same as in the classification for surface soils (Histosol, Spodosol, Oxisol, Vertisol). If the definition was significantly modified, and soil orders had been separated or joined, new terms were proposed. These names were partially borrowed from the revised Food and Agriculture Organization (FAO) World Soil Map legend (FAO-UNESCO-ISRIC, 1988), like Calcisol, Gypsisol and Gleysol, and partially invented, like Argillisol and Protosol. To specify the classification, the authors proposed a list of modifiers: albic, allophonic, argillic, calcic, carbonaceous, concretionary, dystric, eutric, ferric, fragic, gleyed, gypsic, nodular, ochric, salic, silicic, vertic and vitric.

The system received criticism from two opposite points of view. Dahms and Holiday (1998) noted that it is still rather difficult to separate diagenetic and pedogenetic properties in paleosols. And Buurman (1998) pointed out the loss of information due to elimination of many diagnostic properties which the authors of the classification had considered to be dynamic. He also disparaged the use of identical terminology for surface and buried soils with different definitions, which could cause confusion.

Another attempt to classify buried paleosols was made by Nettleton et al (1998). The authors proposed a complete change of the definition of soil orders for buried soils, and modified the order names, adding a prefix *krypt*. The list of orders included Kryptohistosols, Kryptospodosols, Kryptandisols, Kryptoxisols, Kryptovertisols, Kryptaridisols, Krypteldisols, Kryptevolvisols and Kryptaddendosols. Most of these orders had direct analogies among surface soils, but the definitions of the three latter orders are modified completely. Kryptoeldisols corresponded to Ultisols, and Kryptoevolvisols to Alfisols but, because the determination of base saturation is of no use for buried paleosols, a total reserve of weatherable minerals was used instead of base saturation as a diagnostic criterion. The authors made an interesting attempt to correlate some dynamic soil properties (such as base saturation and soil pH) with the stable ones (total chemical composition and clay content).

Later on Nettleton et al (2000) developed a key for paleosol classification, added new definitions and modifiers, modified the terminology and verified the classification on an example of 12 paleosols all over the world. Currently it is the most complete working system for paleosols classification based on the US Soil Taxonomy. The renewed system uses the prefix *paleo* instead of *krypto*, and intends also to include surface paleosols in the scope of the classification. The resulting list of orders is the following:

- Paleoaddendosols – paleosols without expressed diagnostic horizons, 'other soils';
- Paleoandisols – paleosols having *andic* or *vitric* material;
- Paleoaridisols – paleosols with a horizon cemented with calcium carbonate, gypsum or silica;
- Paleoeldisols – paleosols with a horizon of clay illuviation, poor in weatherable minerals;
- Paleoevolvisols – paleosols with a horizon of clay illuviation, rich in weatherable minerals;

- Paleohistosols – paleosols with an organic layer;
- Paleoinceptisols – paleosols with distinctly altered horizon (clay formation, iron depletion or accumulation etc.);
- Paleomollisols – paleosols with a humus-enriched horizon;
- Paleooxisols – deeply weathered paleosols;
- Paleospodosols – palesols with a horizon of iron, aluminium and humus illuviation;
- Paleovertisols – clayey paleosols with evidence of soil swelling and shrinking.

The following modifiers are terms for being more specific: accretionary, buried (cryptic), complete, truncated, welded, carbonate-enriched, unleached, leached, gleyed, oxidized, residual, alluvial, colluvial, eolian, pyroclastic, extensive and inextensive.

The system is vulnerable to criticism, as is any classification. For example, Paleoaridisols are defined by the presence of petrocalcic, petrogypsic horizons or a duripan because the presence of soft secondary carbonates is a labile property. However, in many cases petrocalcic horizons form in a wide range of climates in soils with shallow water tables and the authors recognize that petrocalcic and petrogypsic horizons may also form under Xeric or Ustic water regimes (Nettleton et al, 2000). It may lead to misinterpretation because the Paleoaridisols order by its name implies certain paleoenvironmental conditions. Regardless, this classification is an important step for unifying the terminology of paleosols.

WRB-based paleosols classification

Buurman (1998) argued that the US Soil Taxonomy with its focus on soil regimes was not the best basis for classifying paleosols and that the WRB might be a better system. However, until recently no attempt was made to adapt the WRB to include paleosol classification. The only available scheme (Krasilnikov and García-Calderón, 2006) proposes the classification of buried soils from the very recent in age to lower Cainozoic. Older lithified soils are not considered. The Holocene buried soils, disregarded by most paleopedologists as paleosols (Bronger and Cutt, 1989; Nettleton and Olson, 1999) are included in the classification because of their importance for archaeological studies and for paleoenvironmental interpretations at high latitudes where important climatic changes have occurred even within the last 10,000–12,000 years. It is valid for both complete and truncated soil profiles, either starting from the depth of more than 200cm from the actual soil surface or, independently of the depth, separated from the actual surface soil by a layer of sediments of 50cm thickness or more. The terminology of the system follows the first edition of the WRB (FAO-ISRIC-ISSS, 1998) adding the prefix *infra* if the definition is similar to that of the WRB for a surface analogy. In the case of a different definition or joining several groups, the name is different. For example, soils with a 'cultural layer', which may correspond to former Technosols, are called Archaeosols.

The group Negrosols include buried Chernozems, Phaeozems, Kastanozems and Umbrisols, because all the criteria for separating these reference groups (soft secondary carbonates and base saturation) are labile and cannot be applied to buried soils. The list of groups in the paleosol classification is the following:

- Archaeosols – paleosols with a 'cultural layer';
- Infraandosols – paleosols with pyroclastic materials and the products of their alteration;
- Infraanthrosols – paleosols with a deep cultivated layer;
- Infraarenosols – sandy paleosols;
- Infracalcisols – paleosols with a cemented horizon of accumulation of calcium carbonate;
- Infracambisols – paleosols with a distinct alteration of soil material (clay or iron accumulation *in situ*);
- Infracryosols – paleosols with cryogenic features;
- Infradurisols – paleosols with a horizon cemented by silica;
- Infraferralsols – deeply weathered paleosols;
- Infrafluvisols – buried alluvial soils;
- Infragleysols – paleosols with evidences of paleogleying;
- Infraglossisols – paleosols with a horizon of clay accumulation, penetrated by specific tongues of *albic* whitish material;
- Infragypsisols – paleosols with a horizon cemented by gypsum;
- Infrahistosols – paleosols with an organic horizon;
- Infraleptosols – shallow paleosols;
- Infralixisols – paleosols with a horizon of clay accumulation, with low activity clays;
- Infraluvisols – paleosols with a horizon of clay accumulation, with high activity clays;
- Infranitisols – weathered paleosols with a well-developed structure;
- Infraplanosols – paleosols with an abrupt textural change;
- Infraplinthosols – paleosols with hydrogenic iron accumulation;
- Infrapodzols – paleosols with a horizon of accumulation of iron, aluminium and humus;
- Infrasolonetz – paleosols with a horizon of clay accumulation of a specific columnar structure;
- Infravertisols – clayey paleosols with evidence of swelling and shrinking;
- Negrosols – paleosoils with a developed dark-humus-enriched horizon;
- Ochrisols – paleosols with poorly developed morphology, 'other soils'.

The original publication (Krasilnikov and García-Calderón, 2006) includes a key for paleosol determination and a description of diagnostic horizons, materials and properties. The classification may be made more specific using the modifiers recommended by the WRB (IUSS Working Group WRB, 2006), but with special caution. The modifiers related to conservative soil properties may be used without limitations but the modifiers reflecting the labile properties (e.g. the presence of carbonates, salts, gleyic or stagnic colour pattern) should be used only if their pedogenic or diagenetic origin can be established. Then, in the first case the modifiers receive a prefix *pedo*, and in the latter case, a prefix *dia*.

References

Ad-hoc-AG Boden (2005) *Bodenkundliche Kartieranleitung*, herausgegeben von der Bundesanstalt für Geowissenschaften und Rohstoffe und den Geologischen Landesämtern in der Bundesrepublik Deutschland, Hannover, 5 Auflage, 438pp (in German)

Bronger, A. and Cutt, J. A. (1989) 'Paleosols: Problems of definition, recognition and interpretation', in A. Bronger and J. A. Cutt (eds) *Paleopedology, Nature and Application of Paleosols, Catena Supplement 16*, Cremlingen, Germany, pp1–7

Buurman, P. (1998) 'Classification of paleosols: A comment', *Quaternary International*, vols 51–52, pp17–20

Dahms, D. E. and Holiday, V. T. (1998) 'Soil taxonomy and paleoenvironmental reconstruction: A critical commentary', *Quaternary International*, vols 51–52, pp109–114

FAO-ISRIC-ISSS (1998) *World Reference Base for Soil Resources*, World Soil Resources Report no 84, UN Food and Agriculture Organization, Rome, 88pp

FAO-UNESCO-ISRIC (1988) *Soil Map of the World, Revised Legend*, World Soil Resources Reports no 60, UN Food and Agriculture Organization, Rome, 92pp

IUSS Working Group WRB (2006) *World Reference Base for Soil Resources*, 2nd edition, World Soil Resources Reports no 103, UN Food and Agriculture Organization, Rome, 128pp

Krasilnikov, P. and García Calderón, N. E. (2006) 'A WRB-based buried paleosol classification', *Quaternary International*, vols 156–157, no 2, pp176–188

Mack, G. H., James, W. C. and Monger, H. C. (1993) 'Classification of paleosols', *Geological Society of America Bulletin*, vol 105, pp129–136

Nettleton, W. D. and Olson, C. G. (1999) 'A paleosol classification with the inclusion of ancient arctic and subarctic region soils', *Chinese Science Bulletin*, vol 44, suppl 1, pp243–254

Nettleton, W. D., Brasher, B. D., Behman, E. C. and Ahrens, R. J. (1998) 'A classification system for buried paleosols', *Quaternary International*, vols 51–52, no 2, pp175–183

Nettleton, W. D., Olson, C. G. and Wysocki, D. A. (2000) 'Paleosols classification: Problems and solutions', *Catena*, vol 41, no 1, pp61–92

Retallack, C. J. (1990) *Soils of the Past: An Introduction to Paleopedology*, Unwin-Hyman, Boston, 396pp

Sedov, S., Solleiro-Rebolledo, E., Gama-Castro, J. E., Vallejo-Gómez, E. and González-Velázquez, A. (2001) 'Buried paleosols of the Nevado de Toluca: An alternative record of Late Quaternary environmental change in central Mexico', *Journal of Quaternary Science*, vol 16, no 4, pp375–389

Soil Survey Staff (1999) *Soil Taxonomy: A Basic System of Soil Classification for Making and Interpreting Soil Surveys*, USDA, Handbook 436, 2nd edition, United States Government Printing Office, Washington DC, 696pp

33
A Review of World Soil Classifications

The scope of world classifications

Among all the existing classifications only three have worldwide coverage: the World Reference Base for Soil Resources (WRB) (IUSS Working Group WRB, 2006), the USDA Soil Taxonomy (Soil Survey Staff, 1999) and the French soil classification (AFES, 1998). De facto, only the first two are used worldwide, and the French classification is only potentially suited for classifying world soils. The older version of the French soil taxonomy (CPCS, 1967) is used more often in the former French colonies than the new classification (AFES, 1998).

Both the WRB and Soil Taxonomy are used worldwide for soil correlation: in technical documents, soil reports and in scientific publications local soil names are usually correlated with the terms of one, or both, of the world classifications. It should be mentioned that the correlation with the US Soil Taxonomy is not always adequate because in most countries precise information on soil moisture regimes is absent; however, they can be approximated from long-term monthly climatic data.

In many countries the US Soil Taxonomy is used for all kinds of activities, including soil inventory, mapping, scientific research and university education. Most Latin American countries (except Mexico, Brazil and Cuba), almost all the Asian countries (except Israel, China and Japan) and even some European countries use the US classification system instead of a national classification. The US Soil Taxonomy is used in about half of the world as a basic system for soil classification.

The WRB was not designed as a full-fledged classification. Its declared objective was to serve as an 'umbrella' system for correlating national soil classifications. However, in some developing countries, for example, in Mexico, Tanzania and Vietnam, it has been successfully applied for medium-scale mapping. The European Union uses the WRB as an umbrella system for soil databases throughout Europe and assists in harmonization of new national soil classifications of the EU member states.

The geographical scope of most national soil classifications is limited to their national borders. As we have seen, traditional soil classifications were designed mainly for agronomical purposes; however, currently society demands more environmental information. Thus, it is interesting to review what soils and soil-like superficial bodies are included in national classifications (Table 33.1). Only the WRB includes a full set of superficial objects, including technogenic substrates, bare rock and underwater sediments. Three soil classifications have almost complete coverage: the French, Austrian and Australian systems. Several countries do not recognize 'exotic' surface sediments as soils, and do not classify soils deeply disturbed by agriculture as a separate class: for example, Canada, Switzerland, Belorussia, Estonia, Israel and Brazil. The majority of classifications separate agricultural 'man-made soils' as a special group; some classifications, like the new Russian system, have a complex hierarchy for agricultural disturbance of soils. Technogenic substrates are recognized as soils in almost half of the national classifications and some countries (e.g. Germany and Russia) have separate classifications for technogenic substrates. In general, most recent classifications have a tendency to include substrates of urban and industrial areas in the scope of their systems. Shallow underwater sediments are regarded as soils only in Germany, Austria, Romania, Australia and Ghana. Even fewer countries, namely France, Austria and Ghana, classify consolidated rocks as soils, at least partly.

As can be expected most countries are testing and evaluating various proposals to incorporate taxa strongly influenced by human activities, and consequently changes are likely to be forthcoming in many national classifications. Currently new research initiatives are underway to broaden the environmental scope of Earth sciences, thus pedology joins others to study the regolith and the vadose zone. Recently the European Union launched an initiative to study 'whole soil-regolith pedology' (Buol, 1994), now termed 'Earth Critical Zone' (ECZ), on a worldwide scale. The proponents desire a taxonomy that embraces soils, regolith and groundwater as an integrated natural body that supports life on Earth. Many parts of the ECZ are rapidly deteriorating due to increasing human impacts on nature, and for these reasons proponents of the ECZ want also to develop a new taxonomy of these combined natural resources.

The structures of soil classifications

The number of taxonomic levels in soil classifications, or the depth of classification (Holman, 1992) varies from two to eight (Table 33.2). The least number of taxonomic levels is in classifications called reference bases, namely the WRB and the new French classification. Usually there are several unstated levels in reference bases. The French classification has an optional collective level, and in the WRB, prefix and suffix qualifiers form two different hierarchical levels. The highest number of taxonomic levels is found in the classifications of Russia, Belorussia and the Czech Republic. The numbers of taxonomic levels and lower taxonomic units (individual soils) do not depend directly on the size of the

Table 33.1 *The representation of soils and soil-like superficial bodies in world classifications*

Superficial bodies	Urban soils	Man-transported materials	Bare rock	Subaquatic soils	Soils deeply transformed by agricultural activities
Classifications					
WRB[1]	X	X	X	X	X
USA[2]	-	-	-	-	Partly
Canada[3]	-	-	-	-	-
France[4]	X	X	X	-	X
United Kingdom[5]	Partly	Partly	-	-	Partly
Germany[6]	-	-	-	X	X
Austria[7]	X	X	Partly	X	X
Switzerland[8]	-	-	-	-	-
The Netherlands[9]	-	-	-	-	X
Poland[10]	X	X	-	-	X
Czech Republic[11]	Partly	Partly	-	-	X
Slovakia[12]	X	X	-	-	X
Hungary[13]	-	-	-	-	-
Romania[14]	X	X	-	X	X
Bulgaria[15]	X	Partly	-	-	X
Former Soviet Union[16]	-	-	-	-	Partly
Russia[17]	-	-	-	-	X
Azerbaijan[18]	X	X	-	-	X
Belorussia[19]	-	-	-	-	-
Estonia[20]	-	-	-	-	-
Latvia[21]	X	X	-	-	X
Lithuania[22]	-	-	-	-	X
Ukraine[23]	-	-	-	-	X
Israel[24]	-	-	-	-	-
China[25]	Partly	Partly	-	-	X
Japan[26]	Partly	Partly	-	-	X
Brazil[27]	-	-	-	-	-
Cuba[28]	-	Partly	-	-	X
Australia[29]	X	X	-	X	X
New Zealand[30]	-	-	-	-	X
Ghana[31]	-	-	Partly	X	-
South Africa[32]	-	Partly	-	-	-

Sources: 1 IUSS Working Group WRB (2006); 2 Soil Survey Staff (1999); 3 Soil Classification Working Group (1998); 4 AFES (1998); 5 Avery (1980); 6 Ad-hoc-AG Boden (2005); 7 Nestroy et al (2000); 8 Arbeitsgruppe 'Bodenklassifikation und Nomenklatur', BGS (2002); 9 De Bakker and Schelling (1966); 10 Polish Society of Soil Science (1989); 11 Němeček et al (2001); 12 Sobocká (2000); 13 Szabolcs (1966); 14 Florea and Munteanu (2000); 15 Ninov (2005); 16 Egorov et al (1977); 17 Shishov et al (2004); 18 Babaev et al (2006); 19 Romanova (2004); 20 Reintam and Köster (2006); 21 Kārkliņš et al (2009); 22 Buivydaite (2002); 23 Polupan et al (2005); 24 Dan and Koyumdjisky (1979); 25 CRG-CST (2001); 26 Fourth Committee for Soil Classification and Nomenclature (2002); 27 EMBRAPA (1999); 28 Instituto de Suelos (1999); 29 Isbell (2002); 30 Hewitt (1998); 31 Adjey-Gyapong and Asiamah (2002); 32 Soil Classification Working Group (1991).

country. The US Soil Taxonomy, for example, has the broadest geographical scope, not only because the US has an extensive and diverse territory, but also because it is a de facto international system that was influenced by many foreign specialists

The ideal structure is not known as all classifications are human constructs, and thus are subject to the limitations of the mind. In theory a simple classification has only three levels, generic, specific and varietal. The generic level corresponds to central concepts (archetypes) the members of which are separated into smaller groups with rather specific sets of characteristics which in turn can be separated into variations of the specific taxa. When soil classifications support soil surveys it is common to have additional levels, first to give an overview of the range of variations using a collective level, and then additional levels to provide sufficient specificity of the variability of soils to produce meaningful field scale maps of important soil features for practical uses.

As Ibáñez et al (2006) stated, the human mind cannot manage a large number of entities at a given moment and has to segregate them into smaller groups to make consistent decisions about them. The number of taxa at a generic level generally corresponds to the size of the country and the state of the knowledge about its soils. The highest number is found in the US, a country with extensive and diverse territory and a high level of soil exploration. Russia, though bigger in territory, has fewer generic taxa because extensive northern territories of the country are still poorly studied. The low number of generic taxa in the WRB may seem strange because this system should cover the entire world. However, the paradox may be understood if one takes into account the historical reasons. The system was made initially as a map legend, and had to be simple. Thus it was artificially reduced, and many groups, in fact, represent collective units rather than different soils. The problem is that the generic level is difficult to find in a reference base system.

All too often overlooked is the protocol to design a soil classification system. Of course it is assumed that the purpose of the scheme is known, as is the domain or population of interest, and the individuals that make up the diversity of the population. An assumed domain is all soils (however defined), so that is not numbered as a level in a hierarchy. Assuming further that three or four levels (categories) may be appropriate, the highest level is defined by an abstract statement about the population that will enable it to be divided into clusters whose properties are thought to be associated with the abstract definition. It has been accepted in modern soil science that environmental factor interactions influence processes in soils, resulting in rather specific morphological properties, thus sets of these features observed in soil profiles are evidence of major soil-forming processes, or of current dynamic processes influencing the morphology and/or behaviour of soil bodies. Each lower level then divides the taxa of the level above it into smaller more homogeneous groups, the definitions of which are less abstract (more specific) than the level above. This process continues to a level where the designers are satisfied that the soil information is sufficient to satisfy the purpose of the classification.

Table 33.2 *Some structural characteristics of the world classifications*

	Number of taxonomic levels	The place of the generic level in the hierarchy	The number of units at the generic level
WRB	2(3)	1	32
USA	6	3	317
Canada	5	2	31
France	2(3)	1(2)	102
United Kingdom	4	2	43
Germany	6	3	55
Austria	5	3	46
Switzerland	7	4	22
The Netherlands	4	4	60
Poland	6	3	35
Czech Republic	8	2	26
Slovakia	7	2	21
Hungary	4	2	37
Romania	7	2	32
Bulgaria	3	1	22
Former Soviet Union	6	1	73
Russia	8	3	227
Azerbaijan	7	3	40
Belorussia	8	2	28
Latvia	3	2	12
Lithuania	4	1	12
Ukraine	6	1	54
Israel	4	1	26
China	4	2(3)	39
Japan	5	2	32
Brazil	6	2(3)	44
Cuba	4	2	36
Australia	6	2	84
New Zealand	6	1	15
Ghana	6	4	35
South Africa	3	1	74

For example, level 1 might be defined as soils with properties reflecting major soil-forming processes, and each class or taxa at that level would have soil properties thought to be associated with that definition. Level 2 might be defined as level 1 taxa soils with properties reflecting secondary processes of soil formation, and the selected properties for each taxa at this level would be associated with secondary features. Level 3 might be defined as level 2 taxa soils, the properties of which are thought to reveal overlapping or intergrading processes, and the properties for each taxa would be selected accordingly. Unfortunately the above rationale is seldom provided in soil classifications so we had to make assumptions about the definitions of the levels, then visualize the kinds of soil profiles that were being represented by each taxa, and on that basis try to make reasonable correlations with the taxa of the WRB as we understand them.

World Reference Base and soil correlation

The WRB was proven to be an effective means of international communication and correlation. In addition the WRB during the last decade has become a melting pot of soil classification ideas. A lot of interesting findings and concepts of national classification systems were absorbed by the WRB, and later, as feedback, were accepted by various national classification systems.

Our experience in correlation was somewhat disappointing. The definitional, structural and diagnostic property differences for taxa in different classifications do not allow correlation of one taxon in a national classification with one taxon of the WRB. Many soil features considered to be important in national classifications are absent in the WRB. This can be easily improved by adding new modifiers to the WRB, but that does not solve the dilemma of correlating conceptually different classifications. For example, some classifications use landscape or moisture regime criteria which are not recognized by the WRB. We found this to be a deadlock as partial membership protocols and acceptance are not yet part of standard operating procedures. This attempt to correlate conceptual taxa from different systems may be considered an exercise in visualization that hopefully will open our eyes to the need for a rigorous universal basic system of soil classification.

References

Ad-hoc-AG Boden (2005) *Bodenkundliche Kartieranleitung*, herausgegeben von der Bundesanstalt für Geowissenschaften und Rohstoffe und den Geologischen Landesämtern in der Bundesrepublik Deutschland, Hannover, 5 Auflage, 438pp

Adjey-Gyapong, T. and Asiamah, R. D. (2002) 'The interim Ghana soil classification system and its relation with the World Reference Base for Soil Resources', Quatorzième reunion du Sous-Comité ouest et centre Africain de corrélation des sols pour la mise en valeur des terres, Abomey, Bénin, 9–13 October 2000, World Soil Resources Reports no 98, UN Food and Agriculture Organization, Rome, pp51–76

AFES (1998) *A Sound Reference Base for Soils* (The 'Referentiel pedologique': text in English), INRA, Paris, 322pp

Arbeitsgruppe 'Bodenklassifikation und Nomenklatur', BGS (2002) *Klassifikation der Böden de Schweiz, Version 30*, Eidgenössische Forschungsanstalt für Agrarökologie und Landbau, Zürich-Reckenholz, 96pp

Avery, B. W. (1980) *Soil Classification for England and Wales (Higher Categories)*, Soil Survey Technical Monograph no 14, Harpenden, 67pp

Babaev, M. P., Dzhafarova, Ch. M. and Gasanov, V. G. (2006) 'Modern Azerbaijani soil classification system', *Eurasian Soil Science*, vol 39, no 11, pp1176–1182

Buivydaite, V. V. (2002) 'Classification of soils of Lithuania based on FAO-UNESCO soil classification system and WRB', *Transactions of the 17th World Congress of Soil Science*, Bangkok, Thailand, 14–21 August 2002, CD, pp2189-1–2189-13

Buol, S. (1994) 'Saprolite–regolith taxonomy', in D. L. Cremeens, R. B. Brown and J. H. Huddleston (eds) *Whole Regolith Pedology*, SSSA Special Publication no 34, Madison, WI, pp119–132

CPCS (1967) *Classification des Sols*, Ecole nationale supérieure agronomique, Grignon, France, 87pp

CRG-CST (2001) *Chinese Soil Taxonomy*, Li Feng (ed), Science Press, Beijing and New York, 203pp

Dan, J. and Koyumdjisky, H. (eds) (1979) *The Classification of Israel Soils*, Committee on Soil Classification in Israel, Special publication no 137, Institute of Soils and Water ARO, Bet Dagan, Israel, 95pp (in Hebrew with English abstract)

De Bakker, H. and Schelling, J. (1966) *Systeem voor bodemklassifikatie voor Nederland*, De hogere niveaus, STIBOKA, Pudoc, Wageningen, 217pp

Egorov, V. V., Fridland, V. M., Ivanova, E. N., Rozov, N. N., Nosin, V. A. and Fraev, T. A. (1977) *Classification and Diagnostics of Soils of USSR*, Kolos Press, Moscow, 221pp (in Russian)

EMBRAPA (1999) *Sistema Brasileiro de Clasificação de Solos*, Embrapa Produção de Informação, Brasília – Embrapa Solos, Rio de Janeiro, 412pp

Florea, N. and Munteanu, I. (2000) *Sistemul Roman de Taxonomie a Solurilor (Romanian System of Soil Taxonomy)*, University 'Al. I. Cuza', Iasi, 107pp

Fourth Committee for Soil Classification and Nomenclature (2002) 'Unified soil classification system of Japan (2nd Approximation)', *Pedologist*, vol 46, no 1, pp36–45 (in Japanese)

Hewitt, A. E. (1998) *New Zealand Soil Classification*, 2nd edition, Maanaki Whenua –Landcare New Zealand Ltd, Dunedin, Landcare Research Science Series no 1, Lincoln, Canterbury, New Zealand, 122p

Holman, E. W. (1992) 'Statistical properties of large published classifications', *Journal of Classification*, vol 9, no 3, pp187–210

Ibáñez, J. J., Arnold, R. and Sánchez Díaz, J. (2006) 'The magical numbers of the USDA Soil Taxonomy: Towards an outline of a theory of natural resource taxonomies', in *Abstracts of the 18th World Congress of Soil Science*, 9–15 July 2006, Philadelphia, PA, p424

Instituto de Suelos (1999) *Nueva Versión de Clasificación Genética de los Suelos de Cuba*, AGRINFOR, Ministerio de la Agricultura, Ciudad de La Habana, Cuba, 64pp

Isbell, R. F. (2002) *Australian Soil Classification*, revised edition, CSIRO Land & Water, Canberra, 144pp

IUSS Working Group WRB (2006) *World Reference Base for Soil Resources*, 2nd edition, World Soil Resources Reports no 103, UN Food and Agriculture Organization, Rome, 128pp

Kārkliņš, A., Gemste, I., Mežals, H., Nikodemus, O. and Skujāns, R. (2009) *Latvijas augšņu noteicējs (Taxonomy of Latvian soils)*, Jelgava, LLU, 240pp (in Latvian, with English summary)

Němeček, J., Macků, J., Vokoun, J., Vavříč, D. and Novák, P. (2001) *Taxonomický klasifikační system půd České Republiky*, ČZU, Prague, 180pp

Nestroy, O., Dannenberg, O. H., English, M., Gessl, A., Herzenberger, E., Kilian, W., Nelhiebel, P., Pecina, E., Pehjamberger, A., Schneider, W. and Wagner, J. (2000) 'Systematische Gliederung der Boden Osterreichs (Osterreichische Bodensystematik 2000)', *Mitteilungen der Osterreichischen Bodenkundlichen Gesellschaft*, vol 60, pp1–104

Ninov, N. (2005) 'Taxonomic list of Bulgarian soils according to the FAO world soil system', *Geography 21*, no 5, pp4–20 (in Bulgarian, English summary)

Polish Society of Soil Science (1989) 'Systematics of Polish Soils', *Roczniki Gleboznani*, vol 40, nos 3–4, 112pp

Polupan, M. I., Solovey, V. B. and Velichko, V. A. (2005) *Classification of Ukrainian Soils*, Agrarna nauka, Kiev, 300pp (in Ukrainian)

Reintam, E. and Köster, T. (2006) 'The role of chemical indicators to correlate some Estonian soils with WRB and Soil Taxonomy criteria', *Geoderma*, vol 136, no 2, pp199–209

Romanova, T. A. (2004) *Diagnostics of Soils in Belarus and their Classification in FAO–WRB System*, Institute for Soil Science and Agrochemistry, National Academy of Sciences of Belarus, Minsk, 428pp (in Russian)

Shishov, L. L., Tonkonogov, V. D., Lebedeva, I. I. and Gerasimova, M. I. (2004) *Classification and Diagnostics of Soils of Russia*, Oykumena, Smolensk, 342pp (in Russian)

Sobocká, J. (ed) (2000) *Morfogenetický klasifikačný systém pôd Slovenska. Bazálna referenčná taxonómia*, Výskumný ústav pôdoznalectva a ochrany pôdy, Bratislava, 74pp

Soil Classification Working Group (1991) *Soil Classification: A Taxonomic System for South Africa*, 2nd (revised) edition, Memoirs on the Agricultural Natural Resources of South Africa no 15, Department of Agricultural Development, Pretoria, 257pp

Soil Classification Working Group (1998) *The Canadian System of Soil Classification*, 3rd edition, Agriculture and Agri-Food Canada Publication 1646, Supply and Services Canada, Ottawa, Ont., 187pp

Soil Survey Staff (1999) *Soil Taxonomy: A Basic System of Soil Classification for Making and Interpreting Soil Surveys*, USDA, Handbook 436, 2nd edition, United States Government Printing Office, Washington DC, 696pp

Szabolcs, I. (ed) (1966) 'Methodology for genetic soil mapping on a farm scale', *Genetikus Talajtérképek*, ser 1, no 9, 112pp (in Hungarian)

Part 3
Folk Soil Classifications

P. Krasilnikov, J. Tabor and R. Arnold

34
Ethnopedology and
Folk Soil Classifications

Ethnopedology

Ethnopedology is a scientific discipline at the border of cultural anthropology and soil science which aims to study folk soil knowledge. On the one hand, it deals with the product of collective creative effort of an ethnic group; on the other hand, it studies the correspondence of indigenous soil knowledge with scientific concepts, and incorporates this knowledge in the practice of soil science. The term was proposed by Williams and Ortiz-Solorio (1981, p336), who compared the application of folk and scientific classification systems for soil mapping in Central Mexico.

Actual ethnopedology includes the study of local myths and rituals related to soil, local soil classifications, local perceptions of soil, its spatial distribution and interactions with other components of a landscape, local land use, management and conservation practices, and integration of local soil knowledge into soil surveys and natural resource conservation practices (Barrera-Bassols and Zinck, 2003). However, the scope of ethnopedology is still under discussion. Some researchers understand ethnopedology in a broad sense, including the study of any vernacular agrarian knowledge, and others argue that ethnopedology should be limited only to folk soil classifications studies (since the term *pedology* is used in soil science to denominate soil genesis, classification and geography research). The majority of scientists consider that the nucleus of ethnopedology is the study of folk soil and land classification systems, but soil mythology, land management and conservation practices should be also included (Tabor, 1990, 1992; Tabor and Hutchinson, 1994).

Vernacular systems are developed by land users and are based on characteristics important to the user. Soils are distinguished by obvious characteristics, such as physical appearance (e.g. colour, texture, landscape position), performance (e.g. production capability, flooding), and accompanying vegetation (Stranski, 1954, 1956, 1957). These distinctions are often based on characteristics important to land management; however, the scientific community has largely ignored them until recently, with anthropologists and geographers being the first to document them (Niemeijer, 1995).

Soil classifications in ancient history

Any human society throughout history classified the objects important for their existence (Holman, 2005). Soil, as an important basis for the development of civilization, also attracted the attention of people. For six millennia humans have tilled, drained and irrigated soils for agriculture (Heiser, 1990). For an even longer period of time soils have been used as a construction material. Four thousand years ago the Chinese were classifying soils according to their productivity and using it as a basis for tax assessment (Ping-Hua Lee, 1921; Finkl, 1982, p1). One of the earliest known soil classification systems in the world can be found in an ancient Chinese book *Yugong* (2500 YBP), where soils of China were classified into three categories and nine classes based on soil colour, texture and hydrologic features (Gong Zitong, 1994). The ancient name for Egypt – *Kemet* – means fertile black alluvial soils; *Deshret* means red desert land. About 3000 YBP different arable soils had different costs in Egypt: '*nemhuna*' soils cost three times more than '*sheta-teni*' soils (Krupenikov, 1981). Theophrastus, an ancient Greek botanist, described clay, sand, stony, salty, swamp, soft and hard soils and their relation to plant cover. In Rome, Marcus Porcius Cato (234–149 BC), in his fundamental book *De Agri Cultura*, described a number of soil types: white clay, red clay, mottled earth (*terra cario sam*) and friable dark earth (*terra pulla*) (Krupenikov, 1981). Mid-American civilizations are also known to have developed soil classifications. At least 45 terms for various soils are documented for pre-Hispanic Aztec culture (Williams, 1975). Thus, all over the world for millennia people were classifying soils according to their appearance and properties.

Not only in ancient times, but even now indigenous soil classifications do exist. In most developed countries they were partially or completely replaced by scientific soil taxonomies. But in developing countries it is still possible to find vernacular systems of soil classification untouched. Most studies on pre-scientific soil classifications have been made in African, Latin American and Asian countries (Barrera-Bassols and Zinck, 2003).

Vernacular vs 'scientific' origin of soil classifications

We mention above a number of ancient soil classifications, although they may not belong to ethnopedological studies. We are not sure of the border between a folk soil taxonomy and an artificial one – that is, whether they were intentional attempts to arrange the knowledge into a rational system. It seems that ancient classifications were all based on indigenous soil classifications, which used common folk words for soil classes. One of the main differences between folk and artificial classification is that the former is not documented; it is not fixed by the rules invented by a closed group. Folk classifications were the living knowledge of the people, while artificial classifications are made by the priests, officials, philosophers or, later, scientists. We prefer describing documented historical classifications as 'pre-scientific' ones belonging in the scope of ethnopedology.

Soil knowledge and soil management practices are tightly linked in agricultural societies, thus the knowledge about soils is widely shared among the people. They developed management and conservation practices together and also worked out a common classification of soils. Classification is a basic human mental activity and language itself is rooted in the classification of the world where each object is given a name, and is grouped with similar objects. The value of soil is recognized in most agricultural cultures of the world from spiritual and mythological levels down to practical knowledge. These names often include soil in the overall picture of the world, as well as providing a necessary communication tool needed for practical purposes. Significant difference in soil knowledge commonly exists among members of a community depending on their age, experience, gender and social status; thus, this knowledge may be regarded as the collective wisdom of a community. The most extensive soil knowledge is found in agricultural societies, whereas knowledge among nomads and hunters is much more limited. Development of indigenous soil knowledge is highly dependent on landscape, land use and cultural history. For example, in Russia more than 150 soil names have been identified for the farming-based European part of the country, while only a few soil names have been identified in northern Siberian nations that depended on hunting and reindeer grazing (Shoba, 2002). The soil names used by Yakutians, Evenks and other native Siberians denote fens and natural sources of salts which are relevant for animal management and hunting.

Early studies of folk soil terminology

Apart from ancient soil classifications, modern studies of soils are also rooted mainly in ethnopedological surveys. In Russia, a systematic survey of folk soil knowledge started in the 16th century when special statistical books were created to evaluate soil resources of the state. These books were prepared by interviewing the peasants about the quality and productivity of their lands and included mainly brief descriptions of soils, like *poor sandy soil, clayey stony soils, fat loams* and so forth. Later, in the 19th century, the survey became more regular, and perennial data were published in a series of books, *Materials on Statistics of Russia*, where a number of local folk soil names for soils were listed. Such material was also used for preparing the first soil maps of Russia (Krasilnikov, 1999). Dokuchaev used the books, *Materials on Statistics of Russia*, for extracting soil names for scientific classification in addition to collecting numerous folk soil names such as *chernozem, solonchak, podzol, solonetz* and *gley* which were incorporated into scientific pedologic literature (Dokuchaev, 1953). In 1915, Lamansky collected more than 200 Russian folk names for soils, and Zaharov published a list of soil names for Georgia and Armenia (Lamansky, 1915; Zaharov, 1915). In the US, Hilgard discussed the need to compare farmers' land classifications with scientific agronomic and soil classifications (Hilgard, 1930). In the 1920s and 1930s, Kubiëna (1953) intensively studied the folk soil terminology of Western Europe and introduced the terms *gyttja*, *dy*, *tangel* and *terra rossa* into scientific literature. In 1925, E. Best compiled an extensive list of soil names used by the Maori in New Zealand (Hewitt, 1992). Bennett and Allison (1928) cited a number of vernacular soil terms in their description of the soils of Cuba. In the 1940s and 1950s, several remarkable ethnopedology studies were made including the comprehensive studies of Bulgarian soil terminology made by Stranski (1954, 1956, 1957), who cited more than 1000 folk soil names. Aubert (1949) made a survey on vernacular soil terms used in Sudan and Senegal, and Calton (1949) described folk soil terminology for Zanzibar Island. Raychaudhri (1958) included a number of folk terms in his paper on the soils of India. Ethnopedology has now become a developed branch of scientific research with hundreds of studies all over the world (Barrera-Bassols and Zinck, 2003).

The use of folk soil terminology in scientific classifications

Indigenous classifications definitely differ from scientific ones. Their structure is either nominal, giving unique names to soils or landscapes, or consists of descriptive names of soils based on their characteristics such as colour (e.g. 'red'), drought tolerance (e.g. 'hot'), fertility (e.g. 'fat') or texture (e.g. 'sandy'). Sandor and Furbee (1996) noted hierarchical structures in some vernacular systems; however, the interpretation that a rule-based hierarchical system in fact exists may have been influenced by the way questions were phrased. Any hierarchical structure may be incidental to a predominately descriptive way of classifying soils. A number of folk soil names deal with 'negative' characteristics that distinguish a usually less-productive soil (excess salinity, stoniness, hardpans, etc.). For example, the name *podzol* originates not from northern Russia, where soils with a bleached upper horizon are widespread, but from middle Russia where the term relates to the poor productivity of these soils (Dokuchaev, 1950). Other Russian folk soil names are also related to some extreme character of the soil: *solonchak*, *usol* and *solonetz* mean saline and/or alkaline soils and *gley* and *zablest* mean soils with excessive moisture. People of different localities sometimes described rather different soils under the same name and the meanings do not correspond with scientific terms. Wilde (1953) noted that folk soil terms 'borrowed' in one region should seldom be used for other regions with different environments. He proposed using local vernacular names for naming soils in newly studied regions. Dokuchaev (1967) noted 'in various areas of Russia, identical kinds of soil are often denoted by totally different names... It is, however, much more frequently the case that totally different soils are called by the same name.' Although farmers generally know their soils well, in many areas the soils are not well correlated between farmers or regions. Sometimes soil characteristics used for vernacular soil classifications are rather surprising. In the case of Haiti, farmers give soils names like 'red' and 'sandy' to denote colour and texture but they also use descriptions like 'fat' or 'hot' to describe soils of high fertility or those that are droughty. Baruya people in New Guinea use a rather detailed soil colour classification because they use soils as pigments for decorating their bodies (Ollier et al, 1971). In fact, we are just beginning to understand the diversity of indigenous soil knowledge. Quite a lot of soil names are perceptive, that is, based not on soil properties, but on vegetation, fertility, relief and such. It is not obvious whether we should consider them as real soil names or not.

The use of folk soil classifications for soil survey and land management

Vernacular soil classifications are not strict like scientific ones; they are mostly descriptive or nominal and have only local importance. Then of what use is it to study them?

Vernacular systems can provide outsiders with a language to communicate with local land users, especially regarding agricultural management and resource tenure. Vernacular systems can also provide technicians and scientists with insight into natural resource management systems that may prove valuable in the inventory and developing of local resources (Tabor, 1992). The soils that are identified by farmers often closely resemble those identified by scientific systems and in some cases the vernacular systems make finer distinctions than would normally be made by soil scientists. For example, the Soninke farmers in drought-prone Senegal, Mauritania and Mali make very fine distinctions between soils with respect to their period and frequency of flooding which would be difficult for a soil surveyor to determine from one site visit (Tabor, 1993). Soil scientists can easily determine what soils are typical of an area by having farmers show them what is typical and even help them map soils. Social and economic aspects of soils can be obtained from interviews of local land users. Learning about vernacular systems of soils can be a non-threatening way of determining social and economic status between families and individuals of a community. Vernacular classification systems also offer a useful vehicle for talking with villagers about agricultural and land tenure issues.

Soil scientists tend to be biased towards the classification systems that they know and commonly separate soils based on the division breaks of their system. This can overly complicate a soil survey, or worse, disregard separations that are important to the farmer. Vernacular systems can provide clues for identifying those soil characteristics that are most limiting to land management and can help a soil scientist identify agricultural interventions that will most economically improve the soil's productivity. This approach to soil surveys can provide better insights into the farming system and in turn can better guide agricultural research (Krasilnikov and Tabor, 2003).

Knowing the local systems of land classification is extremely important in understanding land tenure relationships. Agricultural development projects often disrupt established social and tenure relationships through real or perceived changes in the soil's productivity and land value. Needless disruption in tenure relationships can be avoided if local land classification systems are integrated into the development project's soil and cadastral surveys. These indigenous soil classification systems should be viewed as guiding and complementary to scientifically based systems; however, many soil scientists have ignored this indigenous knowledge, while others consider it as an inadequate substitute for the system that they know. Vernacular systems have the added advantage in that they are widely known by the people of the region. Only the relatively few scientists, technicians and extension agents need to learn the farmer's classification system. They simplify the complexity and continuum of the real world into more easily understandable discrete classes. This simplification is based on criteria that are biased towards its intended use, such as agriculture or engineering. Indigenous knowledge and vernacular systems can be used to help develop scientific classification systems, especially in developing countries with limited resources for agricultural research but with a great need to increase agricultural production and reduce soil degradation. Existing soil surveys of much of the world do not provide sufficient detail to base field-specific recommendations for farmers. Even 1:20,000 surveys can be too small a scale where farmers' fields are small and soil variability is extremely high. Integrating local land classification and its associated information with existing soil surveys will be extremely useful to agricultural extension and research until more detailed soil surveys are available. Soil information collected in this manner will allow soil scientists to better define their scientific classification criteria and use vernacular names that have some meaning to farmers and herders. The location of described soils and the wealth of information collected from the farmers can be recorded in a geographical information system and later be used to help produce accurate maps based on remote-sensing and other data.

Vernacular and scientific systems can be used to identify opportunities for land development by looking for soils with contrasting values as identified by each classification system. For example, vernacular systems along the Senegal River in Mauritania, Senegal and Mali identify an economically important river bank soil, but this soil has been ignored by nearly all soil surveys of the area – in part because it is too small and narrow to draw clearly on most mapping scales (Tabor, 1993). In another example the Zuni Indians of New Mexico, USA have a long history of rainfed agriculture using water conservation techniques and have a vernacular system to differentiate the important characteristics of these soils. Soil scientists described the soils as non-arable, ignoring the fact that the Zuni have been farming the land for hundreds of years. Economic, infrastructural, cultural and social assumptions are part of each folk classification system. Soil scientists can reduce time and increase accuracy in classifying and mapping soils, especially for outsiders, if they use vernacular systems. Local populations have long-term perspectives of soil characteristics as they change with the seasons and as they may be expressed during extremes of climate.

A difficult task in conducting a soil survey is creating a good soil legend that effectively separates soils on their productive capacity and that allows easy identification in the field. Village and farmer interviews allow the scientists to rapidly identify all the soils that are of importance to the farmer, determine each soil's relative productivity and their value for agriculture, forestry and range, and locate typical soils of each type and correlate them to other systems, both scientific and indigenous.

It is reasonable to study indigenous soil classification in order to verify the validity of scientific taxonomies (Militarev, 1993). Scientific classifications based only on statements of science cannot readily be verified without correlation to a real world.

References

Aubert, G. (1949) 'Note on the vernacular names of the soils of the Sudan and Senegal', *Proceedings of the First Commonwealth Conference on Tropical and Sub-Tropical Soils*, Harpenter, 1948, Commonwealth Bureau of Soil Science, Technical Committee, vol 46, pp107–109

Barrera-Bassols, N. and Zinck, J. A. (2003) 'Ethnopedology: A worldwide view on the soil knowledge of local people', *Geoderma*, vol 111, nos 3–4, pp171–195

Bennett, H. H. and Allison, R. V. (1928) *The Soils of Cuba*, Tropical Plant Research Foundation, Washington DC, 410pp

Calton, W. E. (1949) 'A reconnaissance of the soils of Zanzibar protectorate', *Proceedings of the First Commonwealth Conference on Tropical and Sub-Tropical Soils*, Harpenter, 1948, Commonwealth Bureau of Soil Science, Technical Committee, vol 46, pp49–53

Dokuchaev, V. V. (1950) 'On podzol', in V. V. Dokuchaev, *Complete Set of Works*, USSR Academy of Sciences Publishes, Moscow, vol 2, pp248–256 (in Russian)

Dokuchaev, V. V. (1953) 'On the use of the study of local names of Russian soils', in V. V. Dokuchaev, *Complete Set of Works*, USSR Academy of Sciences Publishers, Moscow, vol 7, pp332–340 (in Russian)

Dokuchaev, V. V. (1967) *Selected works of V. V. Dokuchaev*, vol I – *Russian Chernozem*, translated from Russian by N. Kaner, Israel Program of Scientific Translations, Jerusalem, 419pp

Finkl, C. W. Jr (ed) (1982) *Soil Classification*, Hutchinson Ross Publishing Company, Stroudsburg, PV, Benchmark Papers in Soil Science, vol 1, 391pp

Gong Zitong (ed) (1994) *Chinese Soil Taxonomic Classification (First proposal)*, Institute of Soil Science, Academia Sinica, Nanjing, 93pp

Heiser, C. B. Jr, (1990) *Seed to Civilization: The Story of Food*, Harvard University Press, Cambridge, MA, 228pp

Hewitt, A. E. (1992) 'Soil classification in New Zealand: Legacy and lessons', *Australian Journal of Soil Research*, vol 30, no 4, pp843–854

Hilgard, E. W. (1930) *Soils, their Formation, Properties, Composition, and Relation to Climate and Plant Growth in the Humid and Arid Regions*, Macmillan, New York, 593pp

Holman, E. W. (2005) 'Domain-specific and general properties of folk classifications', *Journal of Ethnobotany*, vol 25, no 1, pp71–91

Krasilnikov, P. V. (1999) 'Early studies on folk soil terminology', *Eurasian Soil Science*, vol 32, no 10, pp1147–1150

Krasilnikov, P. V. and Tabor, J. A. (2003) 'Perspectives on utilitarian ethnopedology', *Geoderma*, vol 111, nos 3–4, pp197–215

Krupenikov, I. A. (1981) *History of Soil Science*, Nauka Press, Moscow, 327pp (in Russian)

Kubiěna, W. L. (1953) *Bestimmungsbuch und Systematik der Böden Europas*, Verlag Enke, Stuttgart, 392pp

Lamansky, V. V. (1915) 'An experience of folk soil dictionary', *Pochvovedenie*, no 2, pp61–72 (in Russian)

Militarev, V. Yu. (1993) 'Principles of the theory of classifications in natural sciences', in *The Theory and Methods of Biological Classifications*, Nauka Press, Moscow, pp101–115 (in Russian)

Niemeijer, D. (1995) 'Indigenous soil classifications: Complications and considerations', *Indigenous Knowledge and Development Monitor*, vol 3, no 1, pp1–5

Ollier, C. D., Drover, D. P. and Godelier, M. (1971) 'Soil knowledge amongst the Baruya of Wonenara, New Guinea', *Oceania*, vol 42, no 1, pp33–41

Ping-Hua Lee, M. (1921) 'The economic history of China with special reference to agriculture', *Columbia University Studies in History, Economics, and Public Law*, vol 99, pp1–461

Raychaudhri, S. P. (1958) *Soils of India*, The Indian Council of Agricultural Research, Delhi, revised series, vol 25, 28pp

Sandor, J. A. and Furbee, L. (1996) 'Indigenous knowledge and classification of soils in the Andes of Southern Peru', *Soil Science Society of America Journal*, vol 60, no 5, pp1502–1512

Shoba, S. A. (ed) (2002) *Soil Terminology and Correlation*, 2nd edition, Centre of the Russian Academy of Sciences, Petrozavodsk, 320pp

Stranski, I. (1954) 'Bulgarian folk soil names based on their colour and humidity', *Transactions of Soil Science Institute*, Bulgarian Academy of Sciences Press, Sofia, vol 2, pp281–360 (in Bulgarian)

Stranski, I. (1956) 'Bulgarian folk soil names based on their physical properties', *Transactions of Soil Science Institute*, Bulgarian Academy of Sciences Press, Sofia, vol 3, pp329–410 (in Bulgarian)

Stranski, I. (1957) 'Bulgarian folk soil names based on their miscellaneous features', *Transactions of Soil Science Institute*, Bulgarian Academy of Sciences Press, Sofia, vol 4, pp307–409 (in Bulgarian)

Tabor, J. A. (1990) 'Ethnopedology: Using indigenous knowledge to classify soils', *Arid Lands Newsletter*, vol 30, pp28–29

Tabor, J. A. (1992) 'Ethnopedological surveys: Soil surveys that incorporate local systems of land classification', *Soil Survey Horizons*, vol 33, no 1, pp1–5

Tabor, J. A. (1993) 'Soils of the lower, middle and upper Senegal river valley', in T. K. Park (ed) *Risk and Tenure in Arid Lands. The Political Ecology of Development in the Senegal River Basin*, Arid Lands Development Series, The University of Arizona Press, Tucson and London, pp31–50

Tabor, J. A. and Hutchinson, C. F. (1994) 'Using indigenous knowledge, remote sensing and GIS for sustainable development', *Indigenous Knowledge and Development Monitor*, vol 2, no 1, pp2–6

Wilde, S. A. (1953) 'Soil science and semantics', *Journal of Soil Science*, vol 4, no 4, pp1–4

Williams, B. J. (1975) 'Aztec soil science', *Boletin del Instituto de Geografia*, vol 7, no 2, pp115–120

Williams, B. J. and Ortiz-Solorio, C. A. (1981) 'Middle American folk soil taxonomy', *Annals of Association of American Geographers*, vol 71, no 3, pp335–358

Zaharov, S. A. (1915) 'On soil terminology of native population of Trans-Caucasian region', *Russkiy Pochvoved*, nos 13–14, pp367–373 (in Russian)

35
Folk Soil Terminology, Listed by Regions

Introductory notes

In a previous chapter we discussed the scope and the use of ethnopedology. This chapter presents a number of folk soil names, documented by anthropologists and soil scientists. Unfortunately, the material is not equally complete for all ethnic groups and regions. To some extent it reflects an uneven distribution of soil knowledge among nations. Some ethnic groups developed soil classification systems, and others had only a few soil names in their language (Holman, 2005). A lot of soil terms have disappeared due to cultural erosion (Krasilnikov and Tabor, 2003). The extent of the studies around the world is not equal or uniform (Barrera-Bassols and Zinck, 2003).

We chose a regional presentation of the material because it allows one to make an approximate correlation with international classifications even when the soils are poorly defined, if the soil cover of the region is known. We do not try to present folk soil terms as classification systems: such systems are mainly the results of scientific interpretations. The list of soil names used by countryside people is not structured in a classification format, as the composition and meaning of folk terminology often vary between different villages.

Soil names are listed alphabetically within a region. Where non-Latin alphabet or other characters are used, the names are given in transcription. Synonyms are noted with italics (e.g. belyovaya zemlya – ... See also *seraja zemlya*). If a term has dialectic variants, they are listed with the main term, and noted with bold italics (e.g. beluga – ... Also ***beluha, beluzhina***).

In this list we do not provide any correlation of folk soil names with scientific classifications, although a number of correlations do exist on local levels (e.g. see Williams and Ortiz-Solorio, 1981; Niemeijer and Mazzucato, 2003). A previous attempt to find correspondence between indigenous soil names and the World Reference Base for Soil Resources (WRB) terminology (Shoba, 2002) was not satisfactory for several reasons. First, folk soil terminology and the meaning of the same terms vary not only between the regions, but also within the same village (Niemeijer, 1995; Niemeijer and Mazzucato, 2003). That is why ethnopedological soil surveys, which were popular in the 19th century, were later replaced by 'spade and auger' soil research (Krasilnikov, 1999). Second, proper correlation of most folk soil names with scientific ones is difficult because they are commonly based on single soil properties not included in scientific soil classifications at the highest levels (soil texture, temperature regimes, stoniness), or on perceptive characteristics (productivity in respect to a certain crop). Third, our perception of folk terms is not always adequate: we often consider them as much more precise and descriptive than they are. For example, in Russian geographical literature the Chilean folk soil name *trumao* always referred to specific volcanic soils where the volcanic ash is interlayered with peat horizons. However, in Chile this name is used for any volcanic soil, and the specific meaning was invented by geographers. Thus, any correlation of folk soil terminology with scientific classifications on a global scale would be confusing rather than useful for understanding soil names. The local soil names are briefly explained, but not correlated.

The list of indigenous soil names is far from complete. Unfortunately, many of them are reported in 'grey' literature that is not available in most libraries. For an extensive bibliography of ethnopedological works see a review by Barrera-Bassols and Zinck (2000).

Soil names of Russia, Ukraine and Belorussia

amshara (Belarussia and Pskov, Novgorod, Kalinin regions of Russia) – moss moor, sometimes covered with underdeveloped trees (Murzaev, 1984). Also *amsharina*, *amsharyna* and *omshara*.

aray (Russia, Perm region and Ural mountains, also in use in Komi Republic) – low flooded meadow, covered with harsh unproductive grasses with rare trees of *Alnus* and *Salix* (Murzaev, 1984).

areshnik (Arhangelsk region, Russia) – the name of unfertile gravel limestone soil (Lamanski, 1915).

badaran (eastern Siberia, Russia; Yakut) – mud bog, melting in summer down to 80–90cm, with permafrost (Murzaev, 1984).

bagna (Ukraine, Belarus, southern Russia) – moor, bog, swamp, dirty moist place, often with *Ledum palustre* (Murzaev, 1984). Also *bagno*.

barda (Arhangelsk region, Russia) – the name for infertile gravel and stony soils (Lamanski, 1915).

belaya luda (Kostroma region, Russia) – the name for clay podzolic and stagnogleyic soils (Dokuchaev, 1950a).

belichnyje zemli (Tobolsk region, Russia) – light-coloured variety of forest soils (Lamanski, 1915).

belik (Narym region, Russia) – wet, in places almost white, loose and friable loam, poor in humus (Lamanski, 1915).

belokorok (Arhangelsk region, Russia) – whitish loamy soils (Lamanski, 1915).

belozem (Russia) – a synonym of *podzol*, soils with a bleached eluvial horizon in forest zone of Russia. Dokuchaev (1950b) noted low productivity and poor forestry properties of these soils in comparison with other soils having bleached horizon (also *belyak*). The reasons are unknown now (light texture? Excessive acidity?). Later Dokuchaev brought some confusion to the term *belozem* because he used this name not only for leached forest soils but also for light-coloured soils of desert-steppes zone (*light serozems* in the Soviet classification). Later on Dokuchaev increased the confusion when he later described so-called 'aerial belozems of Erevan' in subtropics of the Transcaucasian region. These were debris-containing soils, with carbonates and gypsum coatings throughout the profile from the surface and this is the meaning used in Russian soil science literature (Gerasimov, 1979).

beluga (Russia) – according to Dokuchaev (1950b) in the Yaroslavski region it was a name for loamy light-grey earths, having colour significantly lighter than that of chernozems. Correlates with light-grey or humus-podzolic soils. In Pskov region it was used for 'clayey podzols' (Lamanski, 1915). Also *beluha, beluzhina, belitsa* and *belyovaya zemlya*.

belun (European Russia) – a synonym *of podzol*, soils with bleached eluvial horizon in forest zone of Russia.

bel (Arhangelsk region, Russia) – a name for soils under swampy wooded growth (Lamanski, 1915).

belyak (Russia) – a synonym of *podzol*, used in a number of regions of European Russia. It is interesting to cite Dokuchaev (1952) who in a private communication to A. N. Engelgardt, wrote that in Smolensk region a peasant 'would never mix *belyak* under ordinary straight, healthy birch forest, with *belozem* (*luda-podzol*) under baklusha – twisted, gnarled, stunted and extremely dense variety of the same tree'. However, for agriculture these soils were regarded as unproductive. Murzaev (1984) cites the dictionary of Dal where the following proverb of peasants from Perm region is presented: 'If one seeds on belyak, he will get white' (i.e. empty). There is a connotation of words: *belyak* has the same root as the Russian word *belyj* – 'white'. It seems that *belyak* was a name for loamy podzolic soils. Lamanski (1915) noted that in Samara region and in western Siberia it was a name for 'podzol-like soils of depressions'.

belyovaya zemlya (Yaroslavl and Nizhni Novgorod regions of Russia) – literally, 'whitish earth'. According to Dokuchaev's description (1950a), it was a name for soils resembling humus-podzolic ones. These soils were divided into *orehovyje* (nut), *sosnovyje* (pine) and *elovyje* (spruce) *belyovye zemli*. The latter two groups seem to differ in texture (pine soils are mostly sandy), while orehovyje (nut) soils were distinguished by their nut-like structure in the upper horizon. Also the term *seraya zemlya* ('grey earth') was used for these soils. In Nizhni Novgorod region this name, as well as *ilovka* and *zolka*, were used for stagnogley soils, differing from chernozems by their low productivity. Dokuchaev (1953) wrote: 'you can often see that over white earth there is a thin layer of peat. No doubt that here, in these low locations, water is stagnated for long in spring after rains; then, though it evaporates, the acid character of the earth remains, preventing chernozem formation.'

belyovyj suglinok (Yaroslavl region, Russia) – loam and clayey soils with light subsurface horizon (Dokuchaev, 1950a). Also *svetloseryj suglinok*.

bobovaya ruda (bean ore) (Russia) – sediments of oxides and hydroxides of iron of hydrogenic origin in bogs and semihydromorphic soils of depressions (Rozanov, 1974).

bolotnaya ruda (bog ore) – the same as *bobovaya ruda*.

borovina (Belorussia, Poland) – a name for dark-coloured soils on outcrops of chalk. Lamanski (1915) also notes that in places sandy soils under pine forests were understood under this name (*bor* in Russian and some other Slavonian languages means 'pine forest'). Later the term was used by Sibirtsev (1951) as a synonym of *rendzina*, and by Kubiëna (1953) for alluvial meadow soils on carbonaceous alluvium. In the Soviet classification (Egorov et al, 1977), the term was used for humus-carbonaceous soils.

borovoy pesok (pine forest sand) (Russia) – weakly coloured in brown colour sandy soil, forming mainly on pine forest terraces in forest or forest-steppe, and more rarely in steppe zones. Coarse sandy analogy of brown forest soils. In Russian classification of 1977 they are characterized as sod-podzolic weakly differentiated soil. First noted as a folk name for the Vyatka region by Dokuchaev (1953).

bublik (Ukraine) – hard stony soil (Murzaev, 1984).

buda (Kaluga region, Russia) – productive soil on burned forest meadow (Lamanski, 1915).

buhovina (Irkutsk region, Russia) – shallow black soil of depressions (Lamanski, 1915).

buzovaya zemlya (Nizhni Novgorod region, Russia) – productive dark-coloured loose soil (Lamanski, 1915).

buzun (Angara forested steppe, Russia) – chernozem-like soils of wet depressions, flooded for a long period (Murzaev, 1984).

chernistche (Russia) – a folk name for dark-coloured clayey soils in Tver and Kostroma region (Dokuchaev, 1950a). The description of these soils allows supposition that this term was used for humus- and mud-gleyed soils. It is also evident from the synonyms, used in Tver district: *gryaz* ('dirt'), *pribolot* ('close to a fen', 'like a fen').

chernoglin (Yaroslavl region, Russia) – clay with mould (Lamanski, 1915), literally 'black clay'.

chernogryaz (Russia) – a folk name for dark-coloured marshy soils near Kaluga and Oryol (Dokuchaev, 1949b), literally 'black dirt'. In properties, similar to meadow-chernozemic soils.

chernozem (Russia, Ukraine, Belorussia) – dark, well-structured, rich in organic matter soil, distinguished by high productivity. Name was used and still is used in folk lexicons not only for denominating certain soils, but also as a name for upper, humus-rich soil horizons, sometimes plough horizons in general. In his early works Dokuchaev also used the term rather broadly, using it both for soil types and for soil horizons. In general, in folk understanding *chernozem* as soil has significantly broader meaning than in scientific pedology, and this name is used significantly broader geographically than the area of distribution of chernozems in scientific understanding. In northern Russian *chernozem* is rather often used for eutrophic peat; in some regions of Karelia this name was used for soils on eluvium and derivates of graphite-like (*shunghite*) shales.

chernozem gorovoy (Chernigov region, Ukraine) – the word originates from the Russian word *gora*, which means a mountain. Dokuchaev (1953) mentioned 'a very unpleasant confusion' caused by this name:

> ... *looking through a statistical report from the Chernigov region, I came across the title gorovoy chernozem that meant 'chernozem of plateau'. I picked up this term and used it in the 15th volume of my 'Materials on the Evaluation of Lands of the Nizhni Novgorod Region', but, fortunately, did not use it while compiling maps. As came out afterwards, when I interviewed real Ukrainians, gorovoy chernozem appeared to be, in fact, totally different from that of dry, high steppe areas. In Ukrainee this word is used to define the upper horizon of chernozemic soil, i.e., the plough layer, which is defined as A in my reports. And this layer is called gorovoy chernozem, regardless of the region where the soil is located; it can be in both dry and elevated steppe, or in lowland.*

chvirets (Ukraine) – moving sands on low terraces of Dnieper (Murzaev, 1984). Also *chvirts* and *shirets*.

dityl (eastern Siberia, Russia; Evenk) – peat soils (Lamanski, 1915).

dubnyazhina (northwestern Russia) – soil with nut-like structure, forming on clays under oak forests; derived from Russian *dub* – 'oak'. Also *dubovitsy*. For more details see *poddubitsy*.

dubovitsa (northwestern Russia) – soil with nut-like structure, forming on clays under oak forests. Also *dubnyazhina*. For more details see *poddubitsy*.

dzurty (Crimea, Ukraine; Turkish) – artificial manured soils on the places of nomad camps (Lamanski, 1915).

gashun (Kalmykia and Buryatiya, Russia, and Mongolia) – solonchak (Murzaev, 1984).

gleba (Belorussia) – soils in general; upper humified layer of soils; soils with a well-humified upper horizon (Murzaev, 1984).

gleevina (Volynsk region, Ukraine) – clayey soil (Lamanski, 1915).

gley (Russia, Ukraine) – a common name for bluish, dove-coloured, whitish paludified horizons in soil, also soils containing these horizons. In various parts of Russia (Dokuchaev, 1950a) the term gley had broader or narrower meaning. In some places the name gley was used for limnoglacial clays; in other places – loamy sand soils with stagnogley horizon.

glina (clay) (Russia, Ukraine and Belorussia) – sediments and soils of heavy texture. Also *gnila, glinnik, glinchak*.

golyak (western Siberia, Russia) – solonetz without vegetation, indicated as a bare place in steppe (Lamanski, 1915).

gorohovataya zemlya (pea-like soil) (Russia) – according to Dokuchaev (1950a), the inhabitants of Lgov and Nezhin used this name for 'darkish' loamy sand soils, close to chernozems, but having low (2.3–3.6 per cent) content of humus.

gromazh (Ukraine, Volynsk region) – according to Dokuchaev (1950a), 'sandy-chalky soil ... it is composed of rather coarse sand and small, not bigger than forest nut, pieces of chalk, and it is extremely productive'.

grud (Belorussia and Ukraine) – uplifted place with rich soils in swampy or upland landscape, where broadleaf or spruce–broadleaf forests grow. The name was used both for characteristics of forest types, and for soils: in the latter case it referred to sod-podzolic or grey forest soils, usually weakly gleyed.

gryaznaya zemlya (dirty earth) (Russia) – according to *Materials on Statistics of Russia* (1859, cit. by Dokuchaev, 1950a), these earths are grey and white loamy sands, occurring as a thin layer on red loam, which impedes atmospheric moisture, and forms a so-called *podmor* or *podmochka* (water-impermeable horizon). Also *podmoristyje zemli*. Seems that these soils are differentiated in texture podzolic and sod-podzolic soils, probably formed on polygenetic sediments.

gryaz (mud, dirt) (Russia) – a folk name for dark-coloured clayey soils in Tver region (Dokuchaev, 1950a). Present in small plots on the outskirts of bogs. Probably, by this term sod- and mud-gleyic soils were understood.

holodnaya pochva (cold soil) (Russia) – 'podzol of depressions' (Lamanski, 1915).

hrustsh (Nizhni Novgorod region, Russia) – gravelly soil (Lamanski, 1915).

hrustshel (Pskov region, Russia) – gravelly soil (Lamanski, 1915).

huchur (eastern Siberia, Russia; Hakasian) – the same as *kudu*.

huzhar (eastern Siberia, Russia; Buryat) – the same as *kudu*.

ikritsa (Oryol region, Russia) – hard, excessively hardened soil (Lamanski, 1915).

ilovataya zemlya (clay soil) (Russia) – according to *Materials on Statistics of Russia* (1859, cit. by Dokuchaev, 1950a), used in Tver and other northern regions for denominating loamy and clayey soils with bleached upper horizons, identical to *svetloserye* or *belyovye suglinki* of Yaroslavski region. Folk synonyms: *pod ilok, nailok*. More probably, a name for texturally differentiated soils: podzolic, sod-podzolic, probably light-grey forest soils.

ilovka (Russia) – noted in *Materials on Statistics of Russia* (1861), cited by Dokuchaev (1950a), as sandy, loamy sand or loamy soil with water-impermeable horizon; it results in low yields on these soils. Described by Sibirtsev in 1908 (Sibirtsev, 1951) as weakly gleyed soils of northwestern Russia on bandy clays. In Soviet classification (Egorov et al, 1977) were included into the subtype of *humus-podzolic soils* (Gagarina et al, 1995).

kameshnik (Vyatka region, Russia) – stony soil, shingle bed (Lamanski, 1915).

kirza (Karelia, Russia) – a layer of frozen earth under melted upper horizons (Murzaev, 1984).

kislitsa (Russia) – soil, where acid grasses grow (Lamanski, 1915).

klyaslaya zemlya (Vyatka region, Russia) – hardened soil (Lamanski, 1915).

krasik (Ural, Tyumen and Sverdlovsk regions, Russia) – clayey unproductive soil of reddish colour (Murzaev, 1984). Also *krasnik*.

krasnuha (Karelia, Leningrad region, Russia) – unproductive clayey red-coloured soil (Murzaev, 1984).

kritsa (Nizhni Novgorod region, Tatarstan, Russia) – cold clayey soil (Lamanski, 1915).

krotovina (mole's burrow) (Nizhni Novgorod region, Russia) – according to Dokuchaev (1953), for peasants this meant not only mole's burrows, where excavated soil-forming material was present, but also any outcrops of parent materials, including that of erosion origin.

kudu (eastern Siberia, Russia; Evenk) – solonchak, used by the animals as a source of salt; 'beast solonchak' (Murzaev, 1984). Also *kuzhur, huzhar* and *huchur*.

kuliga (Russia) – forest soil with admixture of charcoal (Lamanski, 1915).

kuzhur (eastern Siberia, Russia; Tuvin) – the same as *kudu*.

lipkovitsa (western Ukraine) – clayey swampy soil (Lamanski, 1915).

lomuha (Pskov region, Russia) – a name of hard clayey soils (Lamanski, 1915).

luda (Russia) – a folk name of podzolic soils, used in the northeast of European Russia (Dokuchaev, 1950b). Dokuchaev illustrates low productivity of this soil with a folk proverb: 'Where is luda – there is the need'. In Smolensk region, according to Dokuchaev (1950b), this name was used for loamy sand soils, approximately equivalent to the modern understanding of podzol (as Al-Fe-humus soils on light-textured material). In Perm district the name *luda* was used for clayey, cold, grey soils or all blue clay (Murzaev, 1984). The name itself originates from Russian word *ludet* – to harden, because these soils cover with a hard crust when the weather is sunny.

meldovataya zemlya (Russia, Vyatka region) – chernozems with high content of carbonates or on chalk eluvium (Dokuchaev, 1953).

mestshernik (Russia, Ryazan region) – acid soils pine forests, unsuitable for cultivation (Lamanski, 1915).

mokropes (Russia, Pskov region) – slightly clayey sand with stagnant water (Lamanski, 1915). Also *sapun*.

myasiga (Zeysk-Burein plain, Russia) – semi-bog soils, forming in heavy clays under mosses, poplar, underdeveloped vegetation and bushes (Murzaev, 1984).

myasuha (Russia, Pskov region) – clay soil, swampy in wet conditions (Lamanski, 1915).

naglinok (Russia, Samara region) – 'chernozem with clods of clay and debris that are not coloured with humus, or chernozem with a significant admixture of sand' (Dokuchaev, 1950a).

nailkovaya pochva (Russia, Moscow region) – a name of redeposited soils (Lamanski, 1915).

nailok (Russia) – according to *Materials on Statistics of Russia* (1859, cit. by Dokuchaev, 1950a), the name is used in Tver and other northern regions for naming loamy and clayey soils with a bleached upper horizon, similar to *svetloseryj* or *belyovyj suglinok* of Yaroslavski region. Folk synonyms: *ilovataya zemlya, podilok*. More probably, the name means soils differentiated in texture (podzolic, sod-podzolic, probably light-grey forest soils).

naplavnyje zemli (Russia) – soils of the foot of slopes; colluvial soils (Lamanski, 1915).

nyasha (Siberia, Russia) – dirty-grey clayey material, remaining in the place of drying salted and eutrophic lakes (Lamanski, 1915).

oglinok (Russia) – exposed to the surface along the banks of rivers and in hilly landscapes, red-coloured moraine loam with boulders in Yaroslavl region (Dokuchaev, 1950a). The name seems to mean eroded humus-podzolic soil.

okost (Russia, Poltava region) – productive solonetz, solonetzic chernozem (Lamanski, 1915).

oles (Russia and Belorussia) – swampy minerotrophic bog with growths of *Alnus*, usually on the foot of a slope, where groundwater is trickling (Murzaev, 1984). Also *ols*.

orehovaya zemlya (nut-like soil) (Russia) – according to Dokuchaev (1950a) 'both here [in Nizhni Novgorod region] and in other places of Russia the name *orehovaya zemlya* is used for *soil* or *subsoil of podzolic* or *ash* colour, easily falling apart in balls or irregular polygons, usually smaller than a little nut'. Also Dokuchaev noted that in Yaroslavl region 'orehovaya' is a name for moderately dry *belyovaya zemlya* (humus-podzolic or grey forest soils), which is the best comparison with *sosnovaya* and *elovaya belyovaya* soils. It is obvious that in contrast to the latter names *orehovaya* soil is so-called due to its structure, not vegetation.

oreshnik (Russia, Vyatka region) – soil originated from weathered shales (Lamanski, 1915).

paglinok (Russia, Nizhni Novgorod region) – according to *Materials for Statistics of Russia* (1861), cit. by Dokuchaev (1949b), it is 'chernozem of brown or grey colour with clods of clays, with the depth [of the A horizon] 13–22cm'. Probably, the name means clayey and loamy grey forest soils. Also, taking into account that paglinok was described mainly 'along the slopes of rivers and gulleys', one can suggest that it was also a name for eroded chernozems.

pesok (sand) (Russia, Ukraine and Belorussia) – light loose material, composed mainly of big mineral particles, and soils on this material.

plaun (Russia, Archangelsk region) – sticky moist white soil of spruce forests (Lamanski, 1915).

plyvun (Russia) – sand, saturated with water, sandy gley (Lamanski, 1915).

poddubitza (northwestern Russia) – soil with nut-like structure, formed under oak forests in clayey sediments; *dub* means 'oak' in Russian, the word *poddubitsa* literally means 'under oak'. This soil was described by Glinka, who stressed its similarity with grey forest soils. Folk synonyms: *dubnyazhny, dubovitsy*. Later this term was understood more narrowly: the name was used for soils of Volhov alluvial plain under growths of oak and soils of Velikoretsk plain on varved clays. In official classification were referred to as podzolic and brown forest soils (Gagarina et al, 1995).

podilok (Russia) – according to *Materials on Statistics of Russia* (1859, cit. by Dokuchaev, 1950a), used in Tver and other northern regions for loamy and clayey soils with bleached upper horizon, similar to *svetlosery* or *belevy suglinok* of Yaroslavski district. Folk synonyms: *ilovataya zemlya, nailok*. Most probably, the name was used for Albeluvisols or Planosols.

podina (the north of European Russia) – a plot of frozen earth with ground ice, which melts only in autumn or does not melt at all (Murzaev, 1984).

podmoristyje zemli (Russia) – according to *Materials on Statistics of Russia* (1859, cit. by Dokuchaev, 1950a), these earths are grey and white loamy sands, occurring as a thin layer on red loam, which impedes atmospheric moisture, and forms so-called **podmor** or **podmochka** (water-impermeable horizon). Also *gryaznyje zemli*. Seems that these soils are similar to Planosols formed on lithologically different sediments.

podnor (Nizhni Novgorod region, Russia) – fine-grained soils of depressions, where moisture accumulates after rainfall. Upper horizons of these soils are usually bleached, and Dokuchaev (1950c) thought them to be identical with podzols.

podsolonok (western Siberia, Russia) – weakly salted and weakly sodic soils (Lamanski, 1915).

podzol (Russia) – soil, having under the plough horizon some material of whitish, 'ash' colour. In different regions of Russia very different soils were known under this name. The term 'podzol', initially documented in *Materials on Statistics of Russia*, was introduced to the scientific literature by Dokuchaev (1950b). All 'podzols', described by Dokuchaev, are far from the modern understanding of these soils: usually Dokuchaev described under this name Albeluvisols or Stagnosols. The real distribution of podzols in Russia (north-west) was unknown to Dokuchaev; he defined the soils of Finland and north-western Russia as 'stony', and the description of a podzol in Vyborg region was considered by Dokuchaev as an extraordinary fact.

pognoy (eastern Ukraine) – strongly manured earth (Lamanski, 1915).

popel (= pepel – ash; Russia) – a synonym of *podzol*, used by the peasants of Kasimov district (Tatarstan) (Dokuchaev, 1950b). According to Murzaev, in Moldavian Kodra mountains podzolic soils are called **popela**, in western Ukraine – **popelychka**.

popeluha (Russia) – a name of *rendzina* soils in Saratov region (Glinka, 1927). The name is connected either with the fact that in dry condition these soils are dusty, or with the fact that they have colour significantly lighter than chernozems, in places similar to the colour of ash (*popel* or *pepel*

in Russian) (Glinka, 1927). Also *popyluha*. In Voroshilovgrad district
– *pepeluha* (Murzaev, 1984).

potlivaya pochva (Vladimir region, Russia) – heavy textured soils (Lamanski,
1915).

potnaya zemlya (Russia) – always wet, moist soil, mainly due to springs
(Lamanski, 1915).

pribolotok (Novgorod region, Russia) – an earth just after spring flooding of a
river (Lamanski, 1915).

pribolot (Russia) – a folk name for dark-coloured clayey soils in Tver region
(Dokuchaev, 1950a). Met as small outlying areas of bogs. The term seems
to mean humus- and mud-gleyic soils.

pripad (Ukraine) – a depression; also a name of chernozemic sticky clay
(Lamanski, 1915).

pripadlivaya zemlya (Russia) – excessively moistured acid soils, where wheat
easily falls (Lamanski, 1915).

puhlaya zemlya (Russia) – soft, friable soil (Lamanski, 1915).

puhletz (Pskov region, Russia) – soil, having soft and loose compaction of mud
horizon due to silt admixture (Lamanski, 1915). Also *puhlinka*.

puhlyak (Tver region, Russia) – greyish-white soil (Lamanski, 1915).

pushnaya zemlya (Russia) – loose, porous soil (Lamanski, 1915). Also
pushnyak.

pyhun (Russia) – a name for structureless and microstructured soils of
chernozemic and meadow-chernozemic types, used in Siberia (Dokuchaev,
1949b; Rozanov, 1974).

rebrovnik (western Siberia, Russia) – solonchak on a bank of a salted lake with
ridged surface (Lamanski, 1915).

rodyuchij solonetz (western Siberia, Russia) – solonetz-like or solonchak-like
chernozem (Lamanski, 1915).

sapun (Pskov region, Russia) – slightly clayey sand with stagnant water
(Lamanski, 1915). Also *mokropes*.

schyolog (Karelia, Russia) – red clay and soils on such clay (Murzaev, 1984).

seraya zyemlya (grey earth) (Russia) – according to Dokuchaev (1950a), in
Yaroslavl region it was a name for loamy and loamy sand soils, having
lighter colour than chernozems. The name seems to mean grey forest or
sod-podzolic soils. Grey earth was divided by the peasants into grey and
light-grey, and also into *orehovaya*, *sosnovaya* and *elovaya belyovaya
zemlya*. Dokuchaev also cited *Materials on Statistics of Russia*: 'In some
places every soil, if it is not red loam or pure sand, is called grey'.

seropesok (east Ukraine) – podzolized sand (Lamanski, 1915). The term later
was used in soil science as a synonym for chernozem-like soils on coarse
sands.

serovik (Nerchin region, Russia) – grey sandy soil (Lamanski, 1915).

seryak (Perm region, Russia) – grey podzolic loam (Lamanski, 1915).

sharyn (Arhangelsk region, Russia) – forest litter, mud, peat (Murzaev, 1984).

sinyuga (Pskov region, Russia) – blue gleyed clay (Lamanski, 1915).

sipets (Ukraine) – strongly sandy clay (Lamanski, 1915).

solod (Russia) – forested depressions on flat poorly drained watersheds of southern Russia. From Dokuchaev's time is used also for nominating weakly acid residual-salted soil with bleached upper horizon. Glinka (1927) argued against the name 'solod' in soil science, because in folk understanding *solod*, as well as its synonyms *solot*, *osinovye kusty* ('poplar shrubs'), *mokrye kusty* ('wet shrubs'), *baklushi* and *kolki*, were based on the vegetation and topography rather than on soil properties. However the term was accepted in scientific literature and is used in Russian, Canadian and some other soil classifications.

solonchak (southern Russia) – soils of arid and semiarid regions salted from the surface, practically infertile, covered with rare halophitic vegetation. The term is used in many scientific soil classifications of the world. Also *solonistshe*.

solonetz (southern Russia) – unproductive, extremely hard soil with characteristic columnar structure. From early stages of development of soil science the name is used in scientific literature for soils with sodium-enriched exchangeable complex. The term is used in many scientific classifications.

sosnovaya belyovaya zemlya – see for details *belyovaya zemlya*.

sosnyaga (Vologda region, Russia) – sandy soil under pine forest (Lamanski, 1915).

suglina (Russia) – fine sandy, strongly compressed soils (Dokuchaev, 1949b).

suglinok (loam) (Russia, Ukraine, Belorussia) – loose sediments, heavier than sand, but lighter than clays, and soils formed on these sediments. Sometimes is used as a synonym of soils in general. In Nizhni Novgorod region the name was used for brown and grey soils, containing less humus than chernozems (Dokuchaev, 1949b). Also *sugley*.

suhmen (Tomsk region; Russian) – dry clayey soil, where crops are affected by wind (Lamanski, 1915).

suholitsaya zemlya (dry-faced earth) (Nerchin region, Russia) – soil having no groundwater (Lamanski, 1915).

supes (loamy sand) (Russia, Ukraine, Belorussia) – loose sediments, heavier than sand, usually due to greater content of clay and silt particles, also soils formed on these sediments. Also, in Vyatka region, *susupes* (Dokuchaev, 1953).

suzyomok (Karelia, Russia) – black soil with insignificant admixture of sand (Lamanski, 1915). Also *suzem*.

svetloseryj suglinok (light-grey loam) (Yaroslavl region, Russia) – loamy and clayey soil with bleached upper horizon (Dokuchaev, 1950a). Also *belyovyj suglinok*.

talets (Arhangelsk region, Russia) – boggy soil, having springs in the bottom, thus freezing later than other soils (Lamanski, 1915). Also *talitsa*.

trunda (Vyatka region, Russia) – bog soil rich in humus (Lamanski, 1915).

turan (eastern Siberia, Russia; Yakut) – solonchak, usually 'beast solonchak' (Murzaev, 1984). Also *turang*.

tyushklevataya zemlya (Vologda region, Russia) – moist, cloggy earth, inconvenient for cultivation (Lamanski, 1915).

tverdozem (Russia) – hard, compact soil (Lamanski, 1915).

uglinok (Moscow region, Russia) – loamy soil (Lamanski, 1915).

usol (Russia) – in *Materials on Statistics of Russia* (1859, cit. by Dokuchaev, 1950a) it is noted that in Yaroslavl region the name is used for 'such type of chernozem, on the surface of which in hot days salt (?) or saltpetre accumulate in a form of white powder'. Probably means saline chernozem.

usush (Kamchatka, Russia) – dry meadow tundra (Lamanski, 1915).

utaitsa (Perm region, Russia) – whitish podzolic loam (Lamanski, 1915).

vodopojchina (Karelia, Russia) – earth, moistened with water, forcing its way up from below (Lamanski, 1915).

vyaz (central regions of Russia) – spongy, swampy bog (Murzaev, 1984). Also *vyazelitsa, vyazel, vyazilo* and *vyazun.*

yaglaya zyemlya (Russia) – rich in organic matter, productive earth, *chernozem* (Murzaev, 1984).

yedun (Arhangelsk region, Russia) – sandy soil, rapidly soaks up water (Lamanski, 1915).

yelovaya belyovaya zemlya – for details see *belyovaya zemlya.*

zahlest (Russia) – according to *Materials on Statistics of Russia* (1859, cit. by Dokuchaev, 1950a), used in Vladimir region as name of sandy soils, with underlying clay at a depth of 0.35–2.1m. Due to the presence of water-impermeable horizon the soil often turns wet, 'heavy, cold'. Also *zemlya s podklyuchinoy*. Also *zahrest.*

zakletch (Russia) – according to Dokuchaev (1950a), in Nizhni Novgorod region it was a name for chernozem-like soils on dark-brown, 'extremely swampy, strongly marly clay'. Also *zaklet.*

zalezh (Arhangelsk region, Russia) – ploughed earth on a minerotrophic bog (Lamanski, 1915).

zamoristyje zemli (Yaroslavl region, Russia) – strongly moistured soils, where crops can freeze (Lamanski, 1915).

zamochka (Yaroslavl region, Russia) – excessively moistured soils of depressions in regions of occurrence of grey forest and chernozemic soils (Dokuchaev, 1950a). Also *zamorina.*

zatuzhennaya zemlya (Vyatka region, Russia) – hard or hardened soil (Lamanski, 1915).

zemlya zyablaya (cool soil) (Middle Volga region, Russia) – according to Dokuchaev (1950a) 'thin, clayey-sandy, unproductive soil, looking like clayey cement'. Also *zemlya lyadinnaya.*

zemlya lyadinnaya – the same as *zemlya zyablaya.*

zemlya s belinoy (earth with whiteness) (Nizhni Novgorod region, Russia) – a common name for soils, having bleached, non-chernozemic upper horizon (Dokuchaev, 1950a).

zemlya s podklyuchinoy (soil with a spring underneath) (Russia) – according to *Materials on Statistics of Russia* (1859, cit. by Dokuchaev, 1950a), this name was used in Vladimir region of sandy soils, with underlying clay at a depth of 0.35–2.1m. Due to the presence of water-impermeable horizon the soil often turned wet, 'heavy, cold'. Also *zahlest.*

zheltik (Russia) – soils of southern regions of Pskov district on kames and boulder residual-calcareous sands and polygenetic sediments, in which brown illuvial horizon occurs under a humus-enriched horizon (Gagarina et al, 1995). Also were named by various authors *sod-acid poor in humus* and *sod-cryptopodzolic.*

zhem (Russia: Archangelsk region, Karelia) – moist mineral soil with water appearing on the surface (Murzaev, 1984). Also *zhemovatye soils.*

zhestel (Russia, Penza region) – hard earth (Lamanski, 1915).

zhirnaya zemlya (fat earth) (Russia) – soil rich in clay and humus (Lamanski, 1915). Also *zhirnozem.*

zobok (Amur region; Russian) – loamy or loamy sand soils of moderate slopes (Lamanski, 1915).

zola (ash) (non-chernozemic belt of Russia) – hard, dry, unproductive soil (Murzaev, 1984).

zolka (Russia, Middle Volga region) – according to Dokuchaev (1953), a synonym of *podzol.* Dokuchaev illustrates the fact, that these soils were believed to be infertile by the peasants, with a short interview of a local person: 'it is belyak-zolka, no use to cultivate' (Dokuchaev, 1949a).

zoloyed (Orel region, Russia) – unproductive, unsuitable for cultivation earth; more probably, the name is for podzolic soils (Murzaev, 1984).

zolnik (Tver region, Russia) – swamp, bog, boggy soil (Murzaev, 1984); bad grey cold soil (Lamanski, 1915).

zyablaya pochva (Nizhni Novgorod region, Russia) – a name for podzolic-gley soils (Lamanski, 1915).

Soil names of Caucasus

akalo-mitsa (Georgia; Imeretia) – clayey soil (Zakharov, 1915).

bosh (Georgia) – a name for loose solonetzic soils (Lamanski, 1915).

chale (Georgia; Mengrelia) – very productive loamy and loamy sand soils in river valleys (Zakharov, 1915).

chinchiburi (Georgia; Guriya) – clayey heavy light-coloured soil with insignificant content of humus, with clayey impermeable subsoil (Zakharov, 1915).

chirnavos (Georgia; Russian) – a name for chernozems among Russian military settlers (Lamanski, 1915).

chita-diha (Georgia; Mengrelia) – clayey heavy light-coloured soil with insignificant content of humus, with clayey impermeable subsoil (Zakharov, 1915).

dihashko (Georgia; Guriya) – loamy and loamy sand soils, comparatively rich in humus, known for high productivity (Zakharov, 1915).

dobira (Georgia; Guriya) – clayey shallow soil, coloured by humus to dark-brown colour (Zakharov, 1915). In Mengrelia the name is pronounced as *dobera.*

gazha (Azerbaijan) – according to Rozanov (1974), arid or semiarid soil, containing layers of gypsum in profile in the form of fine crystalline powder-like mass. In broader sense the name was used for mud soils with high content of gypsum or carbonates.

hinchkona (Georgia; Mengrelia) – loamy, rich in humus soil with high content of stones (Zakharov, 1915).

hirhati (Georgia; Kahetia) – soils, containing a significant admixture of gravel (Zakharov, 1915).

horhi-mitsa (Georgia; Imeretia) – debris-containing sandy soil (Zakharov, 1915).

hriaki (Georgia; Kahetia) – debris-containing soil (Zakharov, 1915).

karaer (Trans-Caucasian region; Tatar) – dense dark loamy soils (Lamanski, 1915).

kara-shorakyat (Trans-Caucasian region; Tatar) – black solonchak (Lamanski, 1915).

kir (Azerbaijan) – soil, permeated with petroleum (Murzaev, 1984).

kviani-mitsa (Georgia; Imeretia) – stony sandy soil (Zakharov, 1915).

kvinshali-mitsa (Georgia; Imeretia) – sandy soil (Zakharov, 1915). Also *lisi-mitsa*.

kviteli-mitsa (Georgia; Kahetia) – 'yellow soils, zheltozems', clayey soils, mixed with some quantity of calcium carbonate (Zakharov, 1915).

lami (Georgia; Kahetia and Kartalinia) – light-coloured carbonaceous alluvial soils of broad river valleys (Zakharov, 1915).

lamiani-mitsa (Georgia; Kahetia) – soil with high content of clay (Zakharov, 1915).

legaza (Georgia; Guriya) – clayey and loamy shallow soils of red or greyish-yellow colour, containing insignificant quantity of humus, laying on weakly weathered clayey and clayey-lime shale (Zakharov, 1915).

lisi-mitsa (Georgia; Imeretia) – sandy soil (Zakharov, 1915). Also *kvinshali-mitsa*.

magnari (Georgia; Guriya and Mengrelia) – loamy, rich in humus stony soil; forms on slopes, footslopes and in valleys (Zakharov, 1915).

mchate-mitsa (Georgia; Imeretia) – light-textured stony soil (Zakharov, 1915).

mdzime-mitsa (Georgia; Imeretia) – heavy clayey, very deep light soils, occurring on slopes and on the footslopes of mountains, as well as in valleys (Zakharov, 1915).

melchvari (Georgia; Mengrelia) – clayey and loamy shallow soils of red or greyish-yellow colour, containing insignificant quantity of humus, laying on weakly weathered clayey and clayey-lime shale (Zakharov, 1915).

mere (Georgia; Guriya) – very productive loamy and loamy sand soils of river valleys (Zakharov, 1915).

mtis-mitsa (Georgia; Imeretia) – dark-grey loose soil (Zakharov, 1915).

mtredis-peri-mitsa (Georgia; Kahetia) – soil with a significant admixture of lime and fragments of carbonaceous shales (Zakharov, 1915).

mtsire-mitsa (Georgia; Imeretia) – 'worst earth', light shallow loam with clayey-carbonaceous subsoil of cloddy structure; the clods are easily destroyed by fingers (Zakharov, 1915).

oroki (Georgia; Guriya and Mengrelia) – clayey soils (Zakharov, 1915). Also *tsare*.

rike (Georgia; Kartalinia) – grey debris-clayey soil on slopes of valleys (Zakharov, 1915).

sahro (Georgia; Guriya) – loamy, rich in humus soil with high content of stones (Zakharov, 1915).

shavi mitsa (Georgia; Imeretia, Kahetia and Kartalinia) – chernozem-like or mountainous grey forest clayey soil with high content of humus, literally 'black earth' (Zakharov, 1915).

shoraket (Azerbaijan) – solonchak, salted soil (Murzaev, 1984). Also *shoran, shorlag* and *shurezar.*

sila-mitsa (Georgia; Imeretia) – light-textured sandy soil (Zakharov, 1915); in Kartalinia the name is pronounced as *silya-mitsa.*

tetri-mitza (Georgia; Kahetia) – light-coloured marl soil, literally 'white earth' (Zakharov, 1915).

tihiani-mitsa (Georgia; Kahetia) – clayey soil (Zakharov, 1915).

tiri-mitsa (Georgia; Imeretia) – chernozemic clayey, comparatively loose soils (Zakharov, 1915). Also *shavi-mitsa.*

tsare (Georgia; Guriya and Mengrelia) – clayey soils (Zakharov, 1915). Also *oroki.*

tsiteli-etseri (Georgia; Imeretia) – cinnamonic heavy clayey soil (Zakharov, 1915).

tsiteli-mitsa (Georgia; Imeretia) – clayey red-coloured soil, literally 'red earth' (Zakharov, 1915).

ucha-diha (Georgia; Mengrelia) – loamy and loamy sand soils, comparatively rich in humus, distinguished by high productivity (Zakharov, 1915).

yehegnut (Armenia) – a reed bog (Murzaev, 1984).

yetseri (Georgia; Guriya) – podzol-like greyish-yellow loamy or clayey soil with powder structure, where grapevines are cultivated (Zakharov, 1915). Some authors wrote that this name was used for red or greyish-yellow soils with heavy texture with no traces of podzolization. Zakharov (1915) argued against this opinion, because usually these soils were reported to have a bleached upper horizon. In Imeretia the name is pronounced as *etseri*, in Mingrelia – as *entseri.*

zegani (Georgia; Imeretia) – heavy clayey light-coloured soil, rich in humus, laying on slopes of mountains and hills. These soils are used for vine production (Zakharov, 1915).

Soil names of Eastern Europe

ak-toprak (Bulgaria; Turkish) – white unproductive soil (Stranski, 1954). Also *beyaz-toprak.*

alicheva (Bulgaria) – stony soil (Stranski, 1956). Also *koprak* and *tashlyk.*

altyn-tarla (Bulgaria; Turkish) – poor, unproductive soil of yellow colour (Stranski, 1954).

aluga (Bulgaria) – sandy alluvial soils, usually on river bars (Stranski, 1956).

asprohoma (Greece, Bulgaria; Greek) – hydromorphous salted soil (Stranski, 1957).

aurtoprak (Bulgaria; Turkish) – soil difficult for cultivation (Stranski, 1956).

azmak (Bulgaria; Turkish) – permanently excessively moistened peat soil, bog soil (Stranski, 1954). Also *batak* and *gyol*.

babicheva (Bulgaria) – red-coloured light-textured soil, not impeding moisture (Stranski, 1956).

bakarliya (Bulgaria; Turkish) – soil of copper colour (Stranski, 1954).

bardakovitsa (Bulgaria) – heavy clayey soil, used for producing bardatsi – water jugs (Stranski, 1956).

baruga (Bulgaria) – hydromorphic mineral soil under meadow vegetation (Stranski, 1954).

barzitsa (Bulgaria) – soil, where crops ripen earlier (Stranski, 1956).

batak – the same as *azmak*.

belozem (white earth) (Bulgaria) – light, white carbonaceous soils, often salted (Stranski, 1954). Also *belozemna*. Also *byala* prast.

beyaz-toprak – the same as *ak-toprak*.

blana (Bulgaria) – soil, cultivation of which results in the formation of large aggregates, resembling bricks (Stranski, 1956).

blato (Bulgaria) – bog soil (Stranski, 1954). Also *valta, dragovets, maklistshe, mlaka, mochur* and *treskavistshe*.

brashnelik (Bulgaria; Turkish) – powder-like, fine soil on eluvium of solid rock (Stranski, 1956).

brenitsa (Bulgaria) – swampy clayey hydromorphic soil (Stranski, 1954).

buchesta (Bulgaria) – soils, having cloddy structure (Stranski, 1956).

buga (Bulgaria) – well humified soil with evidence of hydromorphism, formed under meadow vegetation; meadow soil (Stranski, 1954).

buhkava (Bulgaria; in Bulgarian means 'fluffy') – loose, easy for cultivation soil (Stranski, 1956). Also *buhna, buhova, buhava, buvkava, buvavitsa* and *buavitsa*.

bulgurliya (Bulgaria; Turkish) – soil, having small grain structure, resembling wheat grains (Stranski, 1956).

buraku (Bulgaria, Romania; Romanian) – dark-coloured soil, chernozem (Stranski, 1954).

byal geren (Bulgaria) – soils, salted from the surface (Stranski, 1957).

byal pesak (white sand) (Bulgaria) – light, sometimes carbonaceous sandy soils poor in humus (Stranski, 1954). Also *byala pestshenitsa*.

byala glina (white clay) (Bulgaria) – light-coloured, usually carbonaceous heavy-textured soils (Stranski, 1954). Also *byala klisavitsa, byala ilovitsa, byala huma, byala kal, byala buovitsa, byala smolnitsa, byala prevlaka*.

byala prast [white earth] (Bulgaria) – light, white carbonaceous soils, often salted (Stranski, 1954). Also *beloprasnitsa*. Also *belozem, belozemna*.

chakyl (Bulgaria; Turkish) – soils with debris. Also *chakalyk* and *chaktoprak* (Stranski, 1956). Also *troshlyak*.

chamur (Bulgaria; from Turkish *çamur* – mud) – sticky, smearing clayey soil (Stranski, 1956). Also *mazna*.

cher-geren – the same as *kara-geren*.

cherna (Bulgaria) – a common name for all dark-coloured soils, both well-drained and hydromorphic (Stranski, 1954). Usually used with other names for closer definition: *chernyava, charnava, cherna zemya, cherna pras, cherno myasto, cherna gnilesta, cherna klisa, chernoklisava, cherna zhila, cherna ilovitsa, cheren geren, cherenpyasak, chernozem.*

cherna vodenitsa – the same as *kara-suluk.*

chervena (Bulgaria) – a common name for red-coloured soils (Stranski, 1954). Usually is used with other definitions: *chervena prast, chervenoprasnitsa, chervenazemya, chervenaglina, chervenagnilesta, chervenka, chervenozem, chervena klisa, cherven kazlach, chervena smolnitsa, chervena kapsida, chervena ilovitsa, chervena ehlovitsa, chervenak, cherven pesak, chervena pestshenitsa, chervena varovita, chervena propuskliva.*

chetin (Bulgaria) – dark-coloured, hardening when dried soils (Stranski, 1956).

chista (Bulgaria) – stoneless soil (Stranski, 1956).

chorbalyk (Bulgaria; Turkish) – sandy, well-drained, easy to cultivate soil (Stranski, 1954).

chukli (Bulgaria) – soils on eluvium of dense carbonaceous rocks (Stranski, 1956).

dargel (Bulgaria) – soils with outcrops of solid rock (Stranski, 1956).

delmelyk (Bulgaria; Turkish) – alluvial soil (Stranski, 1957). Also *dolmalyk.*

dolma-toprak – the same as *delmelyk.*

donesena (Bulgaria) – alluvial soil (Stranski, 1957). Also *mytok, naliv, namol, nanos, napluv, napoy, nariv, naslag, natlak, otlaka* and *rechnitsa.*

dradorak (Bulgaria) – stony soil; name is an imitation of the sound of a plough touching a stone (Stranski, 1956).

dragovets – the same as *blato.*

draskoloto (Bulgaria) – shallow, less than 10cm in depth, soils, with underlying solid rock (Stranski, 1956).

drobena (Bulgaria) – soil, having small grained structure to plough horizon (Stranski, 1956).

dzegrevitsa (Bulgaria; from Turkish *cider* – small shot) – soil, forming small rounded aggregates after rain (Stranski, 1956).

fuchalnik (Bulgaria) – stony soils, while cultivated giving a hissing sound of plough touching stones (Stranski, 1956).

gadzhiktoprak (Bulgaria; from Turkish *gicic* – tickling) – soil containing small stones and detritus (Stranski, 1956).

galbin (Bulgaria) – yellow clayey soil (Stranski, 1954).

galcha (Bulgaria) – soil 'black as a crow' (Stranski, 1954).

gaynache (former Yugoslavia) – a name for degraded cinnamon soils of xerophytic forests, having eluvial-gleyic horizon (Glazovskaya, 1983).

geren (Bulgaria) – salted soils or soils, turning salted when drained (Stranski, 1957). Also *giren, gerenna, gerenyak, gerenyava, gerenliya, gerenlik* and *gerenlek.*

geren-toprak (Bulgaria; Bulgarian, Turkish) – salted soils, flooded with water for a long period (Stranski, 1957).

gernesta (Bulgaria) – humus-carbonaceous soil of the swallow hole regions (Stranski, 1957).

giznalyk (Bulgaria; Turkish) – boggy soil; mineral hydromorphic soil, covered with a thin layer of peat or mould (Stranski, 1954).

gletava (Bulgaria) – clayey, heavy soil, containing carbonates (Stranski, 1956).

gragor (Bulgaria) – sandy or debris soil (Stranski, 1956).

grahchata (Bulgaria) – soil with small cloddy structure, resembling bread crumbs (Stranski, 1956).

gramadlyatsy (Bulgaria) – debris-stony soils (Stranski, 1956).

grancharska kal (Bulgaria) – hard, heavy-textured soil, literally 'potter's mud' (Stranski, 1956).

grashovitsa (Bulgaria) – soil with 'pea' large grained structure (Stranski, 1956).

greu (Bulgaria) – soil of heavy texture (Stranski, 1956).

greznitsa (Bulgaria) – sticky, smearing hydromorphic soil (Stranski, 1954). Also *girizliva*.

grudliva (Bulgaria) – well-structured soil (Stranski, 1956).

gustchernitsy (Bulgaria) – stony infertile soils, where lizards live (in Bulgarian *gustchery* means lizard) (Stranski, 1956).

gyok-toprak (Bulgaria; from Turkish *gok* – blue) – clayey soil of blue colour (Stranski, 1954).

gyol – the same as *azmak*.

haharicha (Bulgaria) – stony soil, giving a sound resembling snoring while cultivated (Stranski, 1956).

halva (Bulgaria; from Turkish name for sweets) – soft, productive, easy to cultivate soil (Stranski, 1956).

hamurkese (Bulgaria; from Turkish *hamur* – paste) – sandy soils with a significant admixture of humus and clays (Stranski, 1956).

hasarlyk (Bulgaria; Turkish) – sandy soil the colour of 'matting' (Stranski, 1954).

horosan (Bulgaria; Persian) – a name of infertile clayey soils with a high content of carbonates used in some regions of Bulgaria (Stranski, 1957).

huhlovitsa (Bulgaria) – red-coloured stony soil (Stranski, 1956).

huma (Bulgaria) – clayey soil (Stranski, 1956). Also *humi, humyak, humenka, humenik, humnik, humova, humesta, umata, umistshe, umenik, umnitsa* and *umlyak*.

il (mud, clay) (Bulgaria) – soil of heavy texture, often hydromorphic soil. Also *ilova, ilovitsa, elova* and *elovitsa*. Often have definition of colour: *cherna ilovitsa, kafyava ilovitsa, byala ilovitsa*, etc. (Stranski, 1956).

inya (Bulgaria) – a name for salted soils, having excretions of salts (resembling rime) on the surface (Stranski, 1957).

kaba (Bulgaria) – loose soil; usually is used for characteristic of chernozem (Stranski, 1956).

kadrika (Bulgaria) – soil where cultivation results in the formation of large aggregates, 'curly' soil (Stranski, 1956).

kafyava (Bulgaria) – brown-coloured soils (Stranski, 1954). Different pronunci-
ations exist; also the name is used with the texture modifiers: *kafenitsa*,
kafenikava, kafyava ilovitsa, kafyava sura, kafyava lezga.

kaldaramesta (Bulgaria) – black cinnamon soil with high content of debris and
stones (Stranski, 1956).

kale (Bulgaria) – extremely hard soil (Stranski, 1956).

kalna (Bulgaria) – sticky, smearing hydromorphic soil; from Bulgarian *kal*
– 'mud' (Stranski, 1954).

kana (Bulgaria) – orange-red soil (Stranski, 1954).

kanchuk (Bulgaria) – stony soil; the name resulted from a characteristic sound
of a spade against a stone (*kapchuk* – in Bulgarian 'drain-pipe') (Stranski,
1956).

kapsida (Bulgaria; Greek) – alluvial soils (Stranski, 1956).

kara (Bulgaria; Turkish) – soil of black colour (Stranski, 1954). Also
karakomsal.

kara-geren (Bulgaria; Turkish, Bulgarian) – well humified productive soils
(Stranski, 1957). Also *cher-geren.*

kara-suluk (Bulgaria; Turkish) – black, sticky clayey soil, badly conducting
moisture (Stranski, 1954). Also *kara-toprak* and *cherna vodenitsa.*

kara-toprak – the same as *kara-suluk.*

karashko (Bulgaria; from Turkish *karasik* – mixed) – a mixture of fine earth
and stones (Stranski, 1956).

karmeztoprak (Bulgaria; from Turkish *kirmizi* – red) – red-coloured soils
(Stranski, 1954). Also *kazlach.*

kartel (Bulgaria) – soil, where ploughing exposes stones on the surface (Stranski,
1956).

kashava (Bulgaria) – soil resembling porridge when wet (Stranski, 1954).

kasnitsa (Bulgaria) – soil, where crops ripen later (Stranski, 1956).

katranitsa (Bulgaria) – black soil (Stranski, 1954).

kayalyk (Bulgaria; Turkish) – stony soils, where fine earth is washed out by
running water (Stranski, 1956).

kaymak (Bulgaria; Turkish) – clayey soil (Stranski, 1956).

kayrak (Bulgaria; Turkish) – sandy soil with admixture of debris and stones
(Stranski, 1956).

kazlach (Bulgaria; from Turkish *kizil* – red) – red-coloured soils (Stranski,
1954). Also *karmeztoprak.*

kekez (Bulgaria) – carbonaceous soil (Stranski, 1957).

kepirtoprak (Bulgaria; Turkish) – clayey soil (Stranski, 1956).

kil-toprak (Bulgaria; from Turkish *kil* – clay) – clayey soil (Stranski, 1956).

kipra (Bulgaria) – soil swelling when moist (Stranski, 1956).

kirecava (Bulgaria, from Turkish *kirec* – lime, carbonates) – a name used
in some regions of Bulgaria for denominating soil with high content of
carbonates (Stranski, 1957).

kirec-toprak – the same as *yoren-toprak.*

kirpich (Bulgaria) – heavy clayey soil, used for manufacturing bricks (Stranski,
1956).

kiselitsa (Bulgaria) – forest dark-coloured acid soil (Stranski, 1957).

kishova (Bulgaria) – strongly hydromorphic dark-coloured soil under meadow vegetation; meadow-bog soil (Stranski, 1954).

klisava (Bulgaria) – sticky, plastic soil; name originates from Bulgarian *klisav hlyab* – 'unperfectly baked bread' (Stranski, 1956).

kokina (Bulgaria; Greek) – red-coloured soil (Stranski, 1954).

komrako (Bulgaria) – poor thin soils, with underlying solid rock (Stranski, 1957).

komsal (Bulgaria; Turkish) – sandy soil. Usually is used with a definition of colour: ***cheren komsal***, *karakomsal*, ***siv komsal***, etc. (Stranski, 1956).

koprak (Bulgaria; Turkish) – stony soil (Stranski, 1956). Also *alicheva* and *tashlak*.

korava (Bulgaria) – productive dark-grey soil with good structure and water-holding capacity (Stranski, 1956).

kravenitsa (Bulgaria) – unproductive soil, 'red as blood', rapidly drying after rain (Stranski, 1954).

kremyechna (Bulgaria) – soil on quartz sands, distinguished by low productivity (Stranski, 1957). Also ***kremenista***, ***kremakliva***, ***kremekliva***, ***kremechak*** and ***kremonyak***.

kreshyak (Bulgaria) – shallow soils, not impeding moisture (Stranski, 1956).

kulesta (Bulgaria; from Turkish *kula* – yellow-red) – ochreous soils, used by local population as pigment (Stranski, 1954).

kumbatak (Bulgaria; Turkish) – alluvial, moist sandy soil (Stranski, 1956).

kumtoprak (Bulgaria; from Turkish *kum* – sand) – sandy soil (Stranski, 1956).

kumyutyuryutoprak (Bulgaria; Turkish) – poor sandy soil, where crops die (Stranski, 1956).

kuruch (Bulgaria; from Turkish *kuru* – dry, weak) – light-coloured clayey-sandy stony soil, usually on alluvial sediments (Stranski, 1954).

kush-boku (Bulgaria; from Turkish *kus* – bird, and *bok* – excrements) – clayey soil with secretions of carbonates (Stranski, 1954).

kyullyutoprak (Bulgaria; from Turkish *kul* – ash) – silty soil, easily eroded by the wind (Stranski, 1956).

lakoviste (Romania) – meadow-chernozemic soils (Rozanov, 1974).

leka (Bulgaria) – easy to cultivate, rather productive soil (Stranski, 1956).

lepkava (Bulgaria) – sticky clayey soil; usually of brownish-black or red colour, in places carbonaceous (Stranski, 1956).

lezga (Bulgaria) – shallow (20–30cm) soil, usually clayey, with underlying solid rock (Stranski, 1957). Also ***lizgavitsa***, ***lizgar***, ***liskavitsa***, ***leskava***, ***liska***, ***liskorak***, ***lyuska*** and ***lyuskavitsa***.

liska (Bulgaria) – soil resembling swampy clay when moist, and hard when dry. Also *lezga*. (Stranski, 1956).

lokva (Bulgaria) – swampy, wet soil (Stranski, 1954).

luma (Bulgaria) – soil, producing large aggregates when tilled (Stranski, 1956).

lyuta (Bulgaria) – soil of heavy texture, hardening when dried (Stranski, 1956). Also *lyutic*.

machkan (Bulgaria) – a name originated from Bulgarian *machka* – 'cat'. Stranski (1956) explains (not very convincingly), that there are soils 'black as a cat'.

maklistshe – the same as *blato*.

mala (Bulgaria) – fine clayey alluvial soil (Stranski, 1957).

martlyk (Bulgaria) – soil, having a good drainage and well heated in spring; this soil can be cultivated in March (Stranski, 1956).

mavrohoma (Bulgaria; Greek) – dark-coloured soil, literally 'chernozem, black soil' (Stranski, 1954).

mazna (Bulgaria) – smearing, sticky clayey soil (Stranski, 1956). Also *chamur*.

melyaga (Bulgaria) – soil on eluvium of chalk (Stranski, 1957). Also *melik* and *melinchava*.

mena zemya (Bulgaria) – soft, loose, usually hydromorphic soil (Stranski, 1956). Also *mekito* and *mekash*.

meshana (Bulgaria) – sandy-clayey soil, a mixture of clays and sand (Stranski, 1956).

mlaka – the same as *blato*.

mlyekanya (Bulgaria) – white carbonaceous, often salted soils (Stranski, 1954).

mochur – the same as *blato*.

modra (Bulgaria) – hydromorphic soil of bluish colour (Stranski, 1954).

mokra (Bulgaria) – dark-coloured hydromorphic soil (Stranski, 1954). Also *mokresh* and *mokritsa*.

morogan (Romania) – dark clayey cracky compact soils (Rozanov, 1974).

muhlevina (Bulgaria) – stony soil, unsuitable for cultivation (Stranski, 1954).

mursava (Bulgaria; old Russian) – dark-cinnamonic soil (Stranski, 1954).

muzga (Bulgaria) – pale-pink hydromorphic soil (Stranski, 1954).

mytok – the same as *donesena*.

naaletleme (Bulgaria; from Turkish *nale nâle* – moan) – swampy black soil, hardening in dry conditions (Stranski, 1956).

naliv (Bulgaria) – alluvial soil. Also *namol, nanos, napluv, napoy, nariv, natlag* and *naslag* (Stranski, 1957).

nasap (Bulgaria; from Romanian *nisip* – sand) – sandy soil (Stranski, 1956).

natsepena (Bulgaria) – dark-coloured soil, swelling when moist and cracking when dried (Stranski, 1956).

navlatsy (Bulgaria) – alluvial soil, containing unsorted material, stones, detritus and sand (Stranski, 1957).

negro (Bulgaria, Romania; Romanian) – chernozem (Stranski, 1954).

nyiok (Hungary) – a name for yellow-brown and cinnamon soils of Hungary. Glinka (1927) believed these soils to be 'zheltozems', similar with Yugoslavian *gaynache*.

ohrena (Bulgaria) – soil, coloured ochreous by iron compounds (Stranski, 1957).

orehova (Bulgaria) – soils of dark-cinnamon, nut colour (Stranski, 1954).

ostrak (Bulgaria) – clayey-sandy soil, easy to cultivate (Stranski, 1956).

otlaka – the same as *donesena*.

paklavitsa (Bulgaria) – dark-coloured clayey soil, which hardens when dry and gets unsuitable for cultivation; from ancient Bulgarian ПЬКЛЪ – 'hell, tar' (Stranski, 1956).

pamant albo (Bulgaria, Romania; Romanian) – white carbonaceous soils (Stranski, 1954).

papalyugova (Bulgaria) – sod-carbonaceous soil, *rendzina* (Stranski, 1957).

parhuda (Bulgaria) – silty soil, easily eroded by the wind (Stranski, 1956).

pepeliva (Bulgaria) – soils, resembling ash both in colour and in consistency (Stranski, 1954). A lot of derivates exist: *pepeliva redka, pepeliva zemya, pepeliva pras, pepelista, pepelivka, pepelyak, pepelasha, pepelnitsa, pepelistshe*. Also there are derivates from other synonyms of the word *pepel* – 'ash' in Bulgarian (skrum and smet): *skrumeva* and *smetnitsa*.

pesakliva (Bulgaria) – sandy soil. Also *pesak, peskovita zemlya, pesatsi, pesako* (Stranski, 1956).

pilyafnitsa (Bulgaria) – soil, where a thin crust forms on the surface in dry state (Stranski, 1956).

pitomna (Bulgaria) – soil without stones and debris (Stranski, 1956).

plava (Bulgaria) – brown-grey, 'pale' soil (Stranski, 1954).

pleskavitsa (Bulgaria) – alluvial clayey soil, hardening when dried (Stranski, 1957).

pluvnitsa (Bulgaria) – soil, slowly conducting water, flooded after rains (Stranski, 1956).

podnicharska (Bulgaria) – red clayey soil, used for filling up crevices in a house (Stranski, 1957).

podsushliva (Bulgaria) – stony, light-textured soil, rapidly drying after rain (Stranski, 1954).

podvir (Bulgaria) – hydromorphic soils, forming in places of outlets of springs (Stranski, 1956).

povlek (Bulgaria) – brown soil, holding water on the surface after rainfall; when dried forms a cracking crust (Stranski, 1954).

pradistche (Bulgaria) – periodically flooded soils (Stranski, 1954).

prahkava (Bulgaria) – loose, friable soil (Stranski, 1956).

prashnitsi (Bulgaria) – small grained humus-carbonaceous soils (Stranski, 1956).

prazharitsa (Bulgaria) – poor soil, often red-coloured, where plants dry in hot weather (Stranski, 1956).

prazlak (Bulgaria) – sandy structureless soil (Stranski, 1956).

prigor (Bulgaria) – soils, where plants dry in hot weather (Stranski, 1956).

priplavniva (Bulgaria) – soil, where crops die (Stranski, 1956).

prokevitsa (Bulgaria) – swampy, dirty soil (Stranski, 1954).

propuskliva (Bulgaria) – sandy soil, not holding moisture (Stranski, 1956).

puhkava (Bulgaria) – soft, very loose soil. Stranski (1956) notes that this name in different regions of Bulgaria was used for soils different in texture and genesis.

pukalitsa (Bulgaria) – soil, swelling when moist (Stranski, 1956).

puriyeva (Bulgaria) – sandy carbonaceous soil, usually with high groundwater table (Stranski, 1956).

puzdarliva (Bulgaria) – stony, unproductive pale-grey soil (Stranski, 1954).

razhdiva (Bulgaria) – yellow-brown, 'rust-coloured' eroded soils of heavy texture on loesses (Stranski, 1954).

rechnitsa – the same as *donesena*.

rehavitsa (Bulgaria) – loose soil (Stranski, 1956). Also *ryaol*.

rendzina (Eastern Europe; Polish) – a name for shallow sticky soils on limestone; the name is an imitation of the sound of a plough touching stones. Stranski (1956) proposed another origin of the term, considering it similar to a Bulgarian word *redina*, or *ryadka*. It is difficult to believe it because humus-carbonaceous soils in the limestone debris cannot have the productive and beneficial properties ascribed to *ryadka*. Now the term is widely used in scientific literature.

rihk (Estonia) – outcrops of Silurian limestone, mixed with boulder moraine, in places in the form of small hills (Glinka, 1927). Also soils formed on such outcrops.

rochio (Bulgaria, Romania; Romanian) – soils of yellowish and reddish colours (Stranski, 1954).

ronliva (Bulgaria) – friable soil (Stranski, 1956).

ropulak (Bulgaria) – slope soils on colluvium; distinguished by high debris and stone content (Stranski, 1956).

ryadka (Bulgaria) – easily plugged soil (Stranski, 1956). Also *redina, redka, rednyak* and *ridak*.

sachenliva (Bulgaria; from Turkish *sican* – mouse) – grey soil (Stranski, 1954).

sakyztoprak (Bulgaria; from Turkish *sakyz* – chewing resin) – clayey compact dark-coloured soils (Stranski, 1956).

saltireto (Bulgaria) – soils with boulders (Stranski, 1957).

saratoprak (Bulgaria; Turkish) – yellow soil, turning into sticky mud after rain (Stranski, 1954).

sedinyak (Bulgaria) – hardened soil (Stranski, 1956).

sert (Bulgaria; from Turkish *sert* – hard) – soil, turning hard in dry season (Stranski, 1956).

sharatur (Bulgaria, Romania; Romanian) – salted soil (Stranski, 1957).

sinya (Bulgaria) – soils, having bluish colour; this name was used both for well-drained soils, containing glauconite, and strongly reduced soils, containing vivianite (Stranski, 1954).

sipkava (Bulgaria) – loose, friable soil (Stranski, 1956).

sitna (Bulgaria) – grey, moisture-holding, fine structured soil (Stranski, 1956).

skripavitsa (Bulgaria) – stony soil, where a plough is creaking against stones (Stranski, 1956).

skrumeva – the same as *pepeliva*.

slatina (Bulgaria, Serbia) – hydromorphic (bog) salted soils; in Serbia the term was used in scientific literature (Stranski, 1957).

smetnitsa – the same as *pepeliva*.

smolniza (Romania, Bulgaria, Albania) – clayey cracky compact soils (Rozanov, 1974). In contrast to the common opinion, do not always have black or dark colour: examples are such names as *byala* (white) *smolnitsa* and *chervena* (red) *smolnitsa* (Stranski, 1954).

smonitsa (Austria, former Yugoslavia) – dark clayey cracky compact soils (Rozanov, 1974).

solaneva (Bulgaria) – a common name for salted infertile soils (Stranski, 1957). Also *solena, solanic, solenitsa, solinets and solistshe*.

solan-toprak (Bulgaria; from Turkish *sulanmak* – moisten) – hydromorphic sandy-clayey soil, usually carbonaceous, of light colour (Stranski, 1954).

spilatsy (Bulgaria) – soils on eluvium of sandstones; known for their low productivity (Stranski, 1957).

starnishka (Bulgaria) – loose, easily conducting water soil, containing sand and stones (Stranski, 1956).

stignata (Bulgaria) – soil, difficult to cultivate (Stranski, 1956).

stikliva (Bulgaria) – compact hydromorphic dark-coloured soil (Stranski, 1954).

studena (Bulgaria) – fine earthy, light-textured soil of light-grey colour, easily eroded by the wind (Stranski, 1956).

sura (Bulgaria) – soil of grey colour (Stranski, 1954).

surosinkava (Bulgaria) – soil, having bluish colour; composed of mixture of sand, debris and clays, perfectly drained (Stranski, 1954).

syutlyuk (Bulgaria; from Turkish *süt* – milk) – white carbonaceous, often salted soils (Stranski, 1954).

Szik (Hungary; Szik or Szek – soda) – a local name for dark-coloured alkaline soils, usually *solonetz* or sodic *solonchak*. Later, the term was used by soil scientists of many countries for denominating black clayey solonetz in southern Europe and northern Africa (Rozanov, 1974). Also *Szek*.

tamna (Bulgaria) – dark-coloured soil, chernozem (Stranski, 1954).

tarchinela soil (Bulgaria) – soil of brown colour (Stranski, 1954).

tashlyk (Bulgaria; Turkish) – stony soil (Stranski, 1956). Also *alicheva* and *koprak*.

tatliya (Bulgaria; from Turkish *tatli* – pleasant) – soil, easy to cultivate (Stranski, 1956).

tezhka (Bulgaria) – soil of heavy texture (Stranski, 1956).

tikla (Bulgaria) – soil with platy structure; often eroded soil on eluvium of shales (Stranski, 1956).

titra (Bulgaria) – sandy stony soil, not holding moisture (Stranski, 1956).

tol (Bulgaria) – humus-carbonaceous soils, having morphologically expressed carbonates at a depth of 25–30cm from the surface (Stranski, 1957). Also *tolistshe*.

tospa (Bulgaria) – a name of soil cover, where in depressions there are cinnamon soils, not holding moisture, easily cultivated, while on the uplifts there are eroded soils of yellow and white colour (Stranski, 1956).

trapka (Bulgaria) – soft, plastic soil (Stranski, 1956).

tresavistshe – the same as *blato*.

troshlyak (Bulgaria) – soils with rock debris (Stranski, 1956). Also *chakal*.

tsarnozyom (Bulgaria) – black soil, chernozem (Stranski, 1954).

tsarvena (Bulgaria) – red-coloured soil (Stranski, 1954).

tutkal (Bulgaria) – red-coloured sticky clayey soil (Stranski, 1956).

tutmaniklyava (Bulgaria) – soft, plastic clayey or loamy soil; from Bulgarian *tutmannik* – 'layered cake with curds' (Stranski, 1956).

tuzla (Bulgaria, from Turkish *tuzlu* – salted) – a common name for salted soils in some regions of Bulgaria (Stranski, 1957). Also *tuzlata and tuzluk*.

usuka (Bulgaria) – difficult to cultivate, hard clayey soil (Stranski, 1956).

valta – the same as *blato*.

varovita (Bulgaria) – carbonaceous chernozem (Stranski, 1957). Also *varovitsa, varnitsa, varenitsa, vernitsa, varovitnyak and varovista zemlya.*

vinozem (Bulgaria) – soil of colour of red wine; is characterized by numerous fragments of sandstone on the surface (Stranski, 1954).

vishnitsa (Bulgaria) – soil of cherry colour (Stranski, 1954).

vodnyava (Bulgaria) – heavy clayey soil with low water permeability (Stranski, 1954).

yaka (Bulgaria) – clayey chernozem (Stranski, 1956).

yamen (Bulgaria; Greek) – bog soil (Stranski, 1954).

yazovina (Bulgaria; from Turkish and Persian *jaz* – protected part of a river plain, where thin particles are accumulated) – alluvial soils of heavy texture (Stranski, 1957).

yoren-toprak (Turkey, Bulgaria; from Turkish *yoren* – lime, and *toprak* – earth) – a name for red-coloured carbonaceous soils similar to *terra rossa* (Stranski, 1957).

yotyuryuktoprak (Bulgaria; Turkish) – 'jingling' while cultivated, stony soil (Stranski, 1956).

yurganplak (Bulgaria; Turkish) – soil, thawing later in spring (Stranski, 1956).

zagrivna (Bulgaria) – chernozem or meadow-chernozem soil (Stranski, 1957).

zheltokovenitsa (Bulgaria) – humified yellow-coloured soil, in texture – sandy clay, easily cultivated (Stranski, 1954).

zhelyeznitsa (Bulgaria) – soil, rich in iron (hydr)oxides (Stranski, 1957).

zhilava (Bulgaria) – plastic clayey soil (Stranski, 1956).

zholta (Bulgaria) – yellow soil. Both the name itself and its derivatives are used for different soils: humus-carbonaceous, forest soils, containing carbonaceous concretions, and also any eroded soils, lacking vegetation (Stranski, 1954). Many derivatives of this name exist, some of which have more specific definitions: *zholta pras, zholtaprasna, zholta zemya, zholtozemka, zholta klisavitsa, zholta ilovitsa, zholta smolnichava.*

zhvizdre (Lithuania) – stony unproductive soil (Murzaev, 1984).

zola (Bulgaria) – sandy structureless soil (Stranski, 1956).

Soil names of Western Europe

Alm (Germany) – a name of marly horizons used in Bavaria; these horizons usually lie under humus horizon (or peat) of bog, especially peat-bog soils (Glinka, 1927).

alvar (Sweden, Estonia) – a name for lands, mainly pastures on the outcrops of limestone, fragmentary, covered with debris soils. Usually are covered with *Juniperus* trees.

baragge (Italy) – ochreous podzolic soils of Italian Alps. Also *ferretto* and *groane* (Principi, 1943).

barros (Portugal, Spain) – a local name for dark compacted cracky soils of Vertisols type (Rozanov, 1974).

bollo (Italy) – earthy, stoneless red-coloured soil on eluvium of limestones (Kubiëna, 1953).

calvero (Spain) – skeletic brown forest and light-chestnut soils of dry forests (Krupenikov, 1981).

carr (eastern England) – peat soil with a large quantity of wood in profile (Kubiëna, 1953).

dy (Sweden) – 'bog clay', soils, forming after overgrowing of forest lakes. Later the term was used by German soil science school (Kubiëna, 1953).

ferretto (Italy) – ochrous podzolic soils of Italian Alps. Also *baragge* and *groane* (Principi, 1943).

Geest (Germany) – a name of marshy plains of northern Germany, from Rain to Elba, where peat-gleyic soils on sands, with underlying clayey sediments, form (Ganssen, 1962).

Gifterde (Germany) – the same as *Maibolt*.

groane (Italy) – ochreous podzolic soils of Italian Alps. Also *baragge* and *ferretto* (Principi, 1943).

gyttja (Sweden) – mud, clayey soils of sea and lake banks of Sweden. Form from marine and lacustrine deposits. The term was used by soil scientists (Wilkander et al, 1950; Kubiëna, 1953).

katterkleygronden (The Netherlands) – a historic name of marsh soils containing pyrite (FeS_2) in Holland, where yellow mottles of jarosite ($KFe_3(SO_4)_2(OH)_6$) form after drainage in profile. The productivity of these soils distinctly falls due to sulphuric acid production in the course of pyrite oxidation. The name 'cat clays' is connected with the fact that unexpected death of crops was ascribed to witchcraft, and a cat was usually associated in the Middle Ages with witches (Pons, 1973). The first scientific description of these soils was done by Carl Linné in 1735. Also **katterkley**.

Klee – the same as *Knick*.

Knick (Germany) – a name of a dense, poor in organic matter clayey horizon in marsh soils. Glinka (1927) noted that these sediments are accumulated mainly in the winter period, and alternate with *Schlick*, rich in organic matter layers, forming in summer. Also *Klee*.

Knickboden (Germany, The Netherlands) – a name for old marsh soils of heavy texture, having dense structureless horizons, *Knick* (Glazovskaya, 1983).

korni (Finland) – forest bogs with a dense tree growth of birch or spruce; the soils are defined as peat-mud-gleyic soils of minerotrophic bogs (Glazovskaya, 1983).

letti (Finland) – a name for open bogs with grass vegetation on peat-mud-carbonaceous soils (Glazovskaya, 1983).

Maibolt (Germany) – a local name for acid sulphate soils of marshes. The name originated from word combination: Maibolt – from Maifelder (meadow) and Kobolt (evil fantastic creature) (Pons, 1973). Also *katterkleygronden*.

Marsch (Germany) – a name for moist meadows and marshy depressions near a sea coast, as well as the soils of these depressions; the territory is flooded only during syzigies. The term is widely used in scientific literature, though its mean varies.

Missen (Germany) – eluvial-gleyic soils, forming in Germany under coniferous forests. Distinguished by pale greyish/light blue surface horizon. In German scientific literature described as *Stagnogley* (Ganssen, 1962). Also **Missenboden** and *Molkenboden*.

mjele (Sweden) – a name for pale podzolic-like soils on sorted silty loamy sands in northern Sweden (Rutherford, 1972).

Molkenboden – the same as *Missen*.

neva (Finland) – a name for forestless Sphagnum bogs of Finland on peat ombrotrophic soils (Glazovskaya, 1983).

polder (The Netherlands) – a name for cultivated drained parts of marshes, usually protected by dykes, distinguished by high productivity. The term is widely used in scientific literature.

Roodorn (Germany) – a name for upper horizon of clayey marsh soils used in Groniningen province. This horizon is partially coloured with oxides and hydroxides of iron in reddish colour; it is rich in humus and has weakly acid reaction (Glinka, 1927).

Schlick (Germany) – a name for horizons of marsh soils rich in clay and organic debris. Glinka (1927) notes that these sediments are accumulated mainly in the summer period, and alternate with *Knick,* poor in organic matter clayey layers, forming in winter.

tangel (Austria) – thick layer of litter on soil surface (Kubiëna, 1953).

terra rossa (Italy; Spanish variant – *tierra roja*) – red-coloured weathering crusts of carbonaceous rocks and soils, developed on these crusts. Form under xerophitic wood vegetation. The term is widely used (with different meanings) in soil science.

till (Scotland) – a name for dense stony, bouldered moraines, and soils formed on them. The term is used in geological literature.

turvekangas (Finland) – a name for dry forested peaty massives on peaty-podzolic-gleyic soils (Glazovskaya, 1983).

wad (The Netherlands) – soils of intertidal zone of seas and oceans, rich in organic matter (Kubiëna, 1953). Also **Watt**.

Soil names of Middle Asia and Near East

adyr (Turkic) – a name of hilly-gullied landscapes of Tyan-Shan piedmont and soils of these landscapes (Rozanov, 1974).

akkum (Middle Asia) – literally 'white sands'; in fact means loose sands without vegetation (Superanskaya, 1970).

ala (Turkmenia) – a local name for clayey desert soil with cracking polygonal surface, a synonym of *takyr* (Murzaev, 1984).

arzyk (Uzbekistan) – a term for cemented gypsiferous, partially carbonaceous layers in the lower part of soil profile of hydromorphic (*saz*) soils (Rozanov, 1974).

bozyngen (Turkmenia) – weakly humified loamy soils of deserts with loose surface and close occurrence (12–15cm) of gypsum. In Russian classification (Egorov et al, 1977) distinguished as a genus of highly gypsiferous grey-brown desert soils.

charchin (Turkmenia) – soil, less productive than *kara-upa* (Lamanski, 1915).

chokolak (Middle Asia) – knobby solonchaks (Egorov et al, 1977).

daryalyk (Uzbekistan) – solonchak in the place of a former river bed (Lamanski, 1915).

dasht (Fergana valley, Tajikistan) – a name for debris and gravel soils of deserts (Lamanski, 1915). Also *desht*.

depiz (Turkmenia) – a name for sandy soils of deserts, rich in gypsum and salts, forming near salted lakes on redeposited salt material. In Russian classification (Egorov et al, 1977) distinguished as a genus of sandy desert gypsiferous soils.

dzhaksy-kebir (Turkmenia) – comparatively productive chubby solonchak (Lamanski, 1915).

dzhaman-kebir (Turkmenia) – infertile chubby solonchak (Lamanski, 1915).

gum (Turkmenia) – sandy soil, sometimes 'soil in general' (Murzaev, 1984).

hak (Middle Asia) – name, used in Middle Asia for denominating salted muds formed after rain on clay outcrops (Rozanov, 1974). Also *kakh*.

hamra (Israel) – soils of zone of dry forests and shrubs, similar in properties to cinnamonic soils.

hayat (Turkmenia) – a name of garden lands (Lamanski, 1915).

karakum (Middle Asia) – literally 'black sands', means sodded, fixed sandy soils (Superanskaya, 1970).

kara-upa (Turkmenia) – productive soil (Lamanski, 1915).

kawir (Iran) – takyr and takyr-like soils (Rozanov, 1974).

kebir (Turkmenia) – chubby solonchak (Lamanski, 1915). Also *dzhaksy-kebir* and *dzhaman-kebir*.

kempyrtash (Kirgizia) – a local name for thick, up to 1–3m, carbonaceous crusts in profile of arid and semiarid soils (Rozanov, 1974).

kersh (southern Kazakhstan) – a local name for thick, up to 1–3m, carbonaceous crusts in profile of arid and semiarid soils (Rozanov, 1974).

kyr (Turkmenia) – infertile soil, covered with stones (Lamanski, 1915).

lut (Iran) – desert plain of Iranian highland, also clayey-debris soils of these territories (Rozov and Stroganova, 1979).

nari (Palestine) – a name for carbonaceous crusts 0.5–2m in depth near Jerusalem (Glinka, 1927).

nazal (Israel) – degraded (differentiated in texture) brown-red sandy soils, forming on carbonaceous sandstone or on sandy dunes. Horizon B is enriched with clay and has pryzmatic structure (Ravikovich, 1960).

oytak (Turkmenia) – desert soils of depressions, where during rainfall excess moisture exists due to collecting surface waters. These soils have a relatively developed humus (up to 10cm) and gleyed horizons.

saz, sazovaja pochva, saz soil (Middle Asia) – a name, used for soils of plains at the foot of mountains, developing under the effect of groundwaters of 'foot', 'saz' regime. Characterized by the presence of marly layers in the lower part of soil profile.

shakat (Kazakhstan) – desert landscape with infertile stony or solonchak-like soils (Murzaev, 1984). Kubiĕna (1953) noted that *shokat* is a Kirgizian name for solonetz.

shivarzamin (Tadzhikistan) – clayey, water-saturated soil (Murzaev, 1984).

shoh (Middle Asia; Arabian) – a name for carbonaceous marly horizon, formed in *saz* soils (Rozanov, 1974).

shor (Middle Asia) – strongly salted soils, formed on the bottom of dried lakes, mostly covered with a salt crust. Also *sor*.

sor – the same as *shor*.

takyr (Middle Asia) – clayey soils of flat depressions of deserts, having hard surface, cracking into polygons.

tugay (Turkmenia) – flood plains; alluvial soils (Lamanski, 1915).

zey (Turkmenia) – salted bog, formed by irrigation waters (Murzaev, 1984).

Soil names of China and Far East

baijang (China) – light-coloured soils (Gong Zitong, 1989).

baijang-tu ('soils, white as milk', China) – eluvial-gleyic soil of alluvial-lacustrine plains of Amur region in Russia and China (Kornblyum and Zimovets, 1961).

cha (China) – fluvial soils (Gong Zitong, 1989).

chaohuanglu (China) – cultivated meadow 'pale' soils (Kanno, 1978).

chaohuangyan (China)– cultivated meadow-burozem soils (Kanno, 1978).

chaoni (China) – flooded paddy soils of southern China, having good drainage (Kanno, 1978).

gobi (ancient Mongolian) – debris-gravel desert. Usually is contrasted with *shamo* (Murzaev, 1984).

he-tu (China) – a name used in northern China for denominating chernozemic and meadow-chernozemic soil with prolonged seasonal frost and high groundwater table (Rozanov, 1974).

heilu-tu (China) – a name for long-cultivated soils of the Loess plateau; for a long time it was believed that they have an artificial layer 25–100cm in depth (Rozov and Stroganova, 1979). Recent studies show that most of these soils have natural accumulation of eolian sediments on the surface.

heini (China) – dark-coloured puddy soils of northern China (Kanno, 1978).

huangjang (China) – a name for eroded soils of the Loess plateau, poor in humus, usually contrasted with productive *heilutu* (Glazovskaya, 1983). According to Kanno (1978) these soils are burozems (brown soils), used in agriculture.

huanglu (China) – cultivated 'pale' (weakly developed) soils (Kanno, 1978).

huangni (China) – surface-gleyed flooded puddy soils of southern China (Kanno, 1978).

huang-tu (China) – a local name for loess soils of northern China, having yellowish, light-rusty and pale colours of cultivated horizon (Sibirtsev, 1951).

kuroboku (Japan) – dark-coloured soils, forming in piedmonts of Central Honsyu and Kyusyu and on Pleistocene terraces. In a broad sense a synonym of *Andosols*, dark-coloured soils of temperate regions formed on volcanic ash. The term is used only in Japanese literature. In some works Andosols and kuroboku are divided (Kato and Matsui, 1979); kuroboku means soil with lower content of volcanic glass, sometimes even not volcanogeneous soils. Kanno (1982) considered kuroboku to be relict Andosols. Matsui (1982) divides kuroboku soils into volcanic and non-volcanic.

kurotsuchi (Japan) – dark-coloured soils. In a broad sense a synonym of *Andosols*, dark coloured soils of temperate regions formed on volcanic ash.

kusari (Japan) – weathered gravel soils of high Pleistocene terraces (Kato and Matsui, 1979).

lengjintian (China) – puddy soils, fed by cold springs (Kanno, 1978).

lou (China) – stratified soils on loesses, affected by prolonged manuring (Gong Zitong, 1989).

maganni (China) – surface-gleyed flooded puddy soils of Middle China (Kanno, 1978).

masa (Japan) – slightly weathered lithogenic soils, Lithosols (Kato and Matsui, 1979).

onji (Japan) – red-coloured moist soils, forming under evergreen subtropical forests on pumice-containing ash loams on Kyusyu and Sikoku islands (Ganssen, 1962).

quingni (China) – puddy gleyic soils of Middle China (Kanno, 1978).

sajong (China) – soils of alluvial plains of eastern China, formed in river valleys, affected by annual flooding. At a small depth they have a horizon of dense irregular carbonaceous and Fe-Mn concretions (Glazovskaya, 1983). Also *shachiang*.

shamo – sandy desert. Usually is contrasted with *gobi* (Murzaev, 1984).

shii (Japan) – a folk term, proposed by Sudzuki and Hachiya (Endo, 1982) for denominating forest soils, developing under chestnut and similar forests of Japan in temperate warm climate (Gerasimov, 1958).

shijang-tu (China) – a name used in central and western China for denominating thick, up to 1–3m, carbonaceous crusts in soil profiles (Rozanov, 1974).

xiantian (China) – salted puddy soils (Kanno, 1978).

youge (China) – puddy gleyic soils of southern China (Kanno, 1978).

yuni (China) – flooded puddy soils of Middle China, having good drainage (Kanno, 1978).

Soil names of south and southeastern Asia

aintel maati (Bangladesh; Bengali) – clay soil (Ali, 2003).

aintel-doash maati (Bangladesh; Bengali) – clay loam soil (Ali, 2003).

bele maati (Bangladesh; Bengali) – sandy soil (Ali, 2003).

bele-doash maati (Bangladesh; Bengali) – sandy loam soil (Ali, 2003).

bhangar (Bengal, India) – a name for soils on comparatively old alluvial sediments of heavy texture, usually having a dense layer with carbonaceous concretions (Raychaudhri, 1958).

bhata (India) – red-coloured soils in western India (Satyanarayana, 1971); according to other sources, near Bombay it is a name for young soils of brown colour formed on clayey alluvium (Raychaudhri, 1958).

chalka (Uttar-Pradesh, India) – red-coloured soils of loamy texture on eluvium of granites and gneisses (Raychaudhri, 1958).

chopan (Bombay, India) – sodic chernozem-like soils on Dekan plateau (Raychaudhri, 1958).

dorsa (India) – brown tropical soils of western India (Satyanarayana, 1971).

goradu (Bombay, India) – old soils of brown colour formed on clayey alluvium (Raychaudhri, 1958).

kallar (northern India) – solonchak with a crust of salts on the surface. Also *pex* (Raychaudhri, 1958).

kankar (India) – dark clayey cracky compact soils, having in the lower part of soil profile a layer of carbonaceous concretions (Rozanov, 1974). Also a name of a dense layer of carbonaceous concretions in the profile. Also *kanhar*.

karail (India) – dark clayey cracky compact soils; the name is used only in the state Uttar-Pradesh (Raychaudhri, 1958).

kari (northern India) – salted peat soils (Raychaudhri, 1958).

karl (northern India) – salted alkaline very deep clayey soils (Raychaudhri, 1958).

khadar (Bengal, India) – a name for soils on young alluvial sediments, usually sandy, without a layer of carbonaceous concretions in the profile (Raychaudhri, 1958).

mar (Bumdelkhand, India) – black compact chernozem-like tropical soils (Raychaudhri, 1958).

matasi (India) – yellow-coloured tropical soils of western India (Satyanarayana, 1971).

pali maati (Bangladesh; Bengali) – silty soil (Ali, 2003).

pali-doash maati (Bangladesh; Bengali) – silt loam soil (Ali, 2003).

parwa (Uttar-Pradesh, India) – brownish-grey soils, having the texture from sandy to clay loam (Raychaudhri, 1958).

rakar (Uttar-Pradesh, India) – mountainous and foothill red-coloured soils, not used in agriculture (Raychaudhri, 1958).

raua (Indonesia) – mangrove soils, remote from the coast and free from excess of moisture (evidently, due to surface uplift), black and loose, suitable for agriculture (Glinka, 1927).

regoor (central India) – black compact chernozem-like tropical soils. The term is used also in Indonesia. Also *regar* and *regada*.

reh (northern India) – solonchak with a crust of salts on the surface. Also *kallar* (Raychaudhri, 1958). Widespread in India and Pakistan on the shore of the Arabian Sea, along the banks of the Indus, and from the gulf of Kutch to Afghanistan (Glinka, 1927). Often contain a horizon of carbonaceous concretions.

sawah (Indonesia, Java) – soils, forming in conditions of excessive moisture on ferrallitic weathered regolith under perennial rice culture. Described by Koenning (cit. by Kanno, 1982). Glinka (1927) pointed to high content of organic matter in these soils (loss on ignition 36 per cent), clayey consistency and the smell of hydrogen sulphide.

tana masan (Malaysia) – acid soil; determined by its taste (Osunade, 1988).

tana payan (Malaysia) – sweet (alkaline) soil; determined by its taste (Osunade, 1988).

tana tawah (Malaysia) – neutral soil; determined by its taste (Osunade, 1988).

tanah balus (Western Java, Indonesia) – fine sand soil of high productivity (Marten and Vityakon, 1986).

tanah berbatu (Western Java, Indonesia) – unproductive soil with high content of fragmental material (Marten and Vityakon, 1986).

tanah beureum (Western Java, Indonesia) – red-coloured unproductive soil (Marten and Vityakon, 1986).

tanah bodas (Western Java, Indonesia) – light-coloured soil of medium productivity (Marten and Vityakon, 1986).

tanah gembur (Western Java, Indonesia) – moist productive soil (Marten and Vityakon, 1986).

tanah gersang (Western Java, Indonesia) – dry unproductive soil (Marten, Vityakon, 1986).

tanah hideung (Western Java, Indonesia) – black productive soil (Marten and Vityakon, 1986).

tanah kasar (Western Java, Indonesia) – coarse sandy unproductive soil (Marten and Vityakon, 1986).

tanah keusik (Western Java, Indonesia) – sandy soil of medium productivity (Marten and Vityakon, 1986).

tanah porang (Western Java, Indonesia) – productive clayey soil (Marten and Vityakon, 1986).

tanah tak berbatu (Western Java, Indonesia) – moderately productive soil without rock fragments (Marten and Vityakon, 1986).

tanah tidak liat (Western Java, Indonesia) – soil, poor in clay, having productivity lower than medium (Marten and Vityakon, 1986).

terai (India) – soils, forming on ancient alluvial sediments of the rivers Ghang and Brahmaputra. Characterized by a pale-brown structureless surface horizon and heavier in texture red-brown carbonaceous compact subsoil. Now almost all these soils are cultivated and are used for rice (Rozov and Stroganova, 1979).

Soil names of America

adobe (North America) – cracky clayey soils of flat alluvial plains of North America. The term was used by Marbut for denominating cracky compacted clayey soils (Rozanov, 1974).

akko (Andes, southern Peru; Quechua) – sandy soils (Sandor and Furbee, 1996).

akko kuchu (Andes, southern Peru; Quechua) – sandy soils of narrow valleys (Sandor and Furbee, 1996).

allin halp'a (Andes, southern Peru; Quechua) – good, productive soils (Sandor and Furbee, 1996).

arcilla (Andes, southern Peru; Spanish) – clayey soils (Sandor and Furbee, 1996). Also *arcilloso*. Since *arcilla* is a common Spanish word, it is used widely in all Latin America.

arenoso (Andes, southern Peru; Spanish) – sandy soils (Sandor and Furbee, 1996). Since *arenoso* is a common Spanish word, it is used widely in all Latin America.

arisco (northeastern Brazil) – sandy soils; soils of light texture, with underlying solid rock (Statishin de Queiroz and Norton, 1992).

baixio (northeastern Brazil) – young alluvial soils. Also *coroa* (Statishin de Queiroz and Norton, 1992).

barro de louça (northeastern Brazil) – soils, having heavy clayey horizon, distinctly differing from overlying horizon in the content of clays (Statishin de Queiroz and Norton, 1992).

barro vermelho (northeastern Brazil) – soils, having red-coloured horizon with features of illuviation clays (Statishin de Queiroz and Norton, 1992).

box lu'um (Mexico, Yucatan; Mayan) – black soils, typical of higher relief areas, with stones of 5–10cm diameter (Bautista et al, 2005).

cak li ch'och' (Guatemala; Quieqchu) – rather productive red-coloured soil (Marten and Vityakon, 1986).

calcario (Andes, southern Peru; Spanish) – carbonaceous soils (Sandor and Furbee, 1996).

caliche (Spanish) – folk name of thick horizons of accumulation of powdery lime in desert soils (Rozanov, 1974). The name is used throughout Latin America, from Mexico to Chile.

cascajal (Andes, southern Peru; Spanish) – stony, gravel-rich soils (Sandor and Furbee, 1996).

cascajo (Cuba; Spanish) – stony soils (Bennett and Allison, 1928).

cau ru li ch'och' (Guatemala; Quieqchu) – unproductive soil with hard surface (Marten and Vityakon, 1986).

cerrado (Brazil; Spanish) – a type of vegetation widespread in Brazil, also related to exhausted soils of ancient surfaces. Now in Brazil these soils are distinguished according to their profile. The term was used in FAO World Soil Map legend on the phase level, but then deleted in 1990.

chahuamai (Peru; Shipobo) – clayey, smearing when moist, 'dirty' soils on non-flooded plots of alluvial plain of river Ukayali and its most important tributaries (Behrens, 1989).

chak lu'um (Mexico, Yucatan; Mayan) – deep red soils (Bautista et al, 2005).

chaki ch'och' (Guatemala; Quieqchu) – dry soil, productive in the rainy season (Marten and Vityakon, 1986).

chaltún (Mexico, Yucatan; Mayan) – red, reddish-brown or black shallow soils of depressions with high amount of stone fragments in karstic areas of Yucatan (Bautista et al, 2005).

chaqua halp'a (Cohabamba, Bolivia) – sandy soils (Zimmerer, 1994).

charanda (Mexico; Michoacán state) – reddish-brown clay (Williams and Ortiz-Solorio, 1981). The name is the same as for local strong alcoholic drink made of sugar cane, somewhat similar in colour to these soils.

charral (Spanish) – Latin American name for xerophitic forests and soils of these landscapes. In a broad sense was used by Wilde (1946) as a term characterizing all soils of dry forests, similar in properties to savanna soils. In the older American classification (Baldwin et al, 1938) these soils were called 'non-carbonaceous brown soils'. Wilde considered as varieties of charral soils such soil types as *terra rossa*, *ferretto* and *finbush*, i.e. understanding the term more in its zonal meaning, as 'soils of xerophytic forests', which is closer to folk understanding of these soils.

chich lu'um (Mexico, Yucatan; Mayan) – soils with abundant gravels (Bautista et al, 2005).

ch'och'ol (Mexico, Yucatan; Mayan) – black soils, typical of higher relief areas, with abundant rock outcrops and coarse fragments of 5cm or more (Bautista et al, 2005).

chorishmai (Peru; Shipobo) – compact, hard soils on non-flooded plots of alluvial plain of river Ukayali and its most important tributaries (Behrens, 1989).

coroa (northeastern Brazil) – young alluvial soils. Also *baixio* (Statishin de Queiroz and Norton, 1992).

cuacab li ch'och' (Guatemala; Quieqchu) – black and brownish-black clayey soils, hardening in dry season (Marten and Vityakon, 1986).

echeri ambakiti (Mexico, Michoacán; Purhépecha) – fertile soil (Zinck and Barrera-Bassols, 2002).

echeri charakirhu (Mexico, Michoacán; Purhépecha) – gravelly soil (Zinck and Barrera-Bassols, 2002).

echeri charanda (Mexico, Michoacán; Purhépecha) – clayey soil (Zinck and Barrera-Bassols, 2002).

echeri charapiti (Mexico, Michoacán; Purhépecha) – reddish soil (Zinck and Barrera-Bassols, 2002).

echeri choperi (Mexico, Michoacán; Purhépecha) – hard soil (Zinck and Barrera-Bassols, 2002).

echeri cuatapiti (Mexico, Michoacán; Purhépecha) – loose soil (Zinck and Barrera-Bassols, 2002).

echeri jauamiti (Mexico, Michoacán; Purhépecha) – deep soil (Zinck and Barrera-Bassols, 2002).

echeri jorhepiti (Mexico, Michoacán; Purhépecha) – warm soil (Zinck and Barrera-Bassols, 2002).

echeri karishiri (Mexico, Michoacán; Purhépecha) – dry soil (Zinck and Barrera-Bassols, 2002).

echeri kurhunda (Mexico, Michoacán; Purhépecha) – multi-layered soil; also alluvial soil (Zinck and Barrera-Bassols, 2002).

echeri kutzari (Mexico, Michoacán; Purhépecha) – sandy soil (Zinck and Barrera-Bassols, 2002).

echeri poksinda (Mexico, Michoacán; Purhépecha) – soil with clods (Zinck and Barrera-Bassols, 2002).

echeri querekua (Mexico, Michoacán; Purhépecha) – sticky soil (Zinck and Barrera-Bassols, 2002).

echeri sahuapiti (Mexico, Michoacán; Purhépecha) – shallow soil (Zinck and Barrera-Bassols, 2002).

echeri spambiti (Mexico, Michoacán; Purhépecha) – yellowish soil (Zinck and Barrera-Bassols, 2002).

echeri terendani (Mexico, Michoacán; Purhépecha) – decomposed litter (Zinck and Barrera-Bassols, 2002).

echeri tshirapiti (Mexico, Michoacán; Purhépecha) – cold soil (Zinck and Barrera-Bassols, 2002).

echeri tsuruani (Mexico, Michoacán; Purhépecha) – single-layered soil; also eroded or washed soil (Zinck and Barrera-Bassols, 2002).

echeri tupuri (Mexico, Michoacán; Purhépecha) – silty or powdery soil (Zinck and Barrera-Bassols, 2002).

echeri turipiti (Mexico, Michoacán; Purhépecha) – dark or black soil (Zinck and Barrera-Bassols, 2002).

echeri uekandirini (Mexico, Michoacán; Purhépecha) – moist soil (Zinck and Barrera-Bassols, 2002).

echeri urapiti (Mexico, Michoacán; Purhépecha) – whitish soil (Zinck and Barrera-Bassols, 2002).

echeri zacapendini (Mexico, Michoacán; Purhépecha) – stony soil (Zinck and Barrera-Bassols, 2002).

greda (Andes, southern Peru; Spanish) – clayey soils (Sandor and Furbee, 1996). Also *gredosa*.

gumbo (Cuba) – clayey dark-coloured carbonaceous, sometimes salted in lower horizons soils with features of slitization, often cracking when dried (Bennett and Allison, 1928). Also *tierra masa*.

ha'ru li ch'och' (Guatemala; Quieqchu) – very productive flooded soil (Marten and Vityakon, 1986).

hay lu'um (Mexico, Yucatan; Mayan) – red and reddish-brown shallow soils of depressions in karstic areas of Yucatan (Bautista et al, 2005).

huachumai (Peru; Shipobo) – soft, loose soils on non-flooded plots of alluvial plain of river Ukayali and its most important tributaries (Behrens, 1989).

k'an li ch'och' (Guatemala; Quieqchu) – rather productive yellow soil (Marten and Vityakon, 1986).

k'ankab (Mexico, Yucatan; Mayan) – red relatively deep (10–50cm) soils of depressions in karstic areas of Yucatan (Bautista et al, 2005).

k'ek li ch'och' (Guatemala; Quieqchu) – productive black soil (Marten and Vityakon, 1986).

k'elli halp'a (Cohabamba, Bolivia) – light-brownish-yellow soil (Zimmerer, 1994).

k'illi halp'a (Andes, southern Peru; Quechua) – yellow-coloured soils and sediments (Sandor and Furbee, 1996).

k'un ru li ch'och' (Guatemala; Quieqchu) – productive soils with a loose surface horizon (Marten and Vityakon, 1986).

kuntayu (Andes, southern Peru; Quechua) – white soils and sediments, mainly of volcanic origin (Sandor and Furbee, 1996).

lima (Cohabamba, Bolivia) – silty soils (Zimmerer, 1994). The name is a common Spanish word and is used widely in all Latin America.

limosa (Andes, southern Peru; Spanish) – silty soils (Sandor and Furbee, 1996). The name is a common Spanish word and is used widely in all Latin America.

llamp'u (Cohabamba, Bolivia) – loamy soils (Zimmerer, 1994).

llank'i (Cohabamba, Bolivia) – clayey soil (Zimmerer, 1994).

llink'i (Andes, southern Peru; Quechua) – clay, clayey soil (Sandor and Furbee, 1996).

llink'i amarillo (Andes, southern Peru; Quechua/Spanish) – clayey soil of yellow colour (Sandor and Furbee, 1996).

llink'i blanko (Andes, southern Peru; Quechua/Spanish) – clayey soil of white colour (Sandor and Furbee, 1996).

llink'i negro (Andes, southern Peru; Quechua/Spanish) – clayey soil of black colour (Sandor and Furbee, 1996).

llink'i rojo (Andes, southern Peru; Quechua/Spanish) – clayey soil of yellow colour (Sandor and Furbee, 1996).

llink'i sallakko (Andes, southern Peru; Quechua) – clayey soil with gravel (Sandor and Furbee, 1996).

llink'i tuqraq (Andes, southern Peru; Quechua) – clay, clayey soil with hard clods (Sandor and Furbee, 1996).

maikon (Peru; Shipobo) – soil (in broad sense), loamy soil on non-flooded plots of alluvial plain of river Ukayali and its most important tributaries (Behrens, 1989).

mapumai (Peru; Shipobo) – clayey soils on non-flooded plots of alluvial plain of river Ukayali and its most important tributaries (Behrens, 1989).

mashimai (Peru; Shipobo) – sandy soils on non-flooded plots of alluvial plain of river Ukayali and its most important tributaries (Behrens, 1989).

massape (northeastern Brazil) – red-coloured and black clayey soils, cracking when dried. Also *coroa* (Statishin de Queiroz and Norton, 1992).

melb ru li ch'och' (Guatemala; Quieqchu) – soils, covered with cracky clayey layer (Marten and Vityakon, 1986).

mero cak li ch'och' (Guatemala; Quieqchu) – very productive reddish-brown ('semi-red') soil (Marten and Vityakon, 1986).

mero k'an li ch'och' (Guatemala; Quieqchu) – rather productive brownish-yellow ('semi-yellow') soil (Marten and Vityakon, 1986).

mero k'ek li ch'och' (Guatemala; Quieqchu) – rather productive brownish-black ('semi-black') soil (Marten and Vityakon, 1986).

mulatto (Cuba) – a local name for humus-saturated tropical soils (Bennett and Allison, 1928). A synonym – *tierra negro-clara*, is cited by the same authors.

mu'ru li ch'och' (Guatemala; Quieqchu) – soils with very loose surface horizon, composed mainly of decayed organic matter (Marten and Vityakon, 1986).

negro (Cuba; Spanish) – black, well humified soils, forming in uplands. Usually have no traces of slitization like Vertisols (Bennett and Allison, 1928).

njore' li ch'och' (Guatemala; Quieqchu) – soils, forming cracks when dried (Marten and Vityakon, 1986).

nut'u akko (Andes, southern Peru; Quechua) – fine sand, loamy sand soil (Sandor and Furbee, 1996).

paramos (Bolivia, Peru, Columbia, Chile; Spanish) – a name of mountainous landscapes of southern Andes, also including the soils. The soils usually are dark-coloured humified profiles, formed on volcanic ash and tuffs.

pec ru li ch'och' (Guatemala; Quieqchu) – soils with stony surface; this group includes soils where stones on the surface have white and blue colour, and covered with wood or shrub vegetation (Marten and Vityakon, 1986).

pedregal (Andes, southern Peru; Spanish) – stony soils with fragments of rocks (Sandor and Furbee, 1996). The name is a common Spanish word and is used widely in all Latin America.

peña (Andes, southern Peru; Spanish) – rock; soils with a duripan, hard cemented horizon (Sandor and Furbee, 1996).

playa (in Spanish – sea beach, flat place) – a name for desert clayey cracky soils like *takyr*, used in Mexico and some southern states of the US (Rozanov, 1974). Mostly used for soils formed on the bottom of dried lakes.

pok ch'och' (Guatemala; Quieqchu) – partially weathered limestone (Marten and Vityakon, 1986).

pradera arenosa (Uruguay; Spanish) – sandy unproductive soils (Bramao and Lemos, 1961).

pradera negra (Uruguay; Spanish) – dark-coloured clayey, usually compact soils (Bramao and Lemos, 1961).

puka halp'a (Cohabamba, Bolivia) – red-coloured soil (Zimmerer, 1994).

pus lu'um (Mexico, Yucatan; Mayan) – soft black earth (Bautista et al, 2005).

qaqa (Andes, southern Peru; Quechua) – rock; weakly developed soil on rock (Sandor and Furbee, 1996).

qhilli (Andes, southern Peru; Quechua) – soil, dusty when dry (Sandor and Furbee, 1996).

q'ulp'a (Andes, southern Peru; Quechua) – a special type of soils, used as food by local population; it is believed that soil is used as a sorbent for phytotoxins abundant in local food (Sandor and Furbee, 1996).

rumiyakk (Andes, southern Peru; Quechua) – rock or soil, hard as rock (Sandor and Furbee, 1996).

sab ru li ch'och' (Guatemala; Quieqchu) – bog, permanently excessively moist soil (Marten and Vityakon, 1986).

sactun ch'och' (Guatemala; Quieqchu) – yellow and grey clayey soils (Marten and Vityakon, 1986).

sak li ch'och' (Guatemala; Quieqchu) – white soil, if loose, relatively productive; if hard, practically useless (Marten and Vityakon, 1986).

samahi'ru li ch'och' (Guatemala; Quieqchu) – productive sandy and loamy soils (Marten and Vityakon, 1986).

seb ru li ch'och' (Guatemala; Quieqchu) – yellow and red clayey soils, hardening when dried (Marten and Vityakon, 1986).

sonsocuite (Nicaragua; Spanish) – clayey dark-coloured soil with features of slitization (Dudal and Bramao, 1965).

sulul li ch'och' (Guatemala; Quieqchu) – imperfectly drained soil with mud upper horizon (Marten and Vityakon, 1986).

tepetate (Mexico; Nahuatl) – cultivated soil of uplands, where large hard aggregates were broken in the course of cultivation (Williams, 1975). In other regions of Mexico the name is used for outcrops of relatively soft sedimentary rocks or cemented soil horizons exposed by erosion (Williams and Ortiz-Solorio, 1981). In scientific literature the term is used for cemented layers in volcanogenic soils and sediments or as a synonym for any indurated soil layer.

tequisquitl (Mexico; Nahuatl) – infertile soils of nitritic salinization. The name is preserved by Nahuatl Indians; it was used in the developed classification of soils of the Aztecs (Williams, 1975). Also *tequixquitlalli*.

terra branca (Brazil; Portuguese – white earth) – a name of leached and forested ferralitic soils of Brazil, having light surface horizons (Glinka, 1927); soils, having heavy clayey horizon, distinctly different from overlying layer in clay content, bleached (Statishin de Queiroz and Norton, 1992).

terra catanduva (Brazil; Portuguese) – a name for clayey tropical soils, rich in iron (though the iron oxides content is less than in ferralitic soils like *terra roxa*) (Glinka, 1927).

terra roxa (Brazil; Portuguese) – purple-red ferralitic soils, forming in the regoliths of basalts of the Brazilian plateau (Rozov and Stroganova, 1979). These soils are characterized by granular structure of the upper horizon (Bramao and Lemos, 1961).

tierra amarilla (Mexico, Guatemala; Spanish) – yellow soils (Williams and Ortiz-Solorio, 1981).

tierra arenosa (Mexico; Spanish) – sandy soils (Williams and Ortiz-Solorio, 1981).

tierra arenosa delgada (Mexico; Spanish) – fine-grained sandy soils (Williams and Ortiz-Solorio, 1981).

tierra arenosa gruesa (Mexico; Spanish) – coarse-grained sandy soils (Williams and Ortiz-Solorio, 1981).

tierra baya (Mexico; Spanish) – red-brown sandy loam soils (Bellon and Taylor, 1993).

tierra blanca (Mexico; Spanish) – light (white) soils (Williams and Ortiz-Solorio, 1981).

tierra café (Mexico; Spanish) – chestnut soils (Williams and Ortiz-Solorio, 1981).

tierra café oscura (Mexico; Spanish) – dark-chestnut soils (Williams and Ortiz-Solorio, 1981).

tierra canela (Mexico; Spanish) – cinnamon soils (Williams and Ortiz-Solorio, 1981).

tierra cascajo (Cuba; Spanish) – loamy soil with reddish-brown illuvial horizon, containing fragments of quartz and limestone 5–10cm in diameter (Bennett and Allison, 1928). Also *tierra de cascajo bermejo.*

tierra cascajosa (Mexico; Spanish) – carbonaceous, strongly stony loamy sand soils, having low productivity (Bellon and Taylor, 1993).

tierra colorada (Mexico; Spanish) – red-coloured sandy loamy soils (Bellon and Taylor, 1993).

tierra colorada-arenosa (Mexico; Spanish) – red-coloured loamy sand and sandy loam soils (Bellon and Taylor, 1993).

tierra de barro (Mexico; Spanish) – clayey soils (Williams and Ortiz-Solorio, 1981).

tierra de barro amarillo (Mexico; Spanish) – yellow clayey soils (Williams and Ortiz-Solorio, 1981).

tierra de barro blanco (Mexico; Spanish) – white clayey soils (Williams and Ortiz-Solorio, 1981).

tierra de barro colorado (Mexico; Spanish) – red clayey soils (Williams and Ortiz-Solorio, 1981).

tierra de barro pardo (Mexico; Spanish) – dark clayey soils (Williams and Ortiz-Solorio, 1981).

tierra masa (Cuba; Spanish) – clayey dark-coloured carbonaceous, sometimes salted in lower horizons soil with features of slitization, often cracking when dried (Bennett and Allison, 1928). Also *gumbo.*

tierra migajón (Mexico; Spanish) – plastic soils (Williams and Ortiz-Solorio, 1981).

tierra negra (Latin America; Spanish) – in Mexico – black loamy soil with high content of organic matter, distinguished by high productivity (Bellon and Taylor, 1993). In some other countries – clayey dark-coloured soils with features of slitization (Rozov and Stroganova, 1979).

tierra negra-clara (Cuba; Spanish, literally: black-white earth; it means 'earth of colour of mulatto skin, of a person whose parents were negro and white') – a local name of humus-saturated tropical soils (Bennett and Allison, 1928). Also *mulatto.*

tierra parda (Mexico; Spanish) – brown soils (Williams and Ortiz-Solorio, 1981).

tierra preta (Brazil, Peru; Spanish) – a South American name for ancient cultivated soils with a thick plough horizon, left by the ancient Indian civilizations (Rozov and Stroganova, 1979).

tierra tupida (Andes, southern Peru; Spanish) – thick, dark-coloured productive soils (Sandor and Furbee, 1996).

trumao (Chile, Argentina; Spanish) – in Russian geographic literature it is a name for volcanic layered ash soils, usually having ochreous layers

(Rozov and Stroganova, 1979). According to the information of Chilean pedologists the name is used for any volcanic soil.

tsek'el (Mexico, Yucatan; Mayan) – black soils with very little fine earth, typical of higher relief areas; bedrock outcrops take the form of promontories (Bautista et al, 2005).

tullu halp'a (Cohabamba, Bolivia) – 'bony soil', soil, where fragments of underlying rocks are exposed by erosion (Zimmerer, 1994).

t'upuri (Mexico; Otomi) – loam with a significant content of fine sand (Williams and Ortiz-Solorio, 1981).

tzakal cak li ch'och' (Guatemala; Quieqchu) – unproductive 'very red' soil (Marten and Vityakon, 1986).

tzatzalum (Guatemala; Quieqchu) – soil with dense roots, usually recently cultivated (Marten and Vityakon, 1986).

ushpa halp'a (Cohabamba, Bolivia) – soil of grey colour (Zimmerer, 1994).

uspa halp'a (Andes, southern Peru; Quechua) – soils, having colour of ash (Sandor and Furbee, 1996).

yana halp'a (Cohabamba, Bolivia) – black soil (Zimmerer, 1994).

yanaq halp'a (Andes, southern Peru; Quechua) – black soils (Sandor and Furbee, 1996).

yeso (North America) – a name for gypsiferous crusts in state of New Mexico (Glinka, 1927). When dry these crusts are dense and hard, but when moist turn soft and water-permeable. The name is a common Spanish word (gypsum); it is widely used in all the arid regions of the continent.

yuraq halp'a (Andes, southern Peru; Quechua) – white soils (Sandor and Furbee, 1996).

zoguitl (Mexico; Nahuatl) – clayey soil. The name is preserved by Nahuatl Indians; it was used in the developed classification of soils of the Aztecs (Williams, 1975).

Soil names of northern Africa

badobe (Sudan) – dark compacted cracky soils (Rozanov, 1974).

bile-dian (Sudan) – loamy clay weakly cracking soils (also *mursi*), having carbonaceous concretions several centimetres in diameter. Also *sorgon* (Aubert, 1949).

boi (Sudan) – clayey soils, not forming cracks when dried (Aubert, 1949).

danga (Sudan) – soils of the texture approximately equivalent to sandy loam (Aubert, 1949).

danga-ble (Sudan) – sandy loamy soils of rusty or red colour (Aubert, 1949).

danga-fing (Sudan) – sandy loamy soils of dark-brown or dark-red colour (Aubert, 1949).

dian (Sudan) – soils of loam and clayey texture, containing not less than one-third sandy material (Aubert, 1949).

dian-pere (Sudan) – clayey compact soils, cracking when dried (Aubert, 1949).

erg (Arabian) – a name for sandy desert in northern Africa, usually situated in a big depression (Rozanov, 1974).

feh (northern Africa) – a name for soils of clayey-stony or sandy deserts (Rozanov, 1974).

mursi (Mali, Sudan) – dark clayey cracky compact soils (Rozanov, 1974).

regh (Algeria) – stony gravel desert in Sahara.

sebkcha (northern Africa) – solonchaks with thick salt surface crusts, turbated with the formation of earthy-salt large clods on the surface (Rozanov, 1974).

seno (Sudan) – sandy soils, containing significant admixture of silty material (Aubert, 1949).

serir (northern Africa) – stony desert of lowland regions of Sahara; their surface is covered with gravel of dense rocks, with underlying compressed sand or sandstone (Rozanov, 1974).

shott (Arabian) – a common name in northern Africa for solonchaks of closed depressions, with a bottom covered with a loose layer of salts, turning into salted lake after rain (Rozanov, 1974).

sorgon (Sudan) – loamy clay weakly cracking soils (also *mursi*), having carbonaceous concretions several centimetres in diameter. Also *bile-dian* (Aubert, 1949).

teen suda (Sudan) – dark clayey cracky compact soils (Rozanov, 1974).

tien-tien (Sudan) – sandy soils (Aubert, 1949).

tirs (Algeria, Tunis, Morocco) – dark clayey cracky compact soils (Rozanov, 1974).

Soil names of West Africa

abata (Nigeria; Yoruba) – clayey, permanently flooded soil (Osunade, 1988). Also *ere*, *erofo* and *potopoto*.

alaadun (Nigeria; Yoruba) – loamy soil (Osunade, 1988).

alaadun dudu (Nigeria; Yoruba) – dark loamy soil (Osunade, 1988).

alaadun pupa (Nigeria; Yoruba) – reddish-brown loamy soil (Osunade, 1988).

amo (Nigeria; Yoruba) – heavy clayey, practically infertile soil (Warren, 1992).

baboatandi (Burkina Faso; Mossi) – fertile soil formed on old termite mounds (Niemeijer and Mazzucato, 2003). Also *tuuli*.

baldiol (Senegal; Pulaar) – cultivated clayey soils along drains, periodically flooded (Tabor, 1993). Also *lakhe*.

balogili (Burkina Faso; Gourmantché) – flood plain soil on land bordering riverbed (Niemeijer and Mazzucato, 2003). Also *buanbalgu*.

bere (Senegal; Bambara) – stony soils, with a great quantity of gravel (Tabor, 1993). Also *koche*.

bibiay dodo (Côte d'Ivoire; Bété) – clayey soil of various colours with shiny minerals, used for health supplement (Birmingham, 2003). Also *sini dodo*.

bissiga (Burkina Faso; Mossi) – sandy soil (Niemeijer and Mazzucato, 2003).

bla dodo (Côte d'Ivoire; Bété) – kaolinitic white soil, used mainly for medicinal purposes and for body decoration (Birmingham, 2003).

blo dodo (Côte d'Ivoire; Bété) – sandy soil with some clay (Birmingham, 2003). Also *soko dodo*.

boali (Burkina Faso; Gourmantché) – sticky clayey soil (Niemeijer and Mazzucato, 2003).

bole (Nigeria; Yoruba) – clayey soil (Osunade, 1988).

bolé (Burkina Faso; Mossi) – sticky clayey soil (Niemeijer and Mazzucato, 2003).

bole alaadun (Nigeria; Yoruba) – heavy loamy soil (Osunade, 1988).

bole dudu (Nigeria; Yoruba) – dark heavy loamy soil (Osunade, 1988).

bole funfun (Nigeria; Yoruba) – light-coloured clayey soil (Osunade, 1988).

bole olokuta (Nigeria; Yoruba) – stony clayey soil (Osunade, 1988).

bole pupa (Nigeria; Yoruba) – reddish-brown clayey soil (Osunade, 1988).

bolle (Burkina Faso; Mossi) – clayey soil (Dialla, 1993).

bossay dodo (Côte d'Ivoire; Bété) – sandy soil (Birmingham, 2003). Also *yoromay dodo*.

bossay kpebeu (Côte d'Ivoire; Bété) – black sandy soil, slightly muddy (Birmingham, 2003).

bossay pépé (Côte d'Ivoire; Bété) – white fine sandy soil (Birmingham, 2003). Also *bossay popo*.

bossay popo (Côte d'Ivoire; Bété) – white fine sandy soil (Birmingham, 2003). Also *bossay pépé*.

bossay zéro (Côte d'Ivoire; Bété) – red sandy soil (Birmingham, 2003).

botogo (Niger; Djerma) – hard, compacted and heavy to work dark brown soil (Osbahr and Allan, 2003).

bugri (Burkina Faso; Mossi) – clayey soft, convenient for cultivation soil (Dialla, 1993).

buoogo (Burkina Faso; Mossi) – loamy soil, usually forming in depressions, close to water (Dialla, 1993).

byisri (Burkina Faso; Mossi) – sandy soil (Dialla, 1993).

bys-miuugu (Burkina Faso; Mossi) – red-coloured sandy soil (Dialla, 1993).

bys-sabille (Burkina Faso; Mossi) – sandy black soil (Dialla, 1993).

changoul (Senegal; Pulaar) – silty alluvial soils, periodically flooded by river waters for a short period of time (Tabor, 1993). Also *ko* and *louid*.

dagre (Burkina Faso; Mossi) – compact clayey soil (Dialla, 1993).

dek (northwestern and central Senegal) – fine sandy soils with high content of silty material (Aubert, 1949).

dek-dior (northwestern and central Senegal) – soils, transitional in texture between *dek* and *dior* (Aubert, 1949).

dergia (Senegal; Hassania) – eroded soils, on the surface of which water-impermeable crust forms (Tabor, 1993). Also *karan karan* and *karawal*.

dieri (northern Senegal) – non-flooded sandy soils (Aubert, 1949), red-coloured, non-carbonaceous, usually connected with products of weathering of Paleogenic sandstones (Rozov and Stroganova, 1979).

diezra dodo (Côte d'Ivoire; Bété) – red sticky upland soils (Birmingham, 2003). Also *doudjra dodo*.

dior (north-western and central Senegal) – coarse sandy soils (Aubert, 1949).

dodo kpebeu (Côte d'Ivoire; Bété) – black soil rich in organic matter, with earthworm castes (Birmingham, 2003). Also *serebli kagbo dodo*.

doudjra dodo (Côte d'Ivoire; Bété) – red sticky upland soils (Birmingham, 2003). Also *diezra dodo*.

dougoukoulo fing (Senegal; Bambara) – hydromorphic sandy or sandy loamy soils on alluvial sediments (Tabor, 1993). In the languages of other people of Senegal, sandy and loamy soils are divided, e.g. *faalo* and *salka*.

faalo (Senegal; Pulaar) – hydromorphic loamy sand or sandy loamy soils on alluvial sediments (Tabor, 1993). Also *folo*, *lou wat* and *dougoukoulo fing*.

fara (Senegal; Bambara) – hydromorphic clayey soils with concretions and grey mottles, also clayey compact soils of different extent of hydromorphism, including alluvial ones (Tabor, 1993). In the languages of the other peoples of Senegal, these soils are divided, e.g. *faro*, *khare* and *kolanga*.

faro (Senegal; Soninke) – hydromorphic clayey soils with concretions and grey mottles (Tabor, 1993). Also *fara* and *legrarra*.

firki (Nigeria) – dark clayey cracky compact soils (Rozanov, 1974).

folo (Senegal; Soninke) – hydromorphic loamy sand or sandy loamy soils on alluvial sediments (Tabor, 1993). Also *faalo*, *lou wat* and *dougoukoulo fing*.

fonde (northern Senegal; Soninke, Pulaar) – periodically flooded loamy soils of the valley of River Senegal (Aubert, 1949).

fouga (Senegal; Bambala) – dry soils of uplands, unsuitable for agriculture; after rain turn into wet mud (Tabor, 1993). Chemical analyses showed a high content of exchangeable sodium in these soils. Also *seybo*.

fuaanu (Burkina Faso; Gourmantché) – ephemeral drainage channel soil (Niemeijer and Mazzucato, 2003).

gangani (Niger; Djerma) – hard, highly compacted red soil (Osbahr and Allan, 2003).

gede (Senegal; Soninke) – shallow soils on outcrops of solid rocks (Tabor, 1993). Also *koulou*.

gnailgay (Côte d'Ivoire; Senufo) – gravelly, mostly red soils (Birmingham, 2003).

gnignam bine (Senegal; Soninke) – short flooded sandy alluvial soils (Tabor, 1993). Also *salka* and *trab khakale*.

gouroumbe (Senegal; Soninke) – strongly eroded soil with features of gulley erosion (Tabor, 1993).

holalde (Senegal; Pulaar) – clayey soils of the valley of River Senegal, flooded from July to October (Aubert, 1949). Tabor (1993) defines these soils as alluvial clayey compact soil with features of hydromorphism, similar to *fara*, *khare* and *kolanga*.

ile du (Nigeria; Yoruba) – dark, rich in humus soil (Warren, 1992).

ile funfun (Nigeria; Yoruba) – bleached, strongly leached soil (Warren, 1992).

ile gamo (Nigeria; Yoruba) – silver-grey clayey soil, rich in residues of mica shales; the soil is considered by local population as unsuitable for agriculture, but is used as a cementing material (Warren, 1992).

ile olokuta (Nigeria; Yoruba) – stony soil (Warren, 1992). Also *wokuta*.

karan karan (Senegal; Soninke, Bambara) – eroded soils, on the surface of which water-impermeable crust forms (Tabor, 1993). Also *dergia* and *karawal*.

karawal (Senegal; Pulaar) – eroded soils, on the surface of which water-impermeable crust forms (Tabor, 1993). Also *karan karan* and *dergia*.

katamangna (Senegal; Hassania) – dry soils of uplands, unsuitable for agriculture (Tabor, 1993). Also *katawal* and *katamangue*.

katamangue (Senegal; Soninke, Bambara) – dry soils of uplands, unsuitable for agriculture (Tabor, 1993). Also *katawal* and *katamangna*.

katawal (Senegal; Pulaar) – dry soils of uplands, unsuitable for agriculture (Tabor, 1993). Also *katamangue* and *katamangna*.

khare (Senegal; Soninke) – alluvial hydromorphic clayey compact soils (Tabor, 1993). Also *holalde* and *fara*.

ko (Senegal; Soninke) – silty alluvial soils, periodically flooded by river waters for a short period of time (Tabor, 1993). Also *changoul* and *louid*.

koche (Senegal; Soninke) – stony soils, with a great quantity of gravel (Tabor, 1993). Also *bere*.

kolanga (Senegal; Soninke) – alluvial clayey compact soil with features of hydromorphism (Tabor, 1993). Also *holalde* and *fara*.

korabanda (Niger; Djerma) – black topsoil horizon in any soil (Osbahr and Allan, 2003).

kossogo (Burkina Faso; Mossi) – ephemeral drainage channel soil (Niemeijer and Mazzucato, 2003).

koulou (Senegal; Bambala) – shallow soils on outcrops of solid rocks (Tabor, 1993). Also *gede*.

kpahi dodo (Côte d'Ivoire; Bété) – red, hard impenetrable laterite soil (Birmingham, 2003).

kpankpatanbuagu (Burkina Faso; Mossi) – saline stony soil (Niemeijer and Mazzucato, 2003). Also **kpamkpagu**.

kpay dodo (Côte d'Ivoire; Bété) – gravelly soils (Birmingham, 2003). Also *tekpay dodo*.

kyongo (Burkina Faso; Mossi) – moist clayey soil, easily cultivated (Dialla, 1993).

labu bi (Niger) – black soil, containing relatively much organic matter and distinguished by high productivity (Lamers et al, 1995).

labu cheri (Niger) – red-coloured soil, distinguished by low productivity, considered by the local people as a product of degradation of *labu biri* and *labu kware* (Lamers et al, 1995).

labu kwarey (Niger) – white, leached soil; the local people consider that it forms as a result of degradation of *labu biri* in the course of cultivation or due to erosion (Lamers et al, 1995).

lakhe (Senegal; Soninke) – cultivated clayey soils along drains, periodically flooded (Tabor, 1993). Also *rakhe* and *baldiol*.

legrarra (Senegal; Hassania) – hydromorphic clayey soils with concretions and grey mottles (Tabor, 1993). Also *fara* and *faro*.

lou wat (Senegal; Hassania) – silty alluvial soil, rather seldom flooded, or non-flooded hydromorphic soil on alluvial loams (Tabor, 1993). Also *wallere*.

louid (Senegal; Hassania) – silty alluvial soils, periodically flooded by river waters for a short period of time (Tabor, 1993). Also *changoul* and *ko*.

magnay dodo (Côte d'Ivoire; Bété) – clay soil, used for pottery (Birmingham, 2003).

makpi dodo (Côte d'Ivoire; Bété) – clay soil without gravel (Birmingham, 2003).

makpi kpebeu (Côte d'Ivoire; Bété) – black clay soil (Birmingham, 2003).

makpi pépé (Côte d'Ivoire; Bété) – white clay soil (Birmingham, 2003).

makpi zéro (Côte d'Ivoire; Bété) – red clay soil (Birmingham, 2003).

naare (Burkina Faso; Mossi) – moist clayey soil, easily cultivated (Dialla, 1993).

narawalle (Senegal; Soninke, Pulaar) – soils of uplands with sandy surface horizon, under which there is an illuvial, clay-enriched horizon (Tabor, 1993).

nirakata (Senegal; Soninke) – soils of uplands with sandy, gravel- and stone-enriched surface horizon, under which there is an illuvial, clay-enriched horizon (Tabor, 1993).

ñinmoali (Burkina Faso; Gourmantché) – soil with impermeable layer (Niemeijer and Mazzucato, 2003). Also *pugu*.

paplay dodo (Côte d'Ivoire; Bété) – black mud (Birmingham, 2003). Also *paplay kpebeu*.

paplay kpebeu (Côte d'Ivoire; Bété) – black mud (Birmingham, 2003). Also *paplay dodo*.

parawal (Senegal; Pulaar) – soils of steep slopes with sandy surface horizon, under which there is an illuvial, clay-enriched horizon (Tabor, 1993). Also *parawalle*.

penpeli (Burkina Faso; Gourmantché) – denuded soil where water does not infiltrate because of soil crusting and hard setting (Niemeijer and Mazzucato, 2003). Also *penpelgu*, *penpeligu* and *pempelgu*.

pugu (Burkina Faso; Gourmantché) – soil with impermeable layer (Niemeijer and Mazzucato, 2003). Also *ñinmoali*.

rassempouéga (Burkina Faso; Gourmantché) – the word signifies 'bald' and refers to soils with little or no vegetation, mostly, but not always, because the soil is shallow (Niemeijer and Mazzucato, 2003).

salka (Senegal; Pulaar) – sandy alluvial soils, flooded for a short period (Tabor, 1993). Also *gnignam bine* and *trab khakale*.

seeno (Senegal; Pulaar) – sandy soils, forming on dunes and on the transported derivates of sandstones and quartzites (Tabor, 1993). Also *singue* and *trab beyda*.

serebli kagbo dodo (Côte d'Ivoire; Bété) – black soil rich in organic matter, with earthworm casts (Birmingham, 2003). Also *dodo kpebeu*.

seybo (Senegal; Soninke) – dry soils of uplands, unsuitable for agriculture; after rain turn into wet mud (Tabor, 1993). Chemical analyses showed a high content of exchangeable sodium in these soils. Also *fouga*.

singue (Senegal; Soninke) – sandy soils, forming on dunes and on the transported derivates of sandstones and quartzites (Tabor, 1993). Also *seeno* and *trab beyda*.

sini dodo (Côte d'Ivoire; Bété) – clayey soil of various colours with shiny minerals, used for health supplement (Birmingham, 2003). Also *bibiay dodo*.

soko dodo (Côte d'Ivoire; Bété) – sandy soil with some clay (Birmingham, 2003). Also *blo dodo*.

taberay (Côte d'Ivoire; Senufo) – very soft, mostly light-coloured soils (Birmingham, 2003).

tanbiima (Burkina Faso; Gourmantché) – sandy soil (Niemeijer and Mazzucato, 2003).

tanbuogu (Burkina Faso; Mossi) – fertile soil formed in depression created by digging up clay for bricks to use in construction (Niemeijer and Mazzucato, 2003). Also *timbengu*.

tancagu (Burkina Faso; Gourmantché) – red gravelly soil (Niemeijer and Mazzucato, 2003). Also **tanpkiaku**, **tacadigu** and **tintancaga**.

tanyeray (Côte d'Ivoire; Senufo) – intensively red upland soils (Birmingham, 2003).

taraa (Nigeria; Yoruba) – soil composed mainly of gravel (Osunade, 1988). Also **taara** (Warren, 1992).

tassi (Niger; Djerma) – loose sandy soil (Osbahr and Allan, 2003).

taworo (Côte d'Ivoire; Senufo) – dark, soft, earthy soils (Birmingham, 2003).

tchin tchin (Senegal; Bambala) – sandy soils, both alluvial and occurring on uplands (Tabor, 1993).

tekpay dodo (Côte d'Ivoire; Bété) – gravelly soils (Birmingham, 2003). Also *kpay dodo*.

tepegay (Côte d'Ivoire; Senufo) – soft, pliable, mostly light-coloured soils (Birmingham, 2003).

timbengu (Burkina Faso; Mossi) – fertile soil formed in depression created by digging up clay for bricks to use in construction (Niemeijer and Mazzucato, 2003). Also *tanbuogu*.

tinboanli (Burkina Faso; Gourmantché) – dark loamy soil (Niemeijer and Mazzucato, 2003). Also **tinboangu** and **tinbuanli**.

tinmoanli (Burkina Faso; Gourmantché) – red loamy soil (Niemeijer and Mazzucato, 2003). Also **tinmoanga** and **tinmuanga**.

trab beyda (Senegal; Hassania) – sandy soils, forming on dunes and on the transported derivates of sandstones and quartzites (Tabor, 1993). Also *seeno* and *singue*.

trab khahale (Senegal; Soninke) – sandy alluvial soils, flooded for a short period (Tabor, 1993). Also *salka* and *gnignam bine*.

tunga (Burkina Faso; Mossi) – mountaineous soil (Dialla, 1993).

tuuli (Burkina Faso; Mossi) – fertile soil formed on old termite mounds (Niemeijer and Mazzucato, 2003). Also *baboatandi*.

vallor (northern Senegal) – sandy soils of the valley of River Senegal, flooded from July to October (Aubert, 1949). Probably, another spelling of *wallere*.

wallere (Senegal; Soninke, Pulaar) – silty alluvial soil, seldom flooded (Tabor, 1993). Also *lou wat*.

wokuta (Nigeria; Yoruba) – stony soil (Osunade, 1988). Also *ile-olokuta*.

wokuta dudu (Nigeria; Yoruba) – dark stony soil (Osunade, 1988).

wokuta pupa (Nigeria; Yoruba) – reddish-brown stony soil (Osunade, 1988).

yangi (Nigeria; Yoruba) – latheritic soil (Osunade, 1988).

yangi olomi (Nigeria; Yoruba) – hydromorphic latheritic soil without a horizon of ferrous concretions (Osunade, 1988).

yanrin (Nigeria; Yoruba) – coarse sandy soil (Osunade, 1988).

yanrin dudu (Nigeria; Yoruba) – dark sandy loamy clay soil (Osunade, 1988).

yanrin funfun (Nigeria; Yoruba) – light-coloured sandy soil (Osunade, 1988).

yanrin ogidi (Nigeria; Yoruba) – light, bleached coarse sandy soil (Warren, 1992).

yanrin pupu (Nigeria; Yoruba) – reddish-brown sandy loamy clay soil (Osunade, 1988).

yoromay dodo (Côte d'Ivoire; Bété) – sandy soil (Birmingham, 2003). Also *bossay dodo*.

zegedga (Burkina Faso; Mossi) – gravelly soil (Niemeijer and Mazzucato, 2003).

zi-bugri (Burkina Faso; Mossi) – very soft, loose soil (Dialla, 1993).

zika (Burkina Faso; Mossi) – latheritic soil (Dialla, 1993).

zi-kotka (Burkina Faso; Mossi) – clayey soil with stagnant water regime (Dialla, 1993).

zi-kugri (Burkina Faso; Mossi) – stony soil (Dialla, 1993).

zi-miuugu (Burkina Faso; Mossi) – red-coloured soil (Dialla, 1993).

zi-naare (Burkina Faso; Mossi) – moist heavy loamy soil (Dialla, 1993).

zi-peele (Burkina Faso; Mossi) – white soil (Dialla, 1993).

zipélé (Burkina Faso; Mossi) – denuded soil (Niemeijer and Mazzucato, 2003).

zi-sabille (Burkina Faso; Mossi) – black soil (Dialla, 1993).

Soil names of eastern Africa

babuni (Tanzania) – black and dark-brown clayey soil with some features of salinization and sodicity (Acres, 1984). Also *ibushi*, *itogolo*, *mchanganyika* and *ngungu*.

chamlimani (Tanzania) – shallow unproductive soils with low water-holding capacity (Acres, 1984). Also *changarawe*, *mashishiwe* and *mashololo*.

changarawe (Tanzania) – shallow unproductive soils with low water-holding capacity (Acres, 1984). Also *chamlimani*, *mashishiwe* and *mashololo*.

dambo (eastern Africa) – narrow moist depressions along river valleys and watersheds in savannas of southern Tanzania, southern Congo, Angola, Zambia, Zimbabwe and Malawi, and soils, forming in these depressions. Usually contrasted with *miombo* soils, forming on upland forested places. Dambo soils are similar to *fley* (*vley*) and *fadamma* soils, they are the tropical analogy of meadow soils. Also **dambos**.

ekundu (eastern Africa; Kiswahili) – thick red-coloured soils (FAO-ISRIC-ISSS, 1998).

fadamma (eastern Africa) – soils, forming in African savannas in depressions with periodical excess moistening (Rozanov, 1974). Also *dambo* and *fley* (*vley*).

fue (Zanzibar; Kiswahili) – light-coloured soils, the upper horizons of which are composed of bleached non-aggregated sandy non-carbonaceous material (Calton, 1949).

ibambasi (Tanzania) – solonchak-like soils, having a hard horizon (plinthite) at a depth less than 50cm from the surface, often hydromorphic (Acres, 1984). Also *itogolo*.

ibushi (Tanzania) – black and dark-brown clayey soil with some features of salinization and sodicity (Acres, 1984). Also *babuni, itogolo, mchanganyika* and *ngungu*.

igalungu (Tanzania) – reddish soils of a texture varying from sandy clay loam to clays, having some features of sodicity (Acres, 1984). Also *ikurusi* and *kikungu*.

ikurusi (Tanzania) – reddish soils of a texture varying from sandy clay loam to clays, having some features of sodicity (Acres, 1984). Also *igalungu* and *kikungu*.

ingong'ho (Tanzania) – extremely shallow stony soils in combination with outcrops of parent rocks or ferrous crusts; the depth of soils is not more than 10cm. Often affected by erosion (Acres, 1984). Also *lugulu*.

ipwisi (Tanzania) – imperfectly drained soils of slopes, having mottles of gleyization in lower horizons; the texture varies from sand to clays (Acres, 1984). Also *lukili, manda* and *kisizye*.

isenga (Tanzania) – sandy, loamy sandy, sometimes sandy loamy soils; usually shallow and stony (Acres, 1984). Also *luseni lyape*.

itogolo (Tanzania) – black and dark-brown clayey soil with some features of salinization and sodicity (Acres, 1984). Also *babuni, ibushi, mchanganyika* and *ngungu*. Differs from the listed soils by the presence of plinthite within 50cm of the surface.

kadondolyo (Tanzania) – hydromorphic soils of depressions both related and not related to river network. Mostly have heavy texture, features of gleyization and slitization (Acres, 1984). Also *mbuga, wapi, lukanda* and *manda*.

kifusi (Zanzibar; Kiswahili) – gravel soil on the eluvium of limestones, mostly having weakly differentiated yellow-brown profile (Calton, 1949).

kikungu (Tanzania) – reddish soils, with the texture varying from sandy clay loam to clays, having some features of sodicity (Acres, 1984). Also *igalungu* and *ikurusi*.

kinamo (Zanzibar; Kiswahili) – weakly developed debris soil, usually on the eluvium of limestones (Calton, 1949).

kinongo (Zanzibar; Kiswahili) – red-coloured soils, usually forming on the eluvium of limestones. Tropical analogy of *terra rossa* soils (Calton, 1949).

kisizye (Tanzania) – imperfectly drained soils of slopes, having mottles of gleyization in lower horizons; the texture varies from sand to clays (Acres, 1984). Also *lukili, manda* and *ipwisi*.

lugulu (Tanzania) – extremely shallow stony soils in combination with outcrops of parent rocks or ferrous crusts; the depth of soils is not more than 10cm. Mostly affected by erosion (Acres, 1984). Also *ingong'ho*.

lukanda (Tanzania) – hydromorphic soils of depressions both related and not related to river network. Mostly have heavy texture, features of gleyization and slitization (Acres, 1984). Also *mbuga, wapi, kadondolyo* and *manda*.

lukili (Tanzania) – imperfectly drained soils of slopes, having mottles of gleyization in lower horizons; the texture varies from sand to clays (Acres, 1984). Also *kisizye, manda* and *ipwisi*.

luseni lyape (Tanzania) – sandy, loamy sandy, in places sandy loamy soils; mostly shallow and stony (Acres, 1984). Also *isenga*.

manda (Tanzania) – imperfectly drained soils of slopes and depressions, having mottles of gleyization in lower horizons; clayey textured, having some features of slitization (Acres, 1984). Also *kisizye, mbuga, lukili* and *ipwisi*.

mashishiwe (Tanzania) – shallow unproductive soil with low water-holding capacity (Acres, 1984). Also *changarawe, chamlimani* and *mashololo*.

mashololo (Tanzania) – shallow unproductive soil with low water-holding capacity (Acres, 1984). Also *changarawe, chamlimani* and *ishiwel*.

mbuga (Tanzania) – hydromorphic soils of depressions both related and not related to river network. Mostly have heavy texture, features of gleyization and of slitization (Acres, 1984). Also *lukanda, wapi, kadondolyo* and *manda*.

mchanga (Zanzibar; Kiswahili) – sandy red-coloured tropical soils (Calton, 1949).

mchanganyika (Tanzania) – black and dark-brown clayey soil with some features of salinization and sodicity (Acres, 1984). Also *babuni, ibushi, itogolo* and *ngungu*.

miombo (eastern Africa) – dry places, covered with xerophytic sparse woods in savannas of southern Tanzania, southern Congo, Angola, Zambia, Zimbabwe and Malawi, and soils, forming in these places. Miombo soils, forming on ancient strongly weathered surfaces, are characterized by light texture (the upper horizons consist mainly of quartz sand), strong leaching, low content of humus and low productivity. In the literature they are described mainly as ferrous tropical leached soils.

muchanga (northern Zambia) – a common name of sandy (from coarse sandy to sandy loams) soils, usually applied with definitions of colour and the content of silt (Kerven et al, 1995).

ngungu (Tanzania) – black and dark-brown clayey soil with some features of salinization and sodicity (Acres, 1984). Also *babuni, ibushi, itogolo* and *mchanganyika.*

nkanka (northern Zambia) – upland red-coloured soils, usually clayey, distinguished by significant productivity for most tropical cultures (Kerven et al, 1995).

wapi (Tanzania) – hydromorphic soils of depressions both related and not related to river network. Mostly have heavy texture, features of gleyization and of slitization (Acres, 1984). Also *lukanda, mbuga, kadondolyo* and *manda.*

yumba (Kenya; Mbeere, Kamba and Meru) – clayey soil, rich in kaolinite. The soil is used by local population for pottery (Tabor, 1992).

Soil names of southern Africa

bukutu (Zimbabwe; Shona) – red clayey soils, usually productive, extremely slippery when wet. Also *jiho* (Nyamaphene, 1983).

bungure (Zimbabwe; Shona) – light sandy soils. Also *shapa* (Nyamaphene, 1983).

chiombwe (Zimbabwe; Shona) – clayey soils, unsuitable for cultivation, but used for pottery. Also *mpunzo* and *rondo* (Nyamaphene, 1983).

chishava (Zimbabwe; Shona) – gravelly red-coloured soils, unsuitable for cultivation, but suitable for construction (Nyamaphene, 1983).

chivavane (Zimbabwe; Shona) – grey sodic, weakly salted soils, on the surface of which water stagnates in rainy season (Nyamaphene, 1983).

dhakiumnyama (Zimbabwe; Shona, Ndebele) – a slang term for black clayey soils, cracking when dried (Nyamaphene, 1983).

finbush (SAR; Afrikaans) – a name of sandy young soils under xerophitic wood vegetation, having upper bleached eluvial horizon and a horizon of aluminium, iron and humus illuviation (Soil Classification Working Group, 1991).

flei (SAR; Afrikaans) – a name for tropical and subtropical analogies of meadow soils. Black, brown or grey soils usually with mottles of gleyization, iron concretions and nodules of calcium carbonate. The soils are weakly acid or neutral (Robinson, 1936). They are similar to the soils *pen, dambo and fadamma*. Rozanov (1974) considered this term to mean black clayey compact soils (Vertisols) with excessive groundwater moistening. Rozov and Stroganova (1979) point to the presence of iron-carbonaceous hardened layers in these soils as their specific feature. Also *vley* and *fley.*

gokoro (Zimbabwe; Shona) – light-grey sodic, strongly salted soils (Nyamaphene, 1983).

gova (Zimbabwe; Shona) – black clayey soils, cracking when dried (Nyamaphene, 1983).

ibumba (Zimbabwe; Ndebele) – clayey soils, unsuitable for cultivation, but used for pottery (Nyamaphene, 1983).

ihlabathi (Zimbabwe; Ndebele) – light sandy soils (Nyamaphene, 1983).

isibomvu (Zimbabwe; Ndebele) – red clayey soils, usually productive, extremely slippery in wet conditions (Nyamaphene, 1983).

isidhaka (Zimbabwe; Ndebele) – black clayey soils, cracking when dried (Nyamaphene, 1983).

isikwaka (Zimbabwe; Ndebele) – grey sodic, weakly salted soils, on the surface of which water stagnates in rainy season (Nyamaphene, 1983).

isimunyu (Zimbabwe; Ndebele) – light-grey sodic, strongly salted soils (Nyamaphene, 1983).

jiho (Zimbabwe; Shona) – red clayey soils, usually productive, extremely slippery when wet. Also *bukutu* (Nyamaphene, 1983).

kaamba (Congo) – dark clayey cracky compact soils (Rozanov, 1974).

karu (Namibia; Gottentote) – red-coloured soils on aeolic sediments of southern Africa (Sibirtsev, 1951).

mpunzo (Zimbabwe; Shona) – clayey soils, unsuitable for cultivation, but used for pottery. Also *chiombwe and rondo* (Nyamaphene, 1983).

ndoronya (Zimbabwe; Shona) – hydromorphic, gleyed thixotropic sandy soils (Nyamaphene, 1983).

rondo (Zimbabwe; Shona) – clayey soils, unsuitable for cultivation, but used for pottery. Also *mpunzo and chiombwe* (Nyamaphene, 1983).

rukangarahve (Zimbabwe; Shona) – shallow, gravelly soil with perfect drainage. Also *tsangarahwe* (Nyamaphene, 1983).

rusekenya (Zimbabwe; Shona) – infertile, deep sandy soils. Also *ruzekete* (Nyamaphene, 1983).

shapa (Zimbabwe; Shona) – light sandy soils. Also *sapa* and *bungure* (Nyamaphene, 1983).

tsangarahve (Zimbabwe; Shona) – shallow, gravelly soil with a perfect drainage. Also *rukangarahwe* (Nyamaphene, 1983).

vley – the same as *flei*.

Soil names of Australia and Oceania

anyata (New Guinea; Baruya) – soils, used in agriculture, having the colour of plough horizon varying from black to dark-reddish-brown (5YR 3/2) and dark-brown (7.5YR 3/3). Have weakly acid reaction; texture impossible to correlate due to thixotropic properties of the upper horizon in these soils. By the local people these soils are considered to be productive (Ollier et al, 1971).

aoai (central mountains of New Guinea) – dense reddish-brown clayey soils (Marten and Vityakon, 1986).

biwaka (New Guinea; Baruya) – greenish-grey (7.5GY 6/1) soil material, collected by the indigenous population in marshy places. Rapidly loses colour while oxidized (Ollier et al, 1971).

butuma (Papua-New Guinea; Trobrian islands) – red-coloured soils of light texture near coral reefs, unsuitable for taro cultivation, but favourable for yams (Marten and Vityakon, 1986).

cheragwaka (New Guinea; Baruya) – a red ochreous (5YR 4/6) soil material; sometimes some treatment is necessary, like burning, to obtain the needed pigment. It is used as a body pigment by girls after their first menstruation, women after childbirth and for initiation of witch doctors (Ollier et al, 1971).

chimuwaka (New Guinea; Baruya) – red-coloured (5YR 4/6) soils on limestone, resembling *terra rossa*. Their reaction is acid; the exchangeable complex is saturated mainly with potassium and sodium. These soils are used by local population for growing taro (Ollier et al, 1971).

dawaka (New Guinea; Baruya) – light-yellowish-brown (10YR 6/4) soil material, turning white after burning or drying; includes the material of some anthills. It is used by women for colouring string (Ollier et al, 1971).

detnalaolye (New Guinea; Baruya) – dark-reddish-brown (5YR 3/4) soil under grass vegetation. The reaction is neutral; in recent years these soils have been cultivated by the local population with European tools (Ollier et al, 1971).

dumya (Papua-New Guinea; Trobrian islands) – clayey bog soils, suitable for taro in dry season, but unsuitable for yams (Marten and Vityakon, 1986).

eogwaka (New Guinea; Baruya) – red-coloured (2.5YR 5/8) clay. It is used by the indigenous population as a pigment for the bodies of children and in the third stage of initiation (Ollier et al, 1971).

galaluva (Papua-New Guinea; Trobrian islands) – black, dry soil of heavy texture (Marten and Vityakon, 1986).

gilgai (Australia) – dark clayey cracky compact soils and connected relief (forming due to the swelling of soil material on the surface).

gwegwaka (New Guinea; Baruya) – light-grey (2.5Y 8/2) clay. It is used by the indigenous population on some stages of initiation, and for colouring shield and body in wartime (Ollier et al, 1971).

hali (Hawaii) – red-coloured soils rich in gybbsite (Gerasimov, 1978).

hibber (Australia) – a name of stony deserts of Australia, covered by siliceous gravel, products of weathering of thick siliceous crusts (Rozov and Stroganova, 1979).

ietchiake (New Guinea; Baruya) – reddish-yellow (7.5YR 6/8) clayey soil. It is used by the local people for cultivating taro and pandanus (Ollier et al, 1971).

ikulukwaka (New Guinea; Baruya) – 'strawberry-pink' clay used as pigment for body in ceremonies (Ollier et al, 1971).

jegynye (New Guinea; Baruya) – topsoil that is too thin or is not black enough to be regarded as *anyata*. The colours range from very dark greyish-brown (10YR 3/2), dark-reddish-brown (5YR 3/3) to dark-brown (7.5YR 3/4). The reaction is weakly acid; the horizon has thixotropic properties (Ollier et al, 1971).

kerematua (New Zealand; Maori) – dense, hard clay (Hewitt, 1992).

kereti (New Zealand; Maori) – clay (Hewitt, 1992).

konegwaka (New Guinea; Baruya) – alluvial soils, including salted ones; usually have a shallow humus or mud horizon, not exceeding 10cm; beneath there is alluvial material of various textures, mostly sandy, gravelly or stony. These soils are used by the local people for cultivation of taro, sugarcane, banana, sweet potatoes and other crops (Ollier et al, 1971).

kumawaka (New Guinea; Baruya) – blue to black clay, usually found in patches beneath the topsoil of swampy sites. The blue colour is lost rapidly on oxidation (Ollier et al, 1971).

kurukwaka (New Guinea; Baruya) – non-carbonaceous weakly acid brown (7.5YR 4/4) soils, used for cultivating taro and sweet potatoes (Ollier et al, 1971).

kwala (Papua-New Guinea; Trobrian islands) – black soils near coral reefs, fairly productive, suitable for cultivation of all crops (Marten and Vityakon, 1986).

malala (Papua-New Guinea; Trobrian islands) – poor stony soils, unsuitable for taro, but suitable for yams (Marten and Vityakon, 1986).

mallee (Australia) – solonetz-like soils (Krupenikov, 1981).

numbuchukwaka (New Guinea; Baruya) – red (2.5YR 4/8) clay, used by the indigenous population as a body pigment at dances, at initiation and for war-paint (Ollier et al, 1971).

numwaka (New Guinea; Baruya) – red ochre (5YR 5/6) soils (Ollier et al, 1971).

one hanahana (New Zealand; Maori) – dark soil, containing gravel and small stones; evidently formed on deluvial or slope sediments (Hewitt, 1992).

one kokopu (New Zealand; Maori) – stony soil or soil with a high gravel content (Hewitt, 1992).

one kopuru (New Zealand; Maori) – excessively moistened soil (Hewitt, 1992).

one mata (New Zealand; Maori) – productive dark-coloured soil (Hewitt, 1992).

one pakirikiri (New Zealand; Maori) – soil, containing gravel (Hewitt, 1992).

one parahuhu (New Zealand; Maori) – alluvial soil (Hewitt, 1992).

one pu (New Zealand; Maori) – sandy soil (Hewitt, 1992).

one punga (New Zealand; Maori) – porous light-textured soil (Hewitt, 1992).

one takataka (New Zealand; Maori) – loose soil (Hewitt, 1992).

one tea (New Zealand; Maori) – white soil, formed of sandy material of volcanic origin (Hewitt, 1992).

one tuotara (New Zealand; Maori) – compact brown soil, productive, but requiring loosening by adding sand or gravel (Hewitt, 1992).

one wawata (New Zealand; Maori) – cloggy soil (Hewitt, 1992).

parakiwai (New Zealand; Maori) – loam (Hewitt, 1992).

pu (New Zealand; Maori) – sand (Hewitt, 1992).

pubuti (central mountains of New Guinea) – chocolate-brown, plastic soil with fine granules structure (Marten and Vityakon, 1986).

sawewo (Papua-New Guinea; Trobrian islands) – soils, forming in sink holes of forested coral reefs, suitable for yams (Marten and Vityakon, 1986).

taioma (New Zealand; Maori) – white pipe clay (Hewitt, 1992).

tugke (central mountains of New Guinea) – dense greenish clayey soils, containing ochreous concretions (Marten and Vityakon, 1986).

tuotara wawata (New Zealand; Maori) – brown loose productive soil, suitable for cultivation (Hewitt, 1992).

uku (New Zealand; Maori) – clayey, sticky clay of white or light blue colour (Hewitt, 1992).

weinjuna (New Guinea; Baruya) – black alluvial soils, composed almost totally of organic matter and carbonates; contain a lot of shells (Ollier et al, 1971).

References

Acres, B. D. (1984) 'Local farmers' experience of soils combined with reconnaissance soil survey for land use planning: An example from Tanzania', *Soil Survey and Land Evaluation*, vol 4, no 3, pp77–86

Ali, A. M. S. (2003) 'Farmers' knowledge of soils and the sustainability of agriculture in a saline water ecosystem in Southwestern Bangladesh', *Geoderma*, vol 111, nos 3–4, pp333–353

Aubert, G. (1949) 'Note on the vernacular names of the soils of the Sudan and Senegal', *Communications of the Bureau of Soil Science*, Proceedings of the First Commonwealth Conference on Tropical and Sub-Tropical Soils, 1948, Harpenter, Technical Communication no 46, pp107–109

Baldwin, M., Kellog, C. E. and Thorp, J. (1938) 'Soil classification', in United States Department of Agriculture, *Soils and Men: Yearbook of Agriculture 1938*, US Government Printing Office, Washington DC, pp979–1001

Barrera-Bassols, N. and Zinck, J. A. (2000) *Ethnopedology in a Worldwide Perspective: An Annotated Bibliography*, International Institute for Aerospace Survey and Earth Sciences (ITC), The Netherlands, ITC Publication no 77, 635pp

Barrera-Bassols, N. and Zinck, J. A. (2003) 'Ethnopedology: A worldwide view on the soil knowledge of local people', *Geoderma*, vol 111, nos 3–4, pp171–195

Bautista, F., Diaz-Garrido, S., Castillo-Gonzalez, M. and Zinck, J. A. (2005) 'Spatial heterogeneity of the soil cover in the Yucatan Karst: Comparison of Mayan, WRB, and Numerical Classifications', *Eurasian Soil Science*, vol 38, suppl 1, pp81–88

Behrens, C. A. (1989) 'The scientific basis for Shipobo soil classification and land use: Changes in soil–plant associations with cash cropping', *American Anthropologist*, vol 91, no 1, pp83–100

Bellon, M. R. and Taylor, J. E. (1993) '"Folk" soil taxonomy and the partial adoption of new seed varieties', *Economic Development and Cultural Change*, vol 41, no 4, pp763–786

Bennett, H. H. and Allison, R. V. (1928) *The Soils of Cuba*, Tropical Plant Research Foundation, Washington DC, 410pp

Birmingham, D. M. (2003) 'Local knowledge of soils: The case of contrast in Côte d'Ivoire', *Geoderma*, vol 111, nos 3–4, pp481–502

Bramao, S. L. and Lemos, P. (1961) 'Soil map of South America', *Transactions of the 7th International Congress of Soil Science*, Madison, 1960, Commissions V and VII, Elsevier, Amsterdam, vol IV, pp1–10

Calton, W. E. (1949) 'A reconnaisance of the soils of Zanzibar protectorate', *Communications of the Bureau of Soil Science*, Proceedings of the First Commonwealth Conference on Tropical and Sub-Tropical Soils, 1948, Harpenter, Technical Communication no 46, pp49–53

Dialla, B. E. (1993) 'The Mossi indigenous soil classification in Burkina Faso', *Indigenous Knowledge and Development Monitor*, vol 1, no 3, pp1–5

Dokuchaev, V. V. (1949a) 'On the relation of the age and altitude of a territory on one hand, and of the character and distribution of chernozems, forest soils and solonetz, on the other hand', *Complete Set of Works*, USSR Academy of Sciences Publishers, Moscow, vol 1, pp378–404 (in Russian)

Dokuchaev, V. V. (1949b) 'Russian chernozem', *Complete Set of Works*, USSR Academy of Sciences Publisher, Moscow, vol 3, pp378–404 (in Russian)

Dokuchaev, V. V. (1950a) 'Mapping of Russian soils', *Complete Set of Works*, USSR Academy of Sciences Publisher, Moscow, vol 2, pp69–241 (in Russian)

Dokuchaev, V. V. (1950b) 'On podzol', *Complete Set of Works*, USSR Academy of Sciences Publisher, Moscow, vol 2, pp248–256 (in Russian)

Dokuchaev, V. V. (1950c) 'Soils, vegetation and climate of Nizhni Novgorod region', *Complete Set of Works*, USSR Academy of Sciences Publisher, Moscow, vol 5, pp248–256 (in Russian)

Dokuchaev, V. V. (1952) 'The place and role of modern soil science in science and life', *Complete Set of Works*, USSR Academy of Sciences Publisher, Moscow, vol 6, pp415–424 (in Russian)

Dokuchaev, V. V. (1953) 'On the use of the study of local names of Russian soils', *Complete Set of Works*, USSR Academy of Sciences Publisher, Moscow, vol 7, pp332–340 (in Russian)

Dudal, R. and Bramao, S. L. (eds) (1965) *Dark Clay Soils of Tropical and Subtropical Regions*, Agricultural Development paper no 83, UN Food and Agriculture Organization, Rome, 162pp

Egorov, V. V., Fridland, V. M., Ivanova, E. N., Rozov, N. N., Nosin, V. A. and Fraev, T. A. (1977) *Classification and Diagnostics of Soils of USSR*, Kolos Press, Moscow, 221pp (in Russian)

Endo, K. (1982) 'Forest soils of temperate warm belt of Japan', in A. M. Ivlev and I. V. Ignatenko (eds) *Soils of the Islands and Oceanic Regions of the Pacific Ocean*, Far-East Research Centre of the Russian Academy of Sciences Publishers, Vladivostok, pp111-116 (in Russian)

FAO-ISRIC-ISSS (1998) *World Reference Base for Soil Resources*, Soil Resources Report no 84, UN Food and Agriculture Organization, Rome, 88pp

Gagarina, E. I., Matinyan, N. N., Schastnaya, L. R. and Kasatkina, G. A. (1995) *Soils and Soil Cover of North-West of Russia*, St Petersburg University Press, St Petersburg, 236pp (in Russian)

Ganssen, R. (1962) *Soil Geography*, Foreign Literature Press, Moscow, 272pp (in Russian)

Gerasimov, I. P. (1958) 'Geographical observations in Japan', *Transactions of the Academy of Sciences of the USSR, Geographical series*, no 2, pp54-63 (in Russian)

Gerasimov, I. P. (1978) 'Soils of Hawaii islands and their genetic interpretation', in S. V. Zonn (ed) *Soil Genesis and Geography in Foreign Countries after Investigations of Soviet Geographers*. Nauka Publishing House, Moscow, pp19–38 (in Russian)

Gerasimov, I. P. (1979) 'Belozems', in I. P. Gerasimov (ed) *Genetic Soil Types of the Transcaucasian Subtropics*, Nauka Publishers, Moscow, pp251-255 (in Russian)

Glazovskaya, M. A. (1983) *Soils of Foreign Countries*, Vyshaya Shkola Publishers, Moscow, 312pp

Glinka, K. D. (1927) *Soil Science*, 3rd edition, Novaya Derevnya Press, Moscow, 580pp (in Russian)

Gong Zitong (1989) 'Review of Chinese soil classifications for the past four decades', *Acta Pedologica Sinica*, vol 26, no 3, pp217–225

Hewitt, A. E. (1992) 'Soil classification in New Zealand: Legacy and lessons', *Australian Journal of Soil Research*, vol 30, no 6, pp843–854

Holman, E. W. (2005) 'Domain-specific and general properties of folk classifications', *Journal of Ethnobiology*, vol 25, no 1, pp71–91

Kanno, I. (1978) '"Chinese Soils" compiled by the Nanking Institute of Soil Science, Academia Sinica', *Pedologist*, vol 22, no 2, pp176–181

Kanno, I. (1982) 'Problems of classification of soils of Japan', in A. M. Ivlev and I. V. Ignatenko (eds) *Soils of the Islands and Oceanic Regions of the Pacific Ocean*, Far-East Research Centre of the Russian Academy of Sciences Publishers, Vladivostok, pp97–110 (in Russian)

Kato, Y., Matsui, T. (1979) 'Some applications of paleopedology in Japan', *Geoderma*, vol 22, no 1, pp45-60

Kerven, C., Dolva, H. and Renna, R. (1995) 'Indigenous soil classification systems in Northern Zambia', in D. M. Warren et al (eds) *The Cultural Dimension of Development: Indigenous Knowledge Systems*, Intermediate Technology Publisher, London, pp82–87

Kornblyum, E. A. and Zimovets, B. A. (1961) 'On the origin of soils with a white horizon on the plains of Priamurye', *Pochvovedenie*, no 6, pp55–66 (In Russian)

Krasilnikov, P. V. (1999) 'Early studies on folk soil terminology', *Eurasian Soil Science*, vol 32, no 10, pp1280–1284

Krasilnikov, P. V. and Tabor, J. A. (2003) 'Perspectives of utilitarian ethnopedology', *Geoderma*, vol 111, nos 3–4, pp197–215

Krupenikov, I. A. (1981) *History of Soil Science*, Nauka Press, Moscow, 327pp (in Russian)

Kubiëna, W. L. (1953) *Bestimmungsbuch und Systematik der Boden Europas*, Verlag Enke, Stuttgart, 392pp

Lamanski, V. V. (1915) 'An experience of folk soil dictionary', *Pochvovedenie*, no 2, pp61–72 (in Russian)

Lamers, J. P. A., Feil, P. R. and Buerhert, N. (1995) 'Spatial crop growth variability in Western Niger: The knowledge of farmers and researchers', *Indigenous Knowledge and Development Monitor*, vol 3, no 3, pp17–19

Marten, G. G. and Vityakon, P. (1986) 'Soil management in traditional agriculture', in Marten, G. G. (ed) *Traditional Agriculture in Southeast Asia*, Westview Press, Boulder, CO, pp199–225

Matsui, T. (1982) 'An approximation to establish a unified comprehensive classification system for Japanese soils', *Soil Science and Plant Nutrition*, vol 28, no 2, pp235–256

Murzaev, E. M. (1984) *A Dictionary of Folk Geographic Terms*, Mysl Press, Moscow, 654pp (in Russian)

Niemeijer, D. (1995) 'Indigenous soil classifications: Complications and considerations', *Indigenous Knowledge and Development Monitor*, vol 3, no 1, pp1–5

Niemeijer, D. and Mazzucato, V. (2003) 'Moving beyond indigenous soil taxonomies: Local theories of soils for sustainable development', *Geoderma*, vol 111, nos 3–4, pp403–424

Nyamaphene, K. W. (1983) 'Traditional systems of soil classification in Zimbabwe', *Zambezia*, vol 11, no 1, pp55–57

Ollier, C. D., Drover, D. P. and Godelier, M. (1971) 'Soil knowledge amongst the Baruya of Wonenara, New Guinea', *Oceania*, vol 42, no 1, pp33–41

Osbahr, H. and Allan, C. (2003) 'Indigenous knowledge of soil fertility management in southwest Niger', *Geoderma*, vol 111, nos 3–4, pp457–479

Osunade, M. A. A. (1988) 'Soil suitability classification by small farmers', *Professional Geographer*, vol 40, no 2, pp194–201

Pons, L. J. (1973) 'Outline of genesis, characteristics, classification and improvement of acid sulphate soils', in H. Dost (ed) *Acid Sulphate Soils*, ILRI, Wageningen, Publishers 18, vol 2, pp3–27

Principi, P. (1943) *I Terreni d'Italia*, Sociedade Anonima Editrice Dante Alighieri, Albrighi, Segati XXI, Genoa, Rome, Naples and Gitta di Castello, 242pp

Ravikovich, S. (1960) *Soils of Israel. Classification of the Soils of Israel*, Rehovot, Israel, 86pp

Raychaudhri, S. P. (1958) *Soils of India*, The Indian Council for Agricultural Research, Delhi, Review Series, no 25, 28pp

Robinson, G. W. (1936) *Soils, their Origin, Constitution and Classification. An Introduction to Pedology*, Thos Murby & Co, London, 442pp

Rozanov, B. G. (1974) *Soil Terminology in Russian and Foreign Languages*, in 2 volumes, Moscow, 483pp and 273pp (in Russian)

Rozov, N. N. and Stroganova, M. N. (1979) *Soil Cover of the World*, Moscow State University Press, Moscow, 290pp (in Russian)

Rutherford, G. K. (1972) 'The properties, distribution and origin of white silt soils in Romerike, Norway', *Meddeleser fra det Norske Skøgforsoksvesen*, vol 30, part 2, no 120, 184pp

Sandor, J. A. and Furbee, L. (1996) 'Indigenous knowledge and classification of soils in the Andes of Southern Peru', *Soil Science Society of America Journal*, vol 60, no 5, pp1502–1512

Satyanarayana, K. V. C. (1971) 'Review of research work done on pedology and soil survey. Western zone', *Soil and Water Research in India in Retrospect and Prospect*, I.C.A.R. Technical Bulletin (Agriculture) no 22, Delhi, pp314–323

Shoba, S. A. (ed) (2002) *Soil Terminology and Correlation*, 2nd edition, Centre of the Russian Academy of Sciences, Petrozavodsk, 320pp

Sibirtsev, N. M. (1951) 'Soil Science', *Selected Works*, Agricultural Literature Press, Moscow, vol 1, pp19–465 (in Russian)

Soil Classification Working Group (1991) *Soil Classification. A Taxonomic System for South Africa*, 2nd (revised) edition, Memoirs on the Agricultural Natural Resources of South Africa no 15, Department of Agricultural Development, Pretoria, 257pp

Statishin de Queiroz, J. and Norton, B. E. (1992) 'An assessment of an indigenous soil classification used in the *caatunga* region of Ceara State, Northeast Brazil', *Agricultural Systems*, vol 39, no 4, pp289–305

Stranski, I. (1954) 'Bulgarian folk soil names based on their colour and humidity', *Transactions of Soil Science Institute*, Bulgarian Academy of Sciences Press, Sofia, vol 2, pp281–360 (in Bulgarian)

Stranski, I. (1956) 'Bulgarian folk soil names based on their physical properties', *Transactions of Soil Science Institute*, Bulgarian Academy of Sciences Press, Sofia, vol 3, pp329–410 (in Bulgarian)

Stranski, I. (1957) 'Bulgarian folk soil names based on their miscellaneous features', *Transactions of Soil Science Institute*, Bulgarian Academy of Sciences Press, Sofia, vol 4, pp307–409 (in Bulgarian)

Superanskaya, A. V. (1970) 'Are colour names of rivers terminologic?' in *Local Geographical Terms,* Mysl Press, Moscow, pp120–127 (in Russian)

Tabor, J. A. (1992) 'Ethnopedological surveys: Soil surveys that incorporate local systems of land classification', *Soil Survey Horizons,* vol 33, no 1, pp1–5

Tabor, J. A. (1993) 'Soils of the lower, middle and upper Senegal river valley', in T. K. Park (ed) *Risk and Tenure in Arid Lands. The Political Ecology of Development in the Senegal River Basin,* Monographs on Arid Land Development, The University of Arizona Press, Tucson and London, pp31–50

Warren, D. M. (1992) *A Preliminary Analysis of Indigenous Soil Classification and Management Systems in Four Ecozones of Nigeria,* Discussion Paper RCMD 92/1, International Institute of Tropical Agriculture and the African Resource Centre for Indigenous Knowledge, Ibadan, 28pp

Wilde, S. A. (1946) *Forest Soils and Forest Growth,* Chronica Botanica Company, Waltham, MA, 226pp

Wilkander, L., Hallgren, G., Brink, N. and Johnson, E. (1950) 'Studies on gyttja soils. 2. Some characteristics of two profiles from Northern Sweden', *Kungliga Lantbrukshogskolans Annaler,* vol 17, pp24–36

Williams, B. J. (1975) 'Aztec soil science', *Boletín del Instituto de Geografía,* UNAM, vol 7, pp115–120

Williams, B. J. and Ortiz-Solorio, C. A. (1981) 'Middle American folk soil taxonomy', *Annals of the Association of American Geographers,* vol 71, no 3, pp335–358

Zakharov, S. A. (1915) 'On soil terminology of indigenous population of Trans-Caucasian region', *Russki Pochvoved,* nos 13–14, pp367–373 (in Russian)

Zimmerer, K. S. (1994) 'Local soil knowledge: Answering basic questions in highland Bolivia', *Journal of Soil and Water Conservation,* vol 49, pp29–34

Zinck, A. and Barrera-Bassols, N. (2002) 'Land moves and behaves: Indigenous discourse on sustainable land management in Mexico', *Transactions of the 17th World Congress of Soil Science,* 14–21 August 2002, Bangkok, Thailand, CD-ROM, paper no 316, 11pp

Index

The index includes all the soil names mentioned in the book. Most of the terms are used in plural form, as they are listed in the text. If the term was used both in plural and singular form, it is listed in the index only in plural form; for example, for *Podzol* see *Podzols*. The terms recommended for use in capital letters, as in the French and Brazilian classifications, are listed separately after an analogue in lower-case letters, if exists: for example, ORGANOSOLS follows *Organosols*. The first letter capitalization is not taken into account: for example, *Gley* and *gley* are regarded as the same name. Special characters not used in English alphabet are placed next to similar English letters: Ñ is regarded as N, Ö is regarded as O etc.

Milton Keynes UK
Ingram Content Group UK Ltd.
UKHW031139141024
449569UK00024B/1223